Atlas of Anatomic Pathology

Series Editor
Liang Cheng

Dennis P. O'Malley

Atlas of Spleen Pathology

Springer

Dennis P. O'Malley, M.D.
Clarient Inc./GE Healthcare
Aliso Viejo
California
USA

Adjunct Associate Professor
Department of Hematopathology
M.D. Anderson Cancer Center/
University of Texas
Houston
Texas
USA

ISBN 978-1-4614-4671-2 ISBN 978-1-4614-4672-9 (eBook)
DOI 10.1007/978-1-4614-4672-9
Springer New York Heidelberg Dordrecht London

Library of Congress Control Number: 2012948389

Printed on acid-free paper

Springer is part of Springer Science+Business Media (www.springer.com)

As with all of my works, I would like to dedicate this work to my wife, Karene. It is through her love and support that I accomplish anything.

Preface

The spleen has always been mysterious in the history of medicine. It has had several characters ascribed to it as a seat of emotions, or even as a site of pumping blood. However, in modern medicine, the spleen, while well understood, still leads to diagnostic challenges. This is likely due to its relative rarity as a diagnostic specimen, and the unusual combinations of pathology that can occur. However, when considered in its component parts, splenic pathology can be relatively straightforward and understood as an interesting mix of physiologic, anatomic, and pathologic components. It is my hope that the images in this atlas, along with their descriptions, will help bring some understanding and appreciation to this "organ of mystery."

Acknowledgment

I would like to acknowledge Attilio Orazi, Richard Neiman, Thomas Dutcher, and Peter Banks, who nurtured my interest in hematopathology, and in splenic pathology, specifically. My special thanks to the many pathologists who have shared gross images of splenic pathology; without their help this atlas would be far less illustrative. Finally, I would like to thank all of my colleagues for their help, insights, and support during the production of this book.

Contents

1 Normal Morphologic and Immunohistochemical Findings and Hyperplasias

Normal architecture is basic to an overall understanding of splenic pathology. In this chapter, normal components of spleen are illustrated. These include normal structures such as red and white pulp, stromal elements, and vascular components. In addition, hyperplasias of splenic components are illustrated. These are important, as the differential diagnosis of many splenic lesions and lymphomas are benign proliferations of the same compartments. Follicular hyperplasia and marginal zone hyperplasia are most commonly encountered and also most closely mimicked by follicular lymphoma and splenic marginal zone lymphoma, respectively. Other more rare hyperplasias include proliferations of T cells and immunoblasts. Finally, the immunohistochemical architecture of the spleen is richly illustrated. This is particularly important in the evaluation of splenic lesions, as understanding of normal patterns of staining must precede the evaluation of stains in pathologic processes. A combination of histologic, immunohistochemical, and gross images is presented.

D.P. O'Malley, *Atlas of Spleen Pathology*, Atlas of Anatomic Pathology,
DOI 10.1007/978-1-4614-4672-9_1, © Springer Science+Business Media New York 2013

1.1 Anatomy and Normal Histology

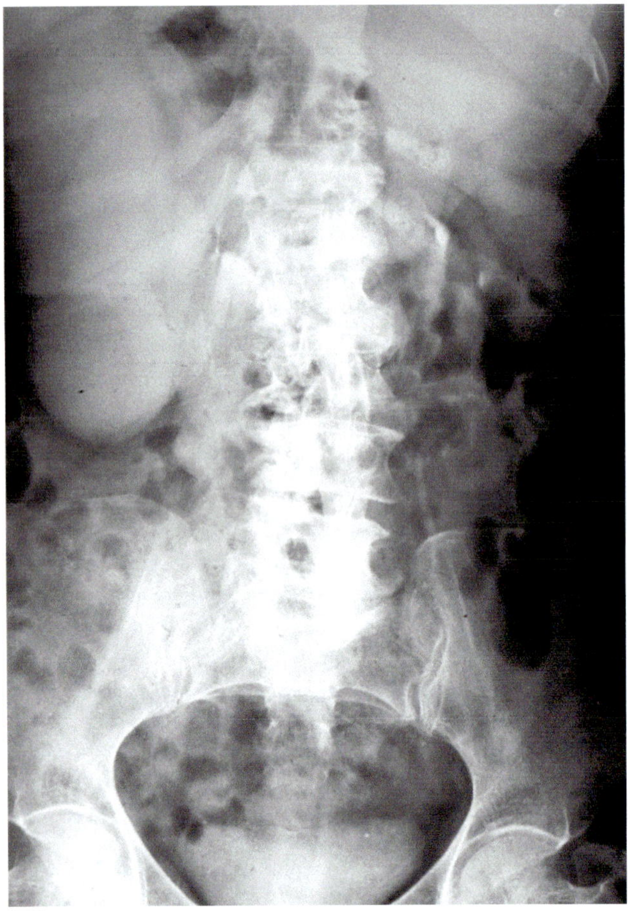

Fig. 1.1 A radiograph of the lower thoracic and abdominal cavity. The spleen is present in the middle left, beneath the ribs. The spleen is seen as a round or bean-shaped organ. It lies adjacent to the stomach and lateral to the pancreas

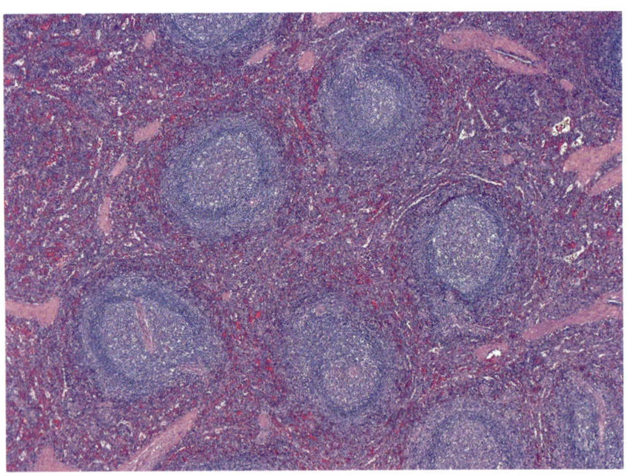

Fig. 1.2 Normal spleen parenchyma. An intermediate magnification of normal spleen parenchyma. In this image, there are appropriate numbers of germinal centers and white pulp admixed with red pulp. The white pulp nodules are in their usual functional form, showing a tripartite arrangement, with a germinal center (pale), a thin mantle zone (dark), and a pale outer marginal zone. Some large caliber vessels (arterioles) are also present. Red pulp shows some patent sinuses, with some circulating red blood cells

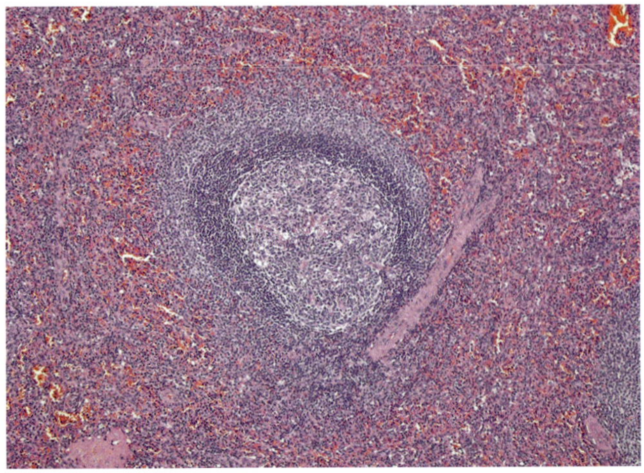

Fig. 1.3 Normal spleen parenchyma. A higher magnification highlighting the features of an individual white pulp nodule. This is composed of a reactive germinal center with mantle zone and marginal zone. Note that there is an adjacent splenic arteriole. White pulp nodules are analogous to "buds" whereas the arterioles are analogous to "branches." The red pulp in this image shows typical features with some patent sinuses and more solid-appearing cords

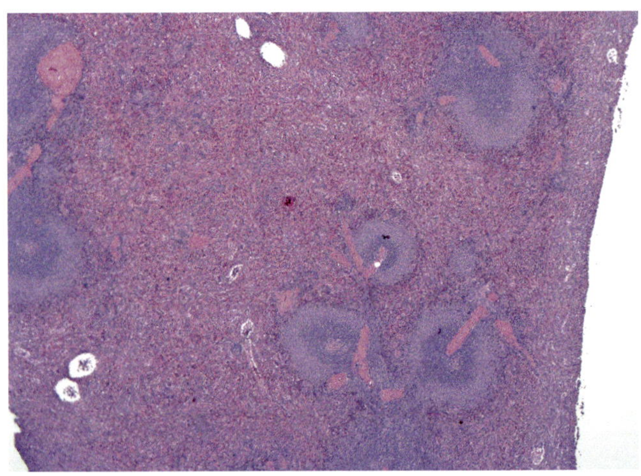

Fig. 1.4 Normal spleen red and white pulp. This low magnification image of spleen shows a typical normal proportion of red pulp (RP) to white pulp (WP). In general, the ratio of RP:WP is about 3–4:1, as in this case

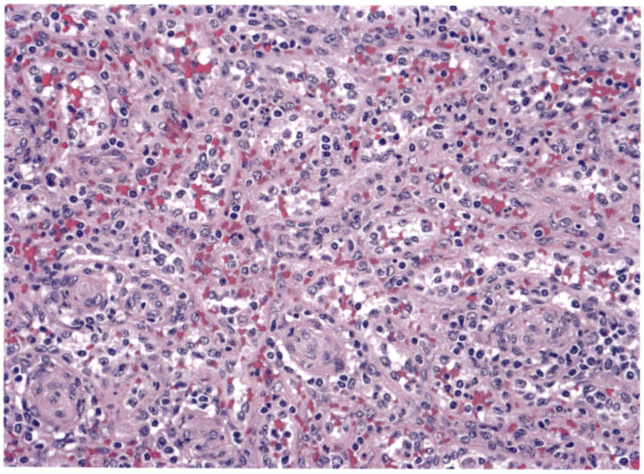

Fig. 1.5 Red pulp. The splenic red pulp is composed of predominantly splenic sinuses and splenic cords. Sinuses are lined by specialized endothelial cells (*eg*, splenic littoral cells). The intervening spaces are splenic cords. They contain stromal elements including splenic macrophages and some occasional lymphocytes

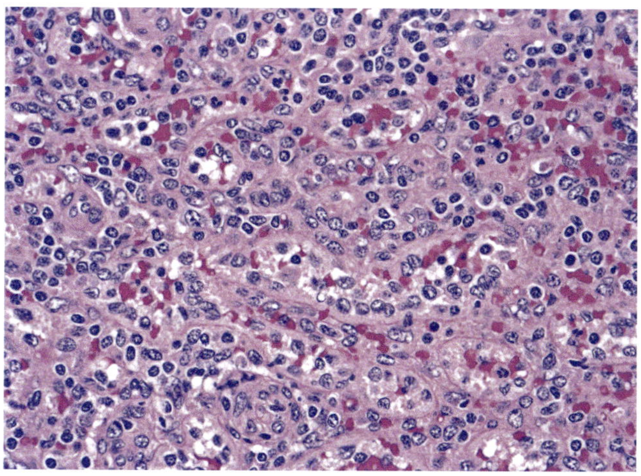

Fig. 1.6 Red pulp. Red pulp is a hypoxic environment. Red blood cells enter into the splenic cords and then must traverse the walls of the sinuses to return to the circulation. If red cells are not deformable, they will be destroyed by splenic macrophages

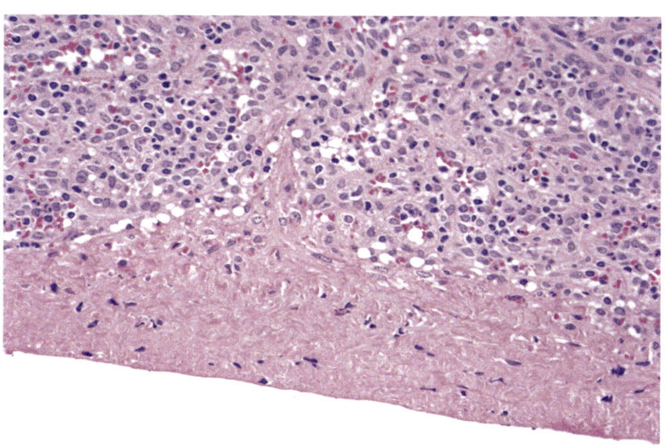

Fig. 1.7 Spleen capsule. In this microscopic image, the spleen capsule is present (*lower*). It is typically paucicellular with only rare, unremarkable spindle-shaped stromal cells present

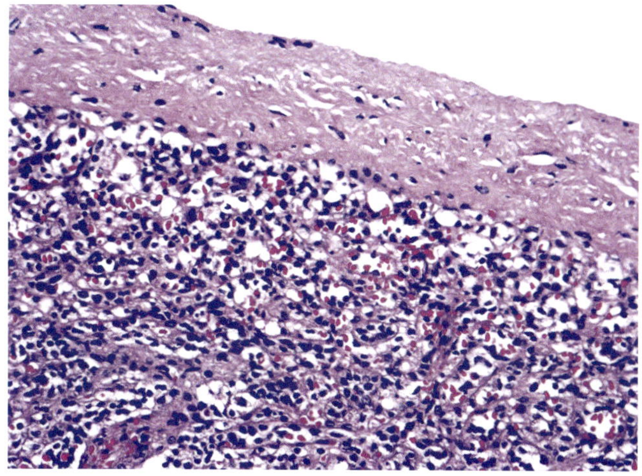

Fig. 1.8 Spleen capsule. Another example of normal spleen capsule (*upper*). The capsule is 50–100 microns thick. The presence of significant cellular infiltrates can indicate pathology

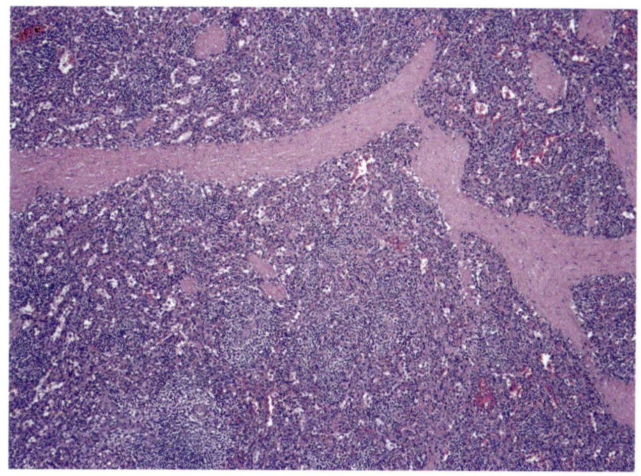

Fig. 1.9 Fibrous trabeculae. Bands of fibrous tissue, contiguous with the capsule, are present in the central portions of the spleen. There are approximately the same thickness as the capsule and provide a framework for vascular and stromal elements in the spleen

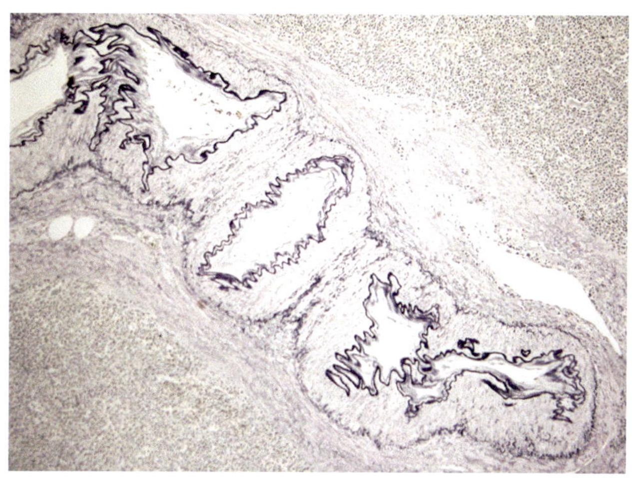

Fig. 1.10 Splenic artery, elastin stain. Elastin stain of normal splenic artery. Note the staining within the artery wall and the notable lack of staining within the splenic red pulp. Elastic fibers are not a normal part of the splenic red pulp architecture

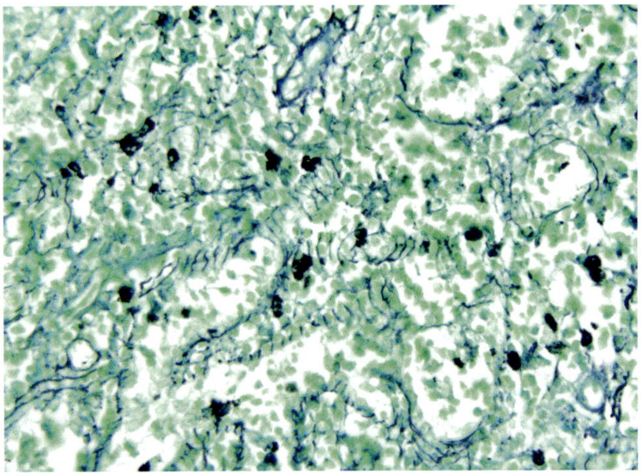

Fig. 1.11 Splenic sinus, Grocott's methenamine silver (GMS) stain. In this GMS stain, the "ring fibers" that wrap the endothelial sinus lining cells are highlighted. These are akin to hoops around the staves of barrel

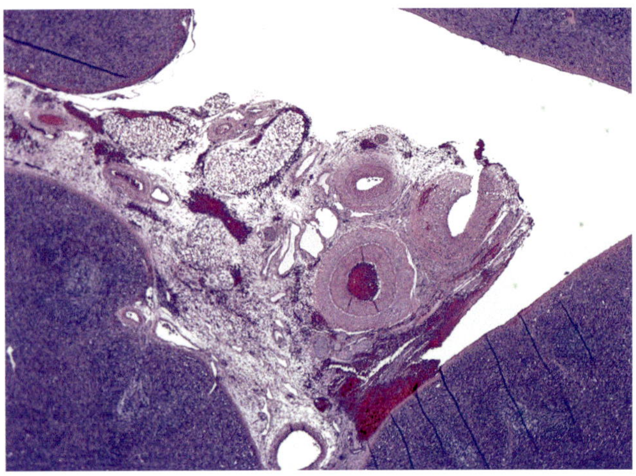

Fig. 1.12 Vessels, splenic hilum. This low magnification image shows vessels in the splenic hilum. Hilar vessels include large caliber arteries and veins, as well as lymphatics. The arteries merge together to form the splenic artery. Only very limited lymphatics are present in the spleen. They are seen only in the adventitial regions of the largest vessels penetrating the spleen. Normal splenic parenchyma does not have lymphatics

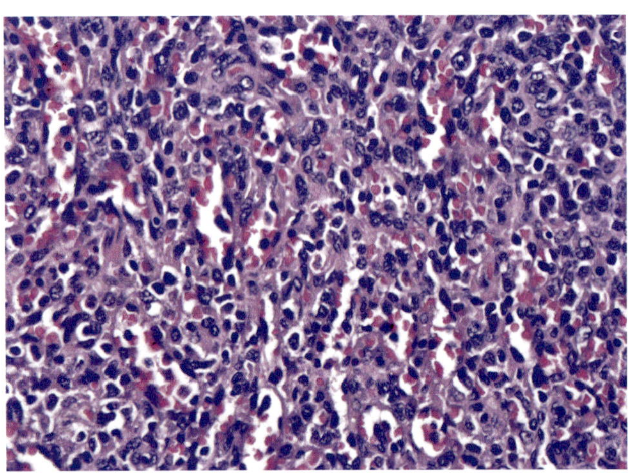

Fig. 1.13 Red pulp sinuses. In this example, the splenic sinuses are patent. Cells within the sinus would likely be seen in the peripheral blood circulation. Other more solid-appearing areas are cords

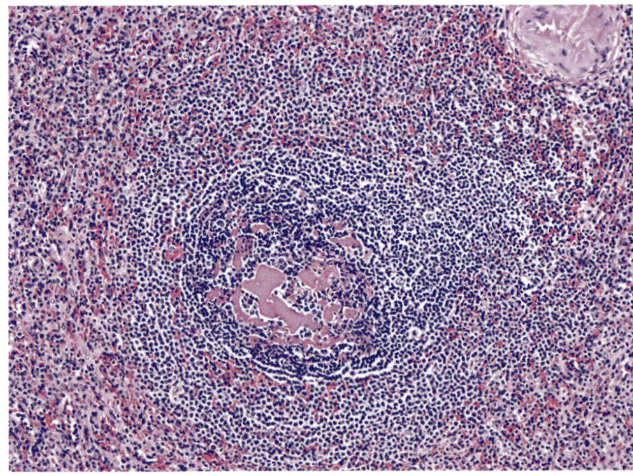

Fig. 1.14 Hyaline deposition, germinal center. In this example of a germinal center, there is hyalinized protein (glassy, pink material) present in the central portion of the germinal center. In some circumstances, this represents immunoglobulin protein deposition; however, in other cases, its exact nature is not known. It is a benign finding and not specific for any diagnosis

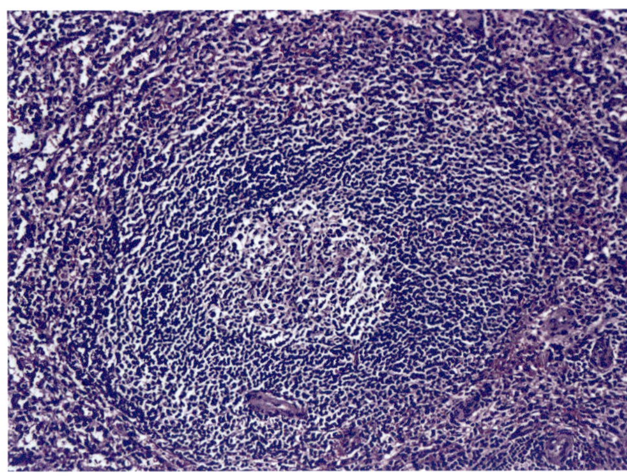

Fig. 1.15 White pulp nodule. This example shows the component of a typical white pulp nodule. The usual follicles have a three-layer structure. The central portion is the germinal center (*eg*, secondary follicle). The second layer is the mantle zone, composed of small, dark-blue, slightly irregular lymphocytes. Finally, the third layer is the marginal zone, composed of mostly small lymphocytes with increased amounts of pale cytoplasm

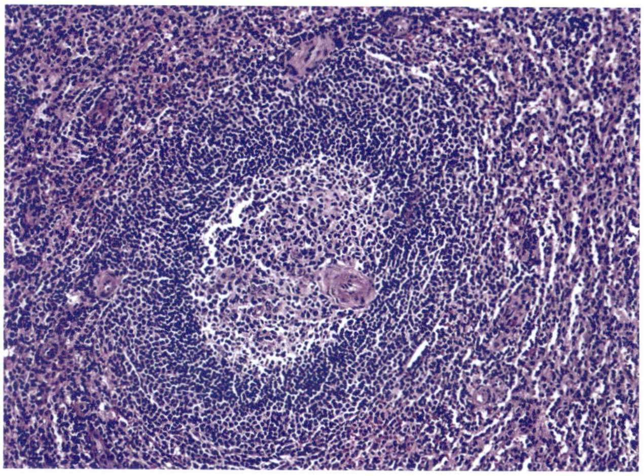

Fig. 1.16 White pulp nodule. As in the previous example, this white pulp nodule is composed of three layers: germinal center, mantle zone, and marginal zone. Note at the periphery is an arteriole in cross section, which continues into the spleen and is surrounded by the periarteriolar lymphoid sheath (PALS) region

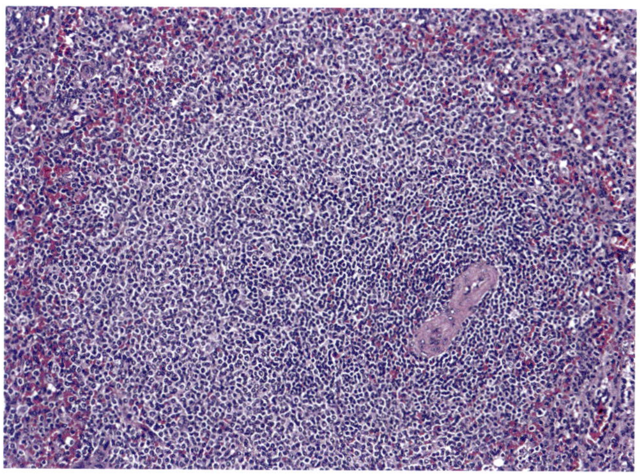

Fig. 1.17 White pulp nodule, primary follicle. In the unreacted spleen, germinal centers have not been formed. In these cases, the white pulp nodules have only two layers, the "mantle cell" core and the marginal zone. These are "primordial" follicles or primary follicles, which have not undergone a germinal center reaction

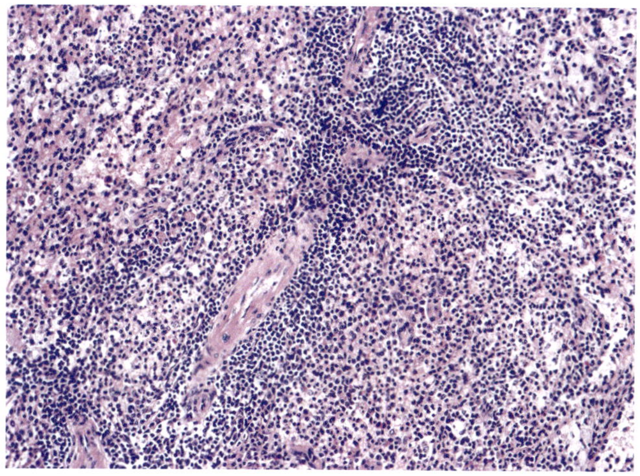

Fig. 1.18 Periarteriolar lymphoid sheath. A longitudinal section of a splenic arteriole with surrounding lymphoid sheath. These lymphoid cells are predominantly T cells, although there are also occasional admixed B cells

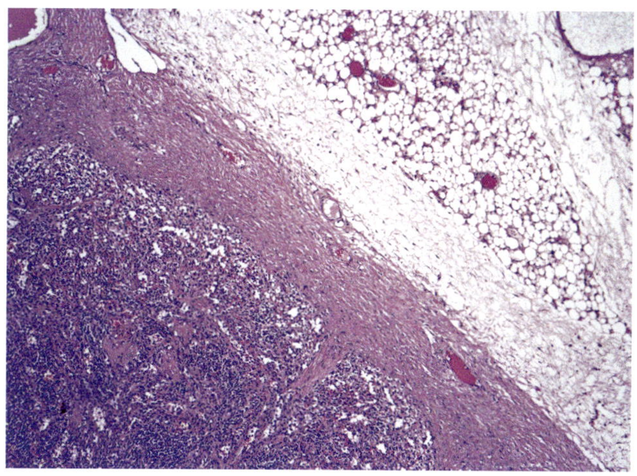

Fig. 1.19 Pediatric spleen. Although not instantly apparent, this is a spleen from a 1-year-old patient. The subtle morphologic clue lies in the perisplenic adipose tissue. There are lipoblasts present, which persist for a short time after birth (2–5 years)

1.2 Hyperplasias

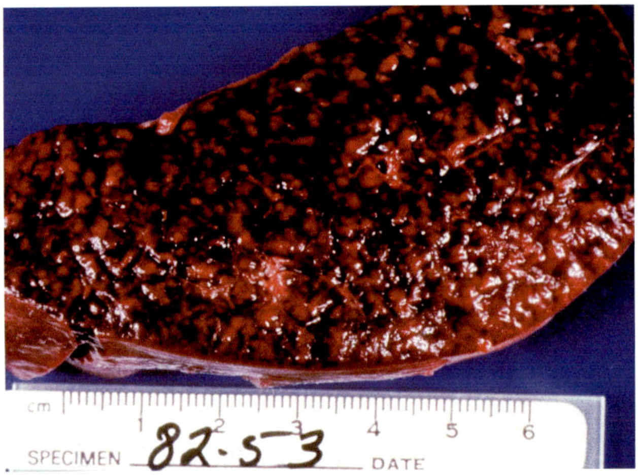

Fig. 1.20 Follicular hyperplasia. This gross image of an enlarged spleen shows an increase in accentuation of white pulp components corresponding to follicular hyperplasia. The accentuation of white pulp nodules is the spleen is said to impart a miliary appearance, which refers to the small white nodules appearing like millet seeds (small grains). Compared to a normal spleen, these white pulp nodules are increased in size and density. The findings are not specific, and could look similar to a lymphoma with involvement of the white pulp. (*Image courtesy of* D. Farhi, Atlanta, GA, USA)

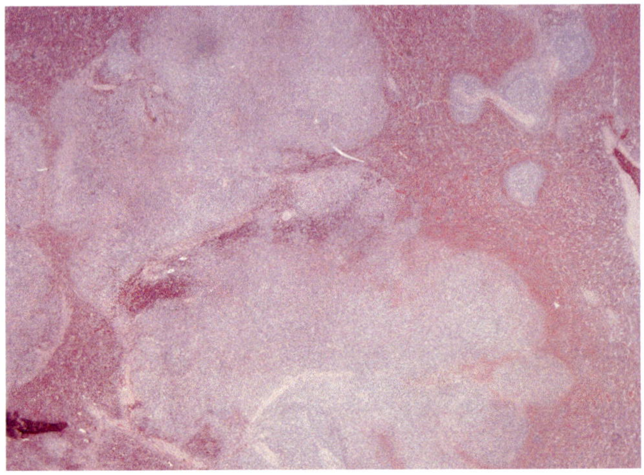

Fig. 1.22 Nodular lymphoid hyperplasia. Nodular (follicular) lymphoid hyperplasia is a rare finding in spleens. It may be grossly visible as an enlarged white pulp nodule, mimicking focal involvement by a lymphoid process or perhaps a granulomatous or inflammatory process. Histologic examination reveals a coalescence of benign reactive germinal centers. These aggregate into nodules, thereby mimicking neoplastic processes. However, the morphologic and immunohistochemical features of the individual follicles and cells are compatible with a reactive process

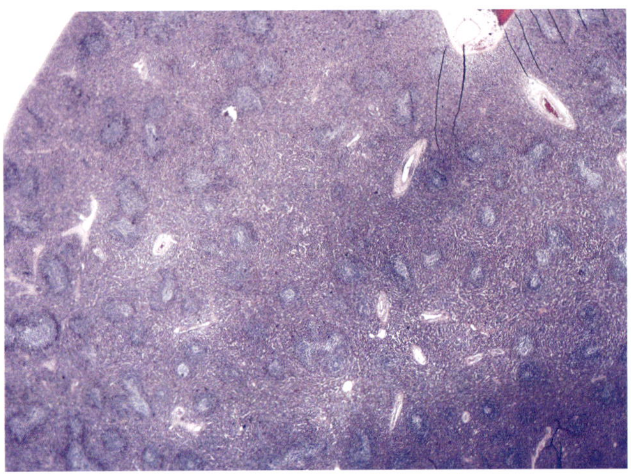

Fig. 1.21 Follicular hyperplasia. In this low magnification image, there is an increase in the proportion of white pulp compared to red pulp, which is seen in follicular hyperplasia. The causes of follicular hyperplasia are broad, and in general are related to immune activation

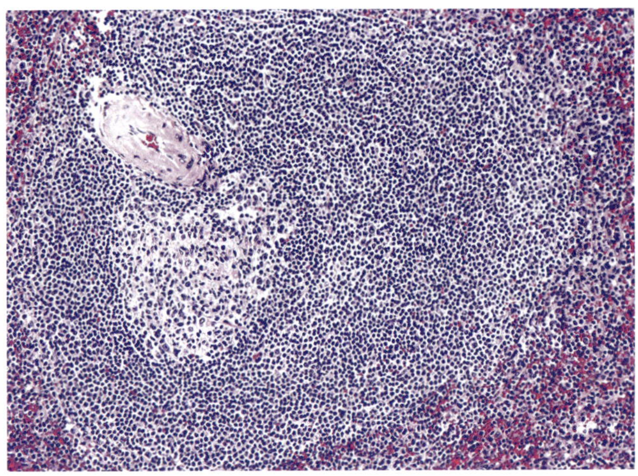

Fig. 1.23 Marginal zone hyperplasia. Hyperplasia of the splenic marginal zone has been arbitrarily defined as a thickening of the marginal zone (third) layer of more than 10 cells. This case is at the borderline of marginal zone hyperplasia, with possibly 10 or fewer cells in the third layer. The remaining features are of a typical reactive follicle

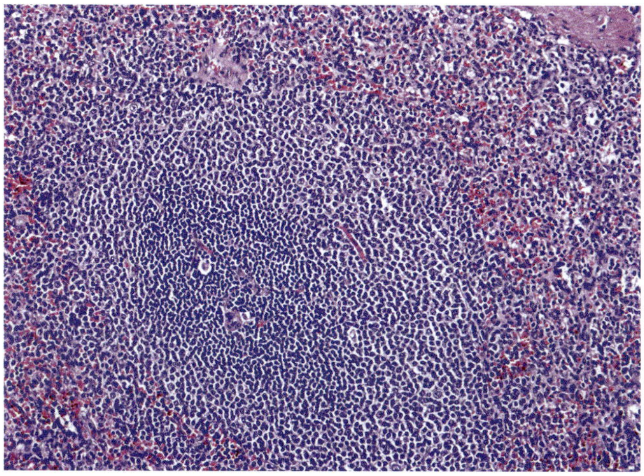

Fig. 1.24 Marginal zone hyperplasia. In this case, there is clear expansion of the outer (pale) layer of the follicular structure. There are more than 10 cells in this layer, which constitutes marginal zone hyperplasia. The cytologic composition is predominantly small and intermediate sized lymphocytes, with round nuclei and increased amounts of pale cytoplasm, imparting their marginal zone appearance. They are distinguished from the "mantle type" cells in the center that are smaller, have dark blue nuclei, and almost no cytoplasm

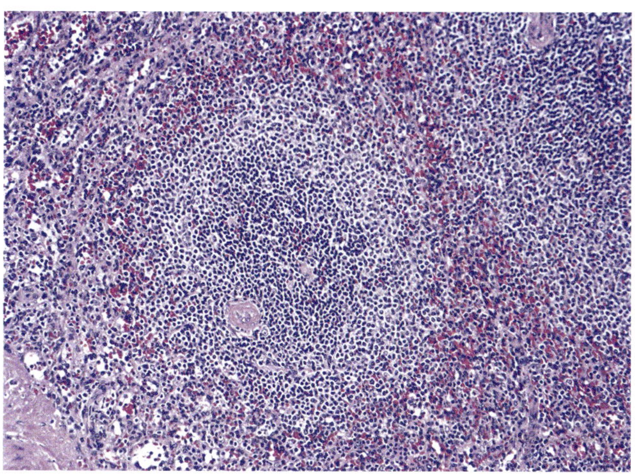

Fig. 1.25 Primary follicle hyperplasia. In this example, there is a primary follicle present. Hyperplasia of primary follicle is a rare finding. It is characterized by an overall increase in white pulp nodules composed of B cells without germinal center formation. The circumstances of primary follicle hyperplasia are most often seen in early immunologic responses, in pediatric spleens and in association with suppression of the normal immune response, such as with steroid therapy

1.3 Normal Immunohistochemistry

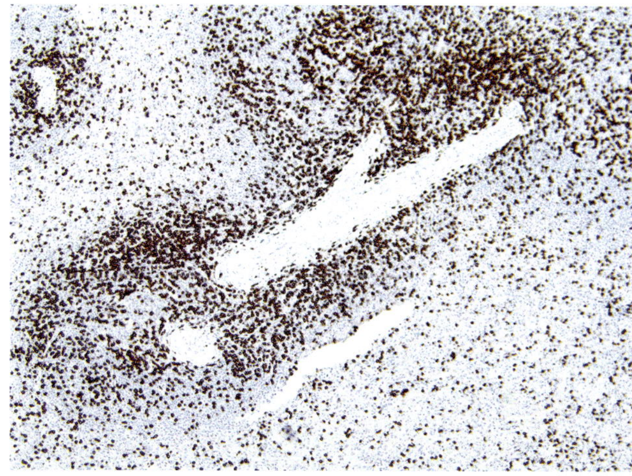

Fig. 1.26 CD3, normal spleen. The distribution of T cells in the normal spleen includes cells within the germinal center, some cells scattered in the mantle and marginal zones, many in the periarteriolar lymphoid sheath, and scattered throughout the white pulp

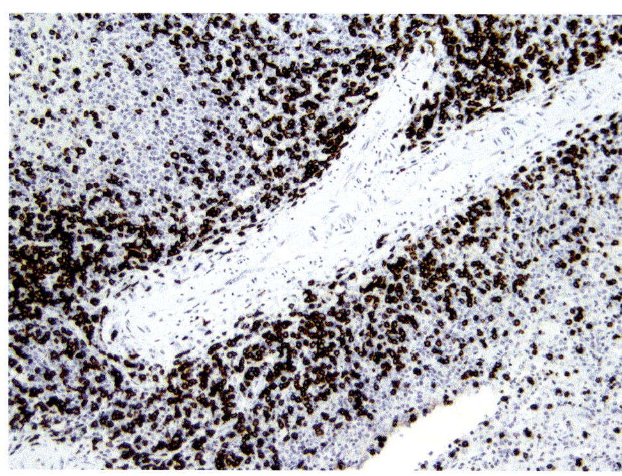

Fig. 1.27 CD3, normal spleen, periarteriolar lymphoid sheath. This CD3 stain highlights CD3-positive T-cells in the periarteriolar lymphoid sheath (PALS) region. In the splenic white pulp, this is a normal distribution of T cells adjacent to arterioles

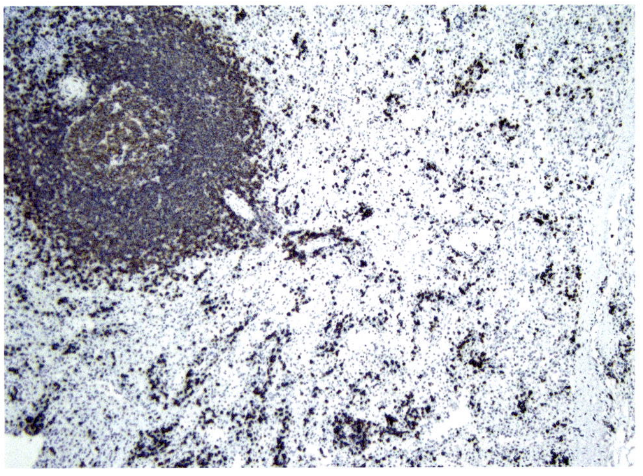

Fig. 1.28 CD20 staining, normal spleen. CD20 stains mature B cells. In this spleen, staining intensity of germinal center compared to the mantle/marginal zone can be appreciated. In addition, there are some scattered clusters and individual B cells within the red pulp

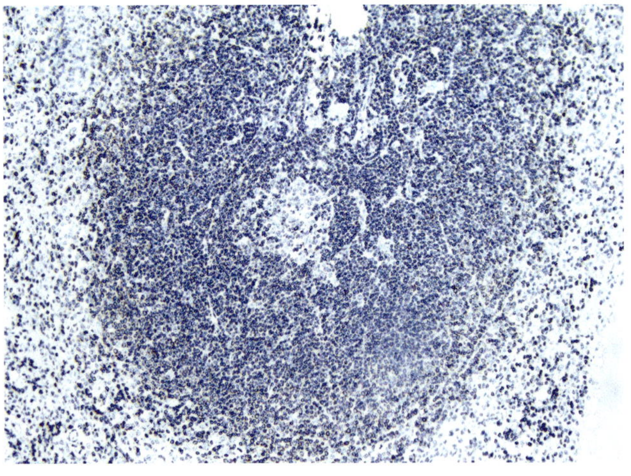

Fig. 1.29 BCL2 staining, normal spleen. Immunohistochemical staining for bcl-2 protein in normal spleen. Note that in white pulp structures, the reactive germinal center is negative, but the normal mantle zones and marginal zones are positive for bcl-2. Scattered T cells in the red and white pulp are also positive

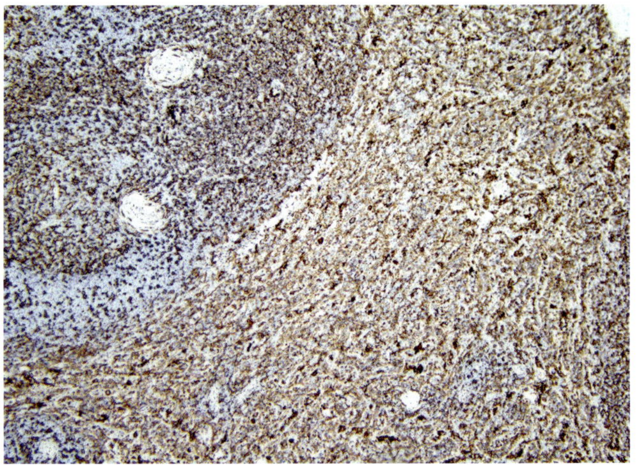

Fig. 1.30 CD4 staining, normal spleen. A CD4 immunohistochemical in spleen highlights helper T cells (within white and red pulp) as well as many macrophages and monocytic cells within the red pulp

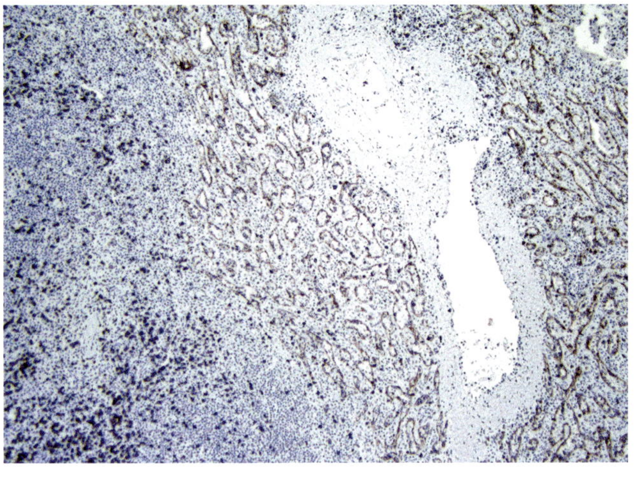

Fig. 1.31 CD8 staining, normal spleen. Aside from cytotoxic T cells, CD8 immunohistochemical stains will highlight splenic littoral (*eg*, sinus lining) cells. A CD8 stain is crucial in delineating the red pulp architecture in the spleen and highlights red pulp pathology when present

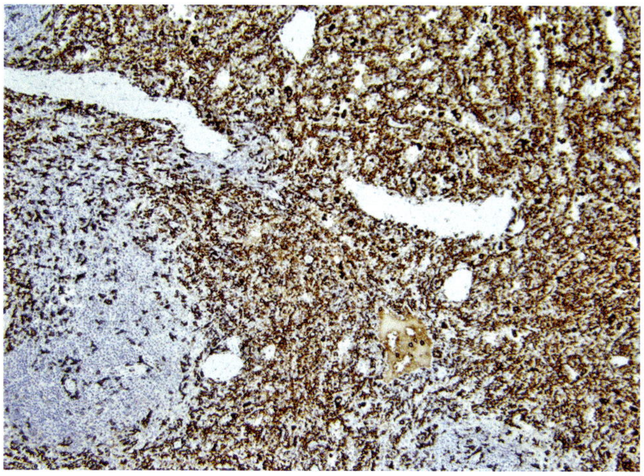

Fig. 1.32 CD163 staining, normal spleen. CD163 highlights macrophages. In the spleen, the cordal regions are strongly positive, and there is some staining in littoral (sinus) cells. There are also some scattered macrophages within white pulp, especially within activated germinal centers

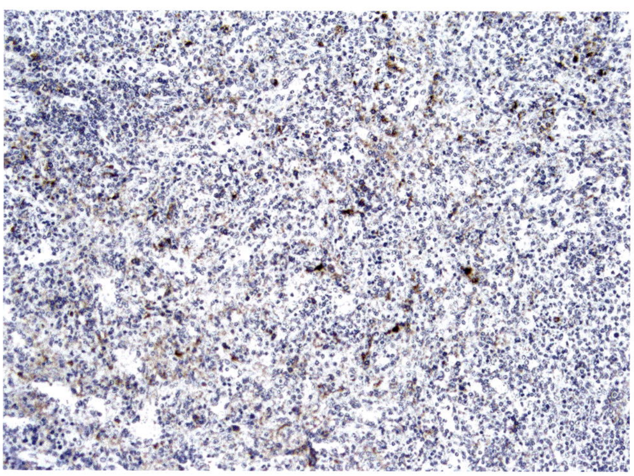

Fig. 1.33 CD42b staining, normal spleen. CD42b stains both megakaryocytes and platelets. Other than rare platelets within the sinuses, no significant staining is seen in the normal spleen. However, rare individual megakaryocytes can be seen in some cases. If there are increases (more than 1/10 HPF), extramedullary hematopoiesis or a neoplastic myeloid proliferation should be considered

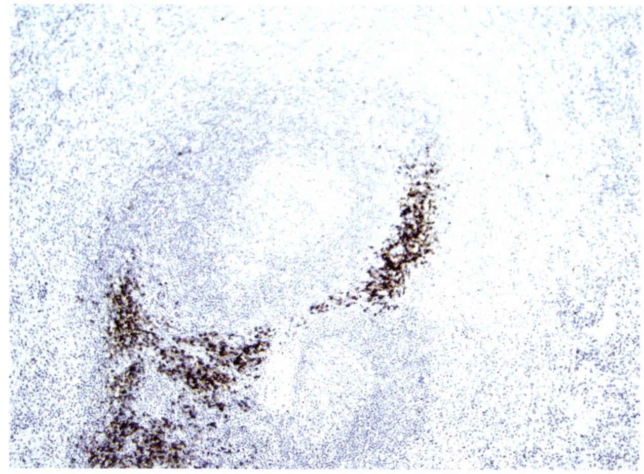

Fig. 1.34 CD123 staining, normal spleen. CD123 is most commonly positive in plasmacytoid dendritic cells. In this case, these are present at the periphery of the marginal zone and PALS regions of white pulp. Some rare B-cell lymphomas express CD123 but in my experience the level of expression can only be detected by flow cytometry and is not positive by immunohistochemistry

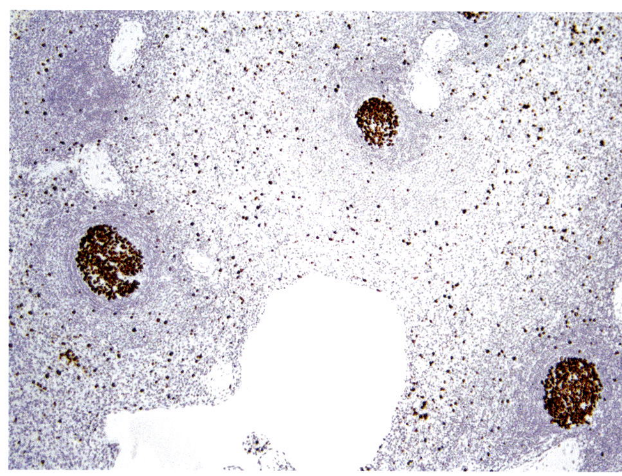

Fig. 1.35 KI-67 staining, normal spleen. Ki-67 is a marker of proliferation. In this image, normal germinal centers (secondary follicles) are highly proliferative. Only rare proliferating cells are identified within the red pulp

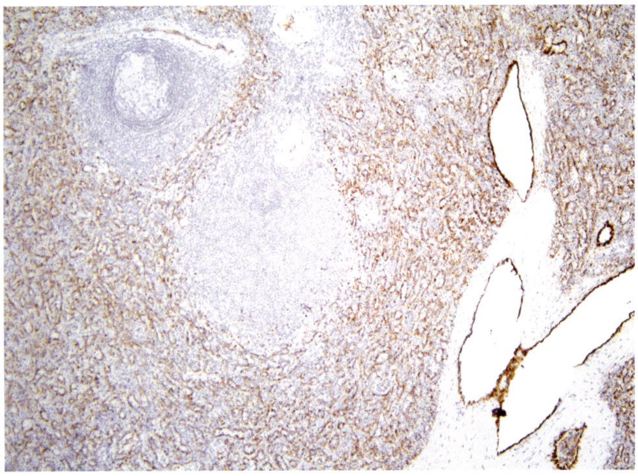

Fig. 1.36 Factor VIII antigen, normal spleen. Factor VIII antigen will stain platelets, megakaryocytes and endothelial elements. In normal spleen, littoral cells (sinus-lining cells) are also positive

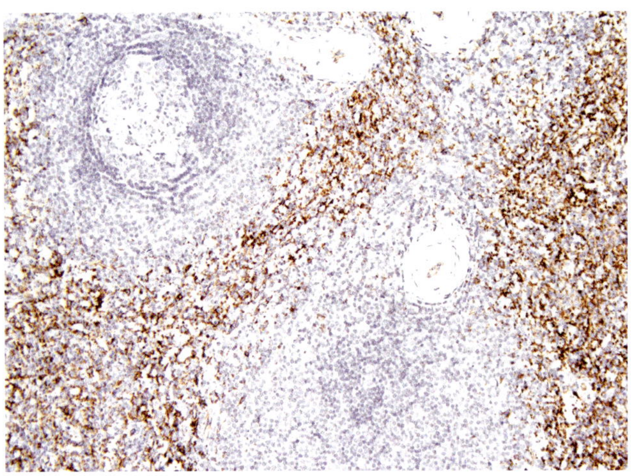

Fig. 1.37 CD31 staining, normal spleen. CD31 stains both endothelial and histiocytic elements. In spleen, there is staining in cordal histiocytes, vascular endothelium, and some staining in littoral cells

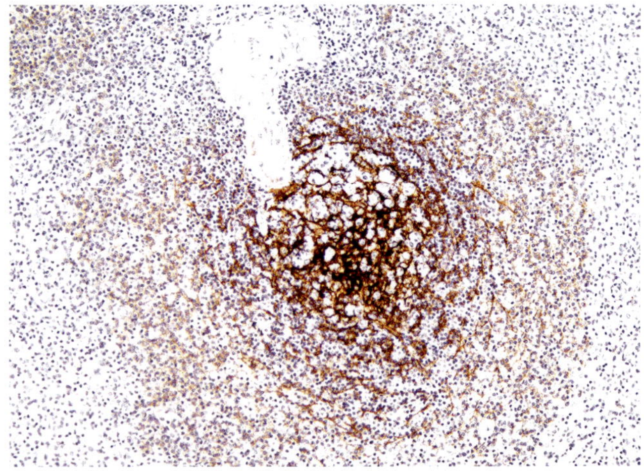

Fig. 1.38 CD21 Staining, normal spleen. CD21 staining in spleen is seen in follicular dendritic cells (FDC) as well as some lymphocytes. Strong reticular staining is seen in the FDC networks of the follicle, as well as some weak and variable staining in lymphocytes of the marginal zone

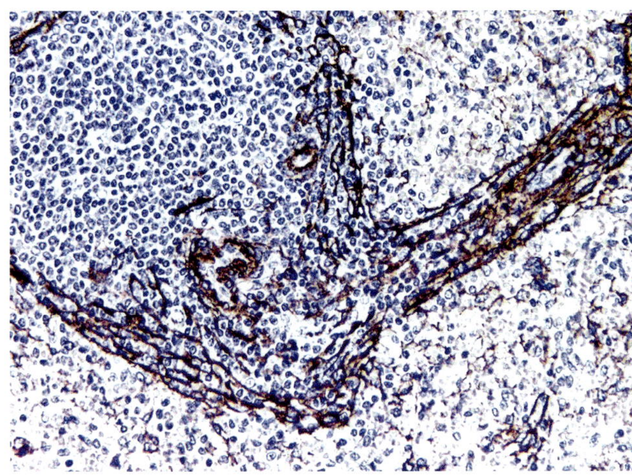

Fig. 1.39 Smooth muscle actin staining, normal spleen. Smooth muscle actin stains myofibroblastic cells in the spleen, including cells that line the margin between splenic marginal zone (white pulp) and red pulp. In some animals, this margin is accentuated, and is a well-defined structure, but it is subtle in humans

Congenital Malformations and Abnormalities of Spleen Location and Histology

2

Congenital abnormalities of the spleen are rarely encountered. However, when present, they may be difficult to interpret due to unfamiliarity. Occasionally, they can mimic serious pathology and should be familiar to pathologists to exclude malignancies. Lesions illustrated in this chapter include accessory spleen, which when present in unusual locations may be a diagnostic problem; intrapancreatic spleen, which is a rare occurrence and probably a variant of accessory spleen; splenosis, which is marked by the presence of multiple, small, fragments of functional splenic tissue within the abdominal cavity; and splenogonadal fusion, a rare lesion that may involve unusual clinical presentations, including left-sided inguinal hernia. Finally, surface grooves present in the spleen are illustrated and their embryologic origin is identified. Both histologic and gross images are presented.

D.P. O'Malley, *Atlas of Spleen Pathology*, Atlas of Anatomic Pathology,
DOI 10.1007/978-1-4614-4672-9_2, © Springer Science+Business Media New York 2013

2.1 Accessory Spleen

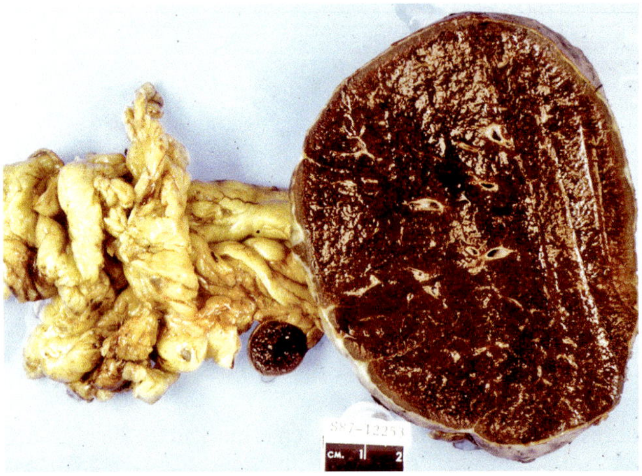

Fig. 2.1 Accessory spleen. A section of spleen with attached pancreas and omentum. In the mass of tissue arising from the hilum, there is a small nodule (approximately 2 cm) of dark red purple tissue. This is an accessory spleen. They are seen in approximately 10 % of the population and are most commonly seen in the splenic hilar region. (*Image courtesy of* D. Farhi, Atlanta, GA, USA.)

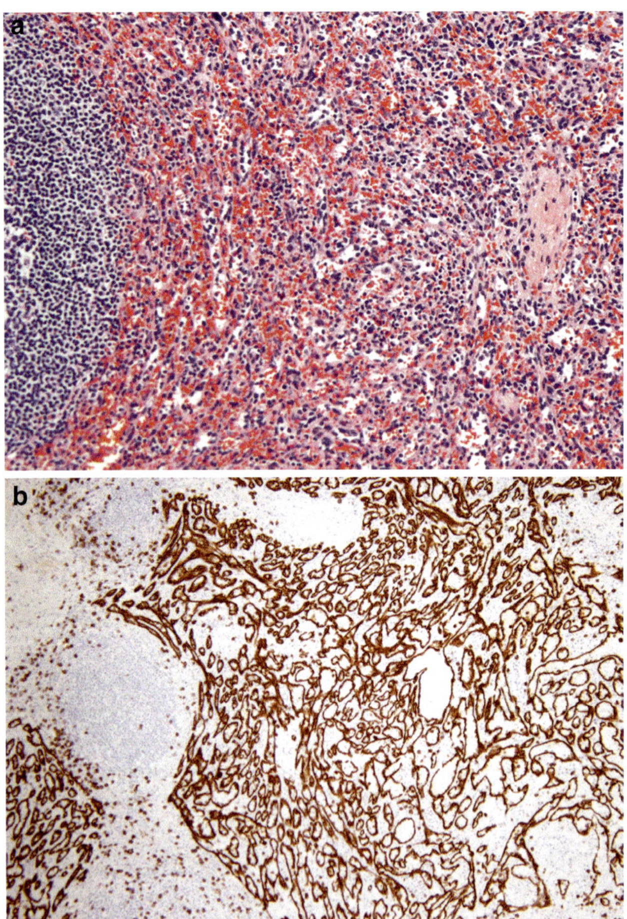

Fig. 2.2 Accessory spleen. This pathologic specimen was submitted as an "enlarged lymph node" with a suspicion of metastatic carcinoma. It was seen adjacent to the pancreas and, as illustrated by histology (**a**) and immunohistochemical stains for CD8 (**b**), is an accessory spleen. CD8 staining highlights the distinctive splenic sinus architecture of the spleen and can be useful in identifying spleen tissue at unusual sites

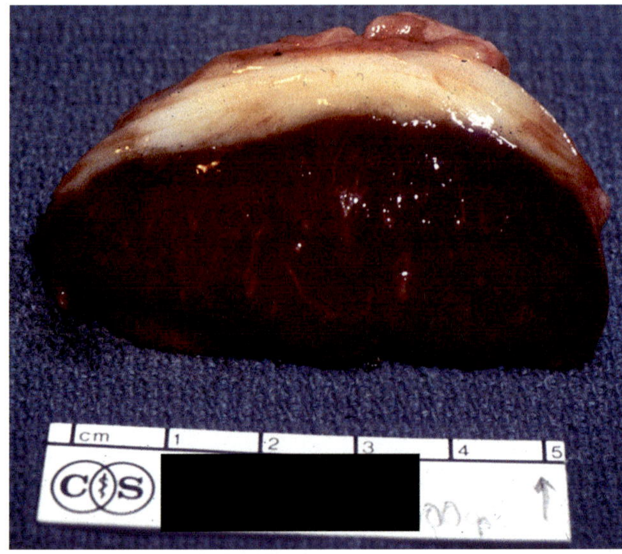

Fig. 2.3 Gross image of a sectioned accessory spleen. In this case, the accessory spleen was approximately 5 cm in diameter. Note that the cut surface has is a deep red color, similar to the spleen. (*Image courtesy of* L. Morgenstern, Los Angeles, CA, USA.)

2.2 Intrapancreatic Spleen

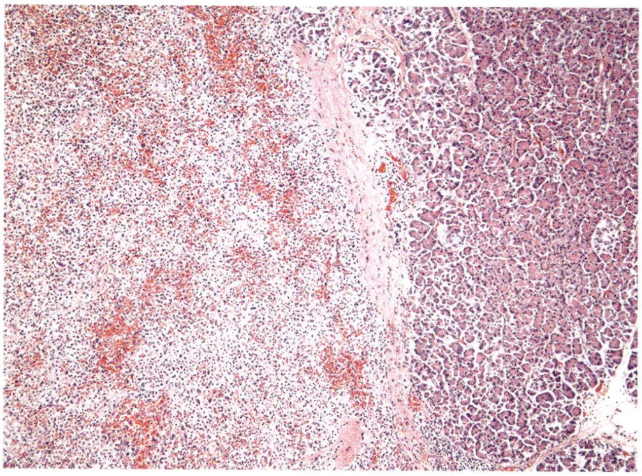

Fig. 2.4 Intrapancreatic spleen is a rare finding associated with embryologic abnormality. Small fragments of spleen tissue are entrapped within a portion, usually distal, of the pancreas (normal pancreatic tissue; *right*). Although these can vary in size, they are usually quite small. In this case, the elements are predominant red pulp components, suggesting that this was a relatively early embryologic event. (*Image courtesy of* R. Mills, Salt Lake City, UT, USA.)

2.3 Polysplenia

Fig. 2.5 Polysplenia. Gross image of polysplenia. In this case, the primordial separate lobes of splenic tissue do not fuse to make a single unified organ. Rather, they are partly fused or remain separated by fibrous bands. Although this anatomic abnormality can be seen associated with normal function, it is often associated with other, severe congenital abnormalities. (*Image courtesy of* D. Farhi, Atlanta, GA, USA.)

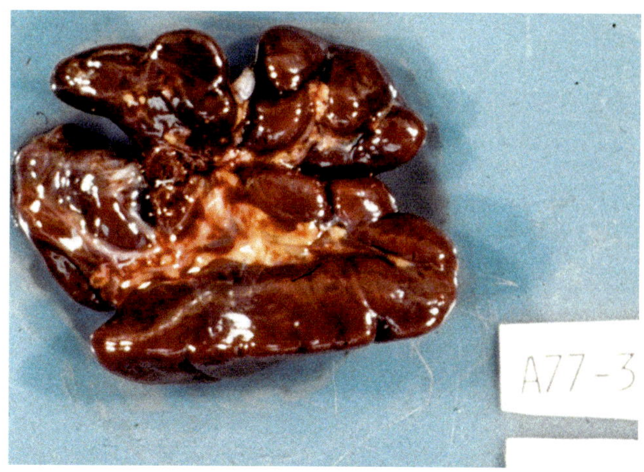

Fig. 2.6 Polysplenia. Another example of a small gross specimen from a fetus with polysplenia. Separated nodules of spleen tissue are present. Each nodule would have its own artery and vein, which eventually merge into the common splenic artery or vein. (*Image courtesy of* D. Farhi, Atlanta, GA, USA.)

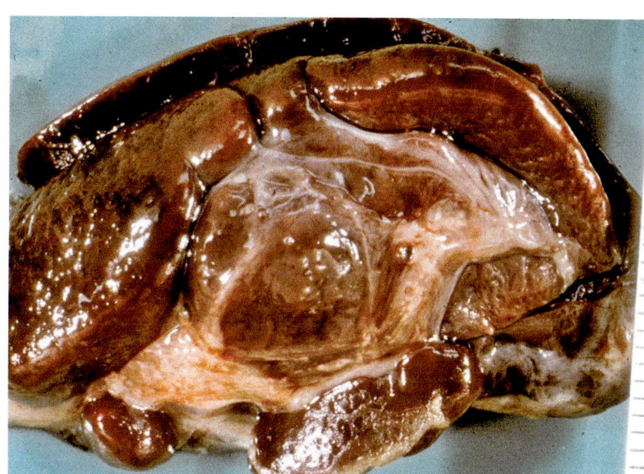

Fig. 2.7 Polysplenia. Still another example of polysplenia. Note that the lobes of the spleen have not completely fused. (*Image courtesy of* D. Farhi, Atlanta, GA, USA.)

2.4 Splenosis

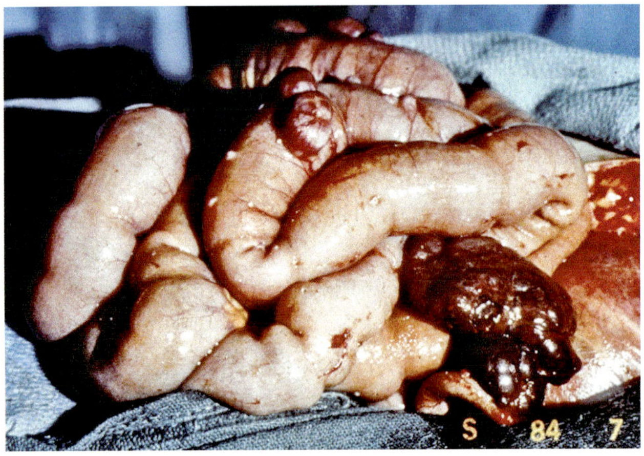

Fig. 2.8 Splenosis. Gross image of splenosis. In this case, there are nodular, dark red tissue fragments attached to segments of bowel. This most often occurs after trauma, when the spleen is ruptured. Small fragments of splenic tissue are seeded throughout the abdominal cavity. Each establishes its own blood supply and functions as a tiny but complete spleen. In cases of splenectomy and pathologic processes, such as hemolytic anemias, these small displaced splenic fragments may enlarge greatly. (*Image courtesy of* D. Farhi, Atlanta, GA, USA.)

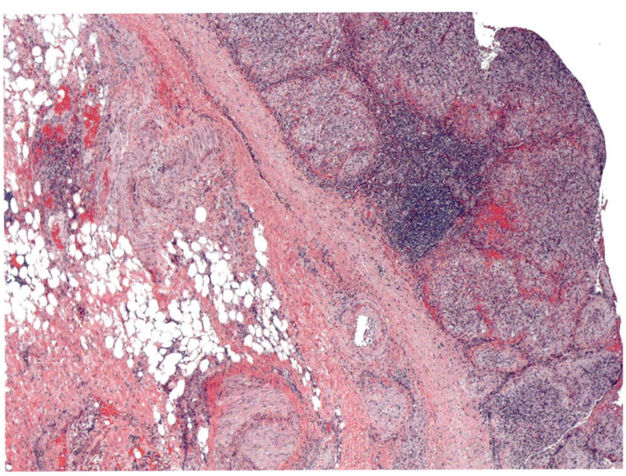

Fig. 2.9 Splenosis. Microscopic image of splenosis. The tissue present is composed of fibroadipose tissue from the peritoneum. At the intraperitoneal surface (*right* and *upper*) are slightly fibrous lymphoid areas. These represent fragments of spleen tissue that were likely "transplanted" to this site following abdominal trauma

2.5 Splenogonadal Fusion

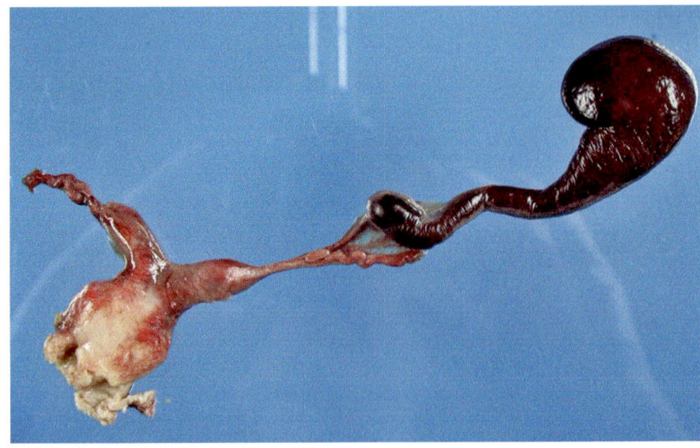

Fig. 2.10 Spleno-ovarian fusion. Gross image of spleno-ovarian (*eg*, splenogonadal) fusion. In this case, there is a physical attachment between the stretched and elongated splenic fragment, which has tethered a portion of the left ovary. In some cases, the fusion may only consist of a thin fibrous band, which may have small entrapped fragments of splenic parenchyma

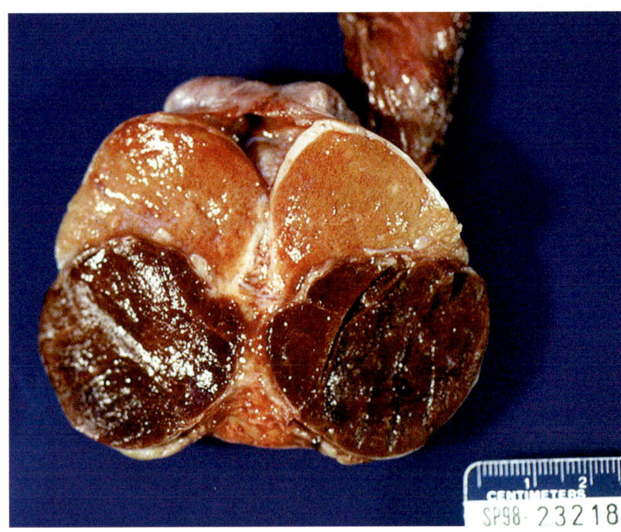

Fig. 2.11 Spleno-gonadal fusion. Ectopic splenic tissue in a sectioned testicle (*lower*). The splenic tissue has the usual deep red color, with normal, somewhat compressed testicular tissue in the upper portion of the sample

2.6 Other Findings

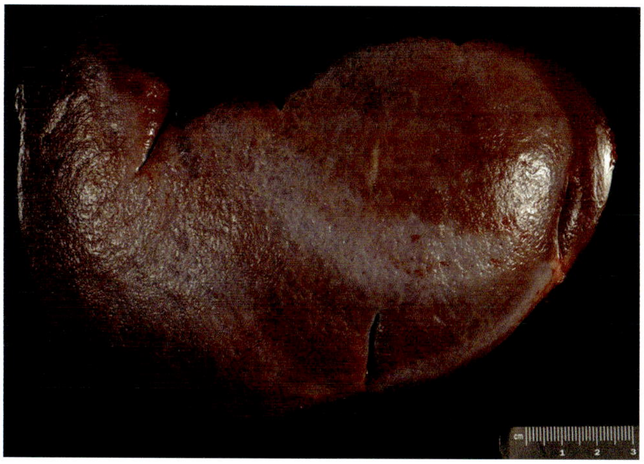

Fig. 2.12 Surface grooves. This spleen is enlarged by a low-grade B-cell lymphoma. However, deep grooves are seen in the medial surface (*upper left*), the upper pole (*far right*), and in the lateral/dorsal surface (*lower, middle*). These grooves are a normal variant, and are residual separations from the original embryologic splenic lobes that are incompletely fused in the adult. They have no specific physiologic consequences. (*Image courtesy of* W. Greaves, Houston, TX, USA.)

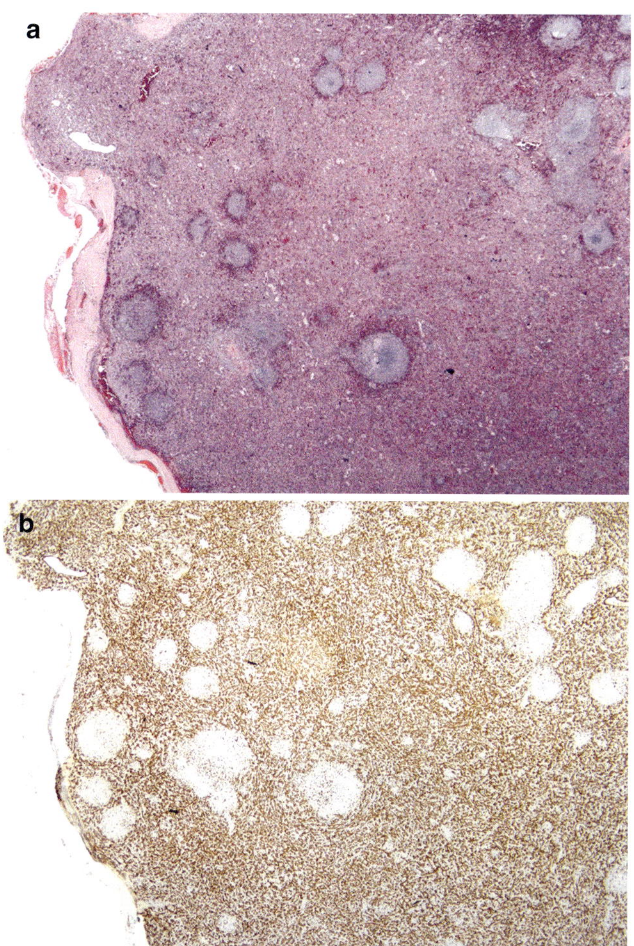

Fig. 2.13 Tissue taken from "left inguinal hernia." In this case, the histology (**a**) and presence of a red pulp architecture staining by CD8 (**b**) show that the "inguinal hernia" was in fact caused by ectopic spleen tissue. In the descent of the testes from the mesonephros, small portions of spleen tissue may be transported. When these fragments are attached, it is splenogonadal fusion; when unattached, these fragments may be present as inguinal lumps, within the inguinal canal, simulating a hernia

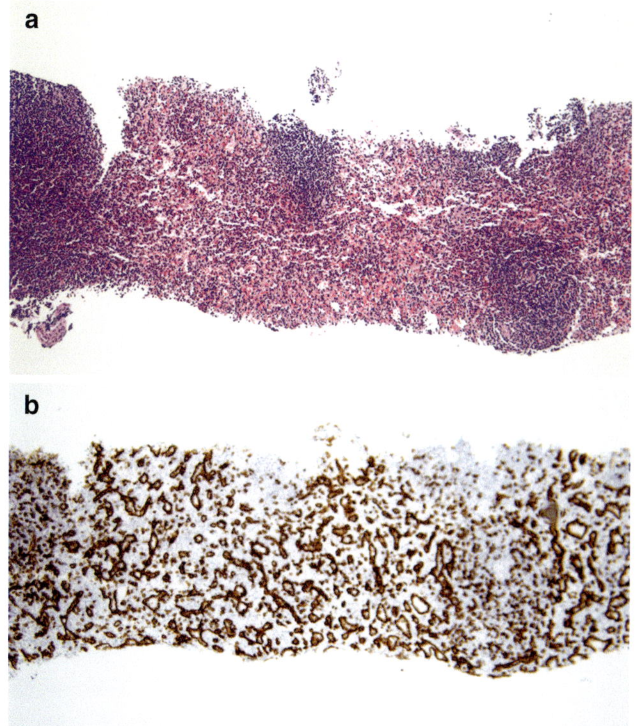

Fig. 2.14 Needle biopsy, lung. A biopsy of the left lower lobe of lung revealed this tissue. In this case, the lung was missed and a sample of normal spleen was taken (**a**). Normal splenic architecture is highlighted by CD8 staining (**b**), which is a useful stain for highlighting splenic red pulp architecture

3 Lymphoid Neoplasms

Perhaps one of the most important topics in splenic pathology is lymphoid neoplasms. A wide variety of lymphomas can be seen in the spleen. These can be divided in several ways including cell derivation (T versus B), pattern of involvement (red pulp, white pulp, mixed, masses), or as predominantly diagnosed in spleen versus manifestations of systemic lymphomas. Lymphomas that predominantly involve the white pulp illustrated are splenic marginal zone lymphoma, chronic lymphocytic leukemia/small lymphocytic lymphoma, lymphoplasmacytic lymphoma, follicular lymphoma, mantle cell lymphoma, and plasma cell disorders. Lymphomas presenting in red pulp that are illustrated include hairy cell leukemia, hairy cell leukemia-variant, splenic diffuse red pulp B-cell lymphoma, B-cell prolymphocytic leukemia/lymphoma, acute lymphoblastic leukemia/lymphoma, hepatosplenic T-cell lymphoma, T-cell large granular lymphocytic leukemia, and other T cell lymphomas. In addition, other lymphomas illustrated include diffuse large B-cell lymphoma and variants, classical Hodgkin lymphoma, and nodular lymphocyte-predominant Hodgkin lymphoma. When appropriate, immunohistochemical staining, gross images and other sites of presentation (peripheral blood, bone marrow) are illustrated.

D.P. O'Malley, *Atlas of Spleen Pathology*, Atlas of Anatomic Pathology,
DOI 10.1007/978-1-4614-4672-9_3,

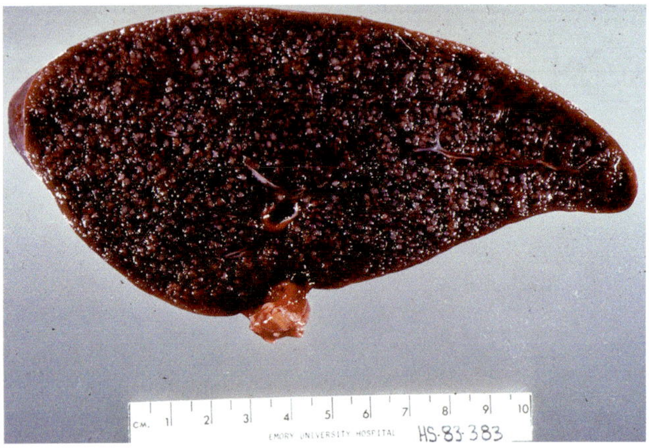

Fig. 3.1 B-cell lymphoma, white pulp. Gross image of sectioned spleen with B-cell lymphoma involving the white pulp. There is a marked accentuation of the normal white pulp architecture, leading to numerous small pale nodules in the dark red background. (*Image courtesy of* D. Farhi, Atlanta, GA)

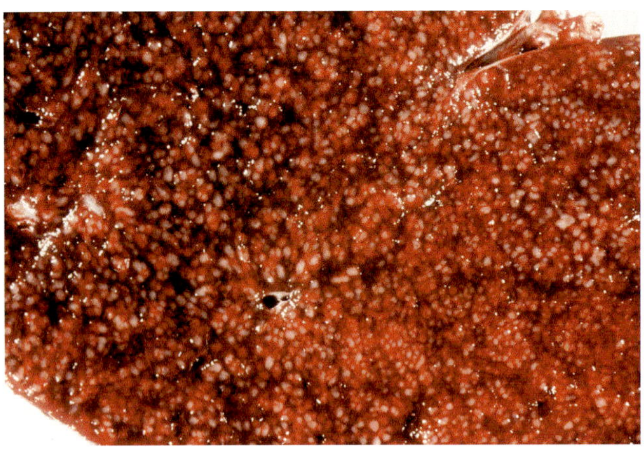

Fig. 3.2 B-cell lymphoma, white pulp. Another gross image of spleen with B-cell lymphoma involving white pulp. Note to overall increase in the number of white pulp nodules. This appearance is called “miliary,” referring to millet seeds

3.1 Splenic Marginal Zone Lymphoma

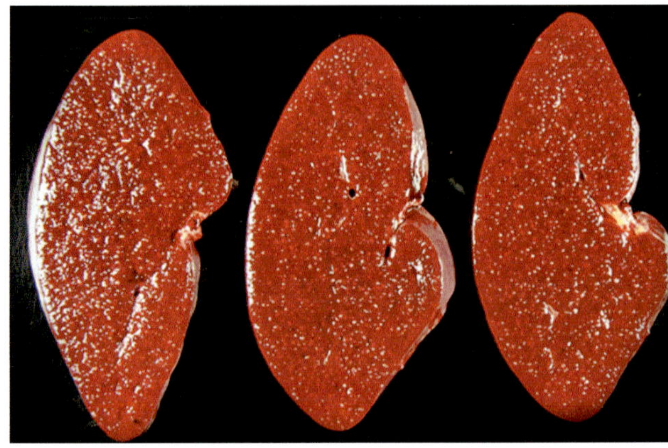

Fig. 3.3 Splenic marginal zone lymphoma. Gross examination of cut spleen with massive involvement by splenic marginal zone lymphoma. Note the accentuation of the splenic white pulp as small white, “miliary” nodules. (*Image courtesy of* A. Sohani, Boston, MA)

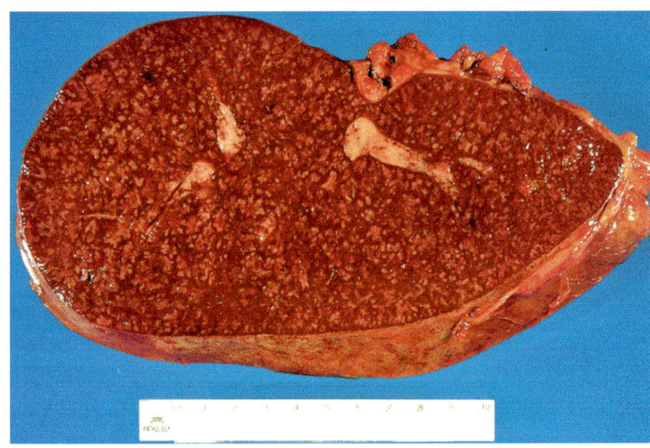

Fig. 3.4 Splenic marginal zone lymphoma. Gross image of 2800 g spleen with splenic marginal zone lymphoma. As in previous case, the white pulp is especially prominent, with a “miliary” pattern. (*Image courtesy of* M. Kamionek, Boston, MA)

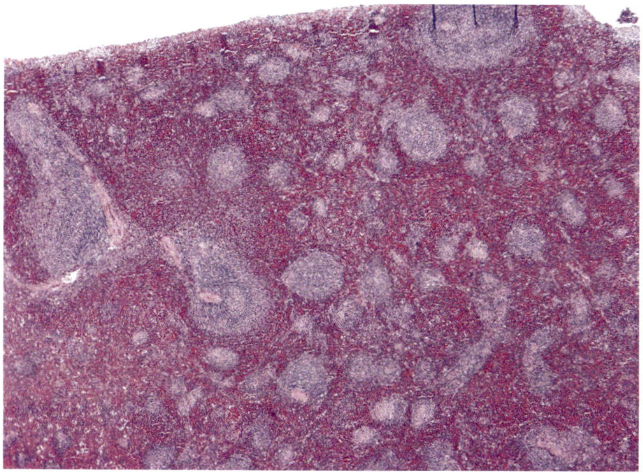

Fig. 3.5 Splenic marginal zone lymphoma. Low magnification of splenic marginal zone lymphoma. There is a marked increase in nodules of white pulp in the spleen. Each of these nodules appear as the miliary nodules seen grossly. Only a portion of these nodules are white pulp; others are nodules of lymphocytes within the red pulp

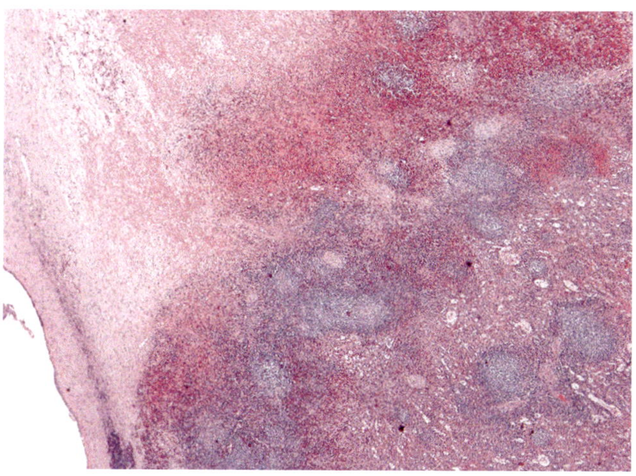

Fig. 3.6 Splenic marginal zone lymphoma. Low-magnification image of splenic marginal zone lymphoma (*lower*) with an area of infarction (*upper*). This spleen ruptured, likely due to overall enlargement, with thinning and friability of the capsule. Although not typical, rupture may be a presenting symptom of low-grade splenic B-cell lymphomas

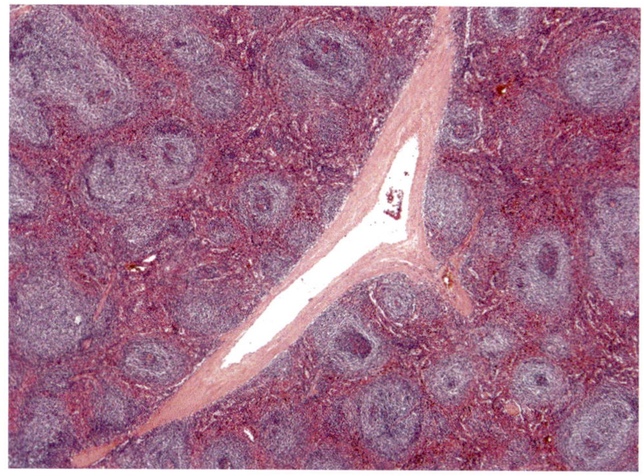

Fig. 3.7 Splenic marginal zone lymphoma. This example of splenic marginal zone lymphoma shows a marked increase in white pulp nodules. Abnormal lymphocytes will increase and track along vascular structures

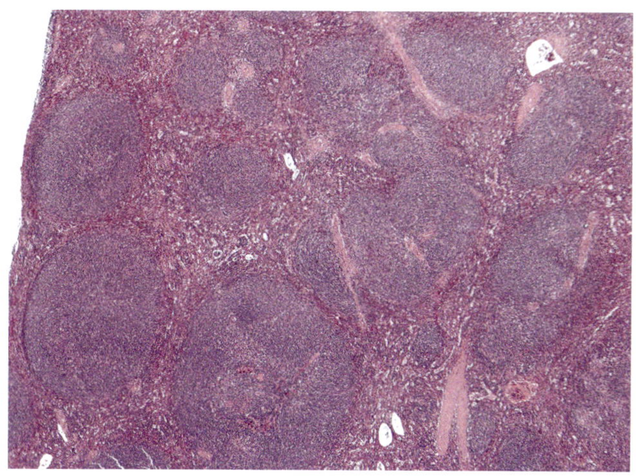

Fig. 3.8 Splenic marginal zone lymphoma. Intermediate magnification of splenic marginal zone lymphoma. In this case, the white pulp nodules are increased in size. Further, their cytologic composition is relatively uniform (a single layer), which can be seen in a variety of splenic white pulp B-cell lymphomas

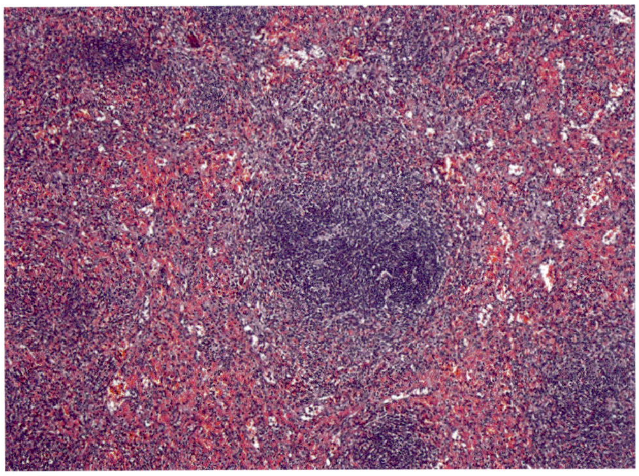

Fig. 3.9 Splenic marginal zone lymphoma. Intermediate magnification of splenic marginal zone lymphoma. Note that the white pulp nodules are disorganized in appearance. In addition, there is an overall increase in lymphocytes within the red pulp

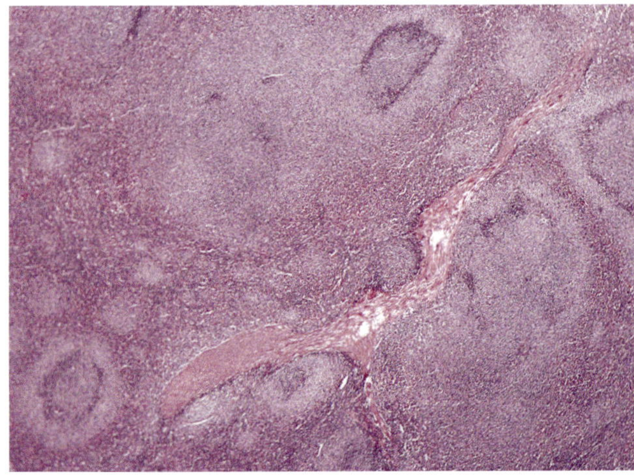

Fig. 3.11 Splenic marginal zone lymphoma. Splenic marginal zone lymphoma showing striking "monocytoid" differentiation. The lymphoma cells are pale compared to the residual mantle zones (*dark blue*) but there is colonization of the reactive follicles by neoplastic lymphocytes

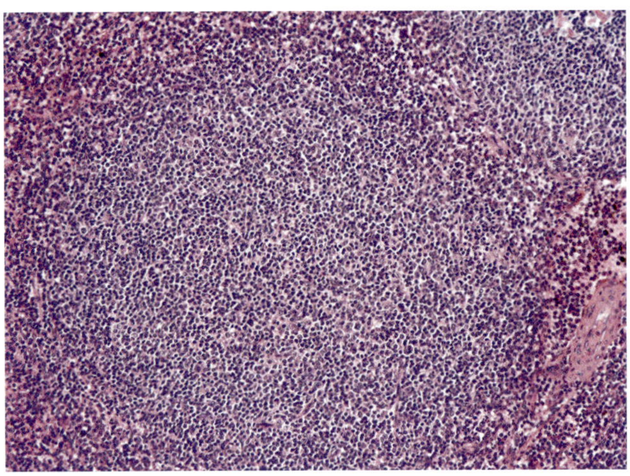

Fig. 3.10 Splenic marginal zone lymphoma. White pulp nodule of splenic marginal zone lymphoma. In this case, there are predominantly small lymphocytes present. However, in the most central portion of the nodule are some scattered larger cells, suggesting colonization of a reactive germinal center by neoplastic lymphoid cells

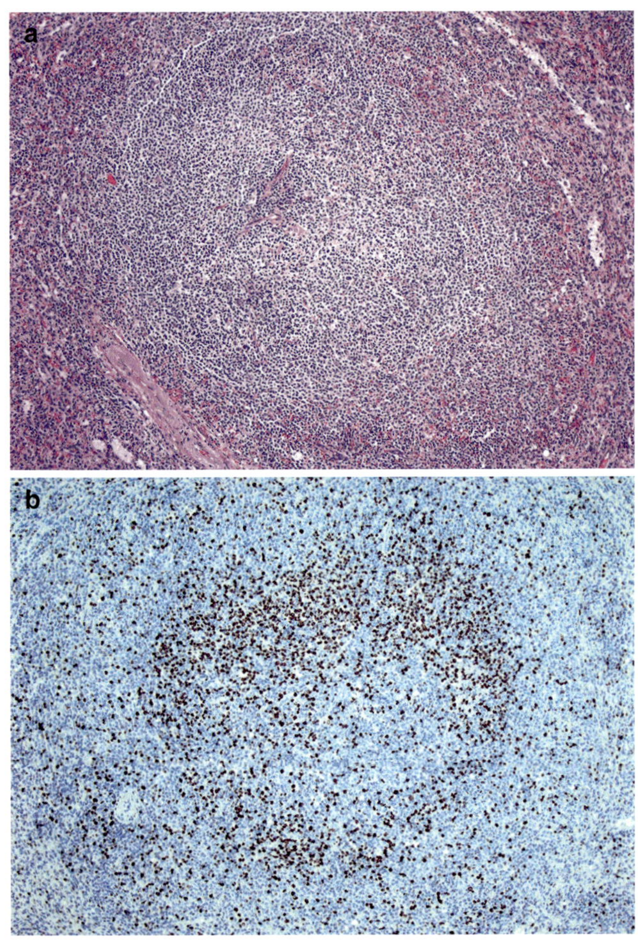

Fig. 3.12 Splenic marginal zone lymphoma. In this example of splenic marginal zone lymphoma, there is colonization of the follicle by lymphoma cells (**a**), which is highlighted by Ki-67 staining (**b**). The highly proliferative normal follicle center cells are displaced by the low proliferation lymphoma cells

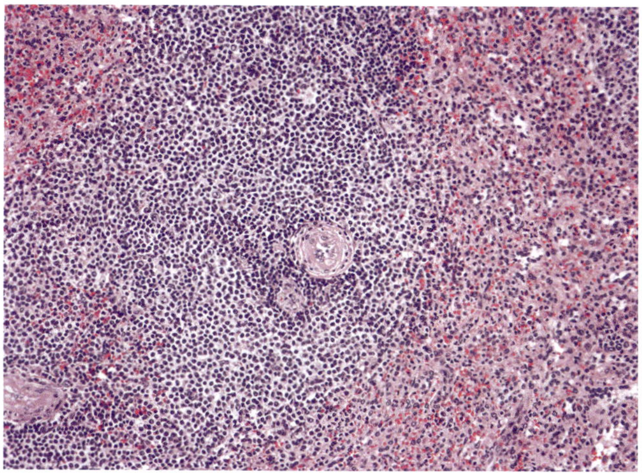

Fig. 3.13 Splenic marginal zone lymphoma. Intermediate magnification showing a small arteriole. Note the abnormal lymphoma cells (increased pale cytoplasm, slightly larger in size) surround the arteriole. Splenic marginal zone cells will often colonize the periarteriolar lymphoid sheath (PALS) region

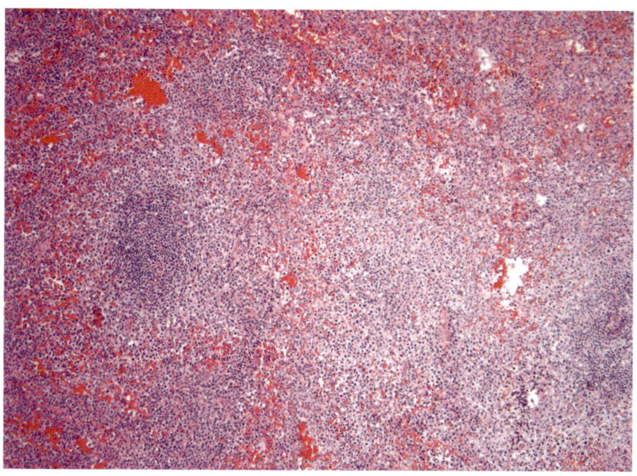

Fig. 3.14 Splenic marginal zone lymphoma with transformation to diffuse large B-cell lymphoma. In this example, there is evidence of low-grade B-cell lymphoma (splenic marginal zone lymphoma) in the center and left portions, while a lymphoma (*right middle*) appears with a diffuse pattern and large cells. This diffuse large B-cell lymphoma likely represents a transformation of the splenic marginal zone lymphoma

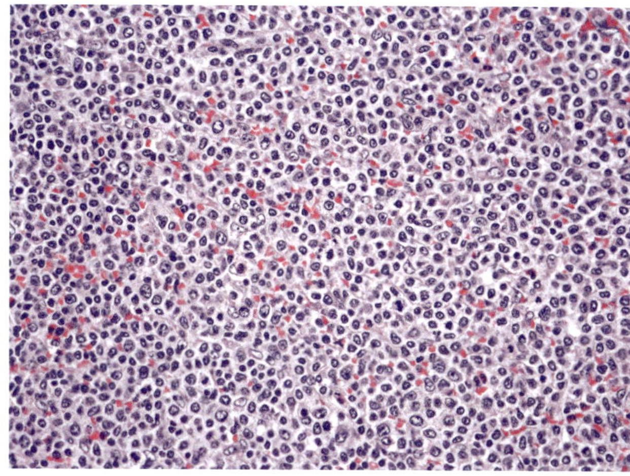

Fig. 3.15 Splenic marginal zone lymphoma. In this example of splenic marginal zone lymphoma, there is evidence of plasmacytoid differentiation. Although rare, plasma cell differentiation can be seen in splenic marginal zone lymphoma. This phenomenon is more common than the presence of lymphoplasmacytic lymphoma in the spleen

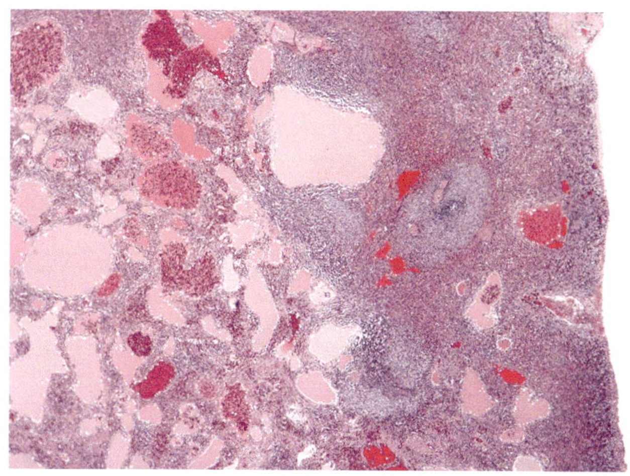

Fig. 3.16 Splenic marginal zone lymphoma with lymphangioma. The presence of more than one diagnosis is not uncommon in spleen resections. In this case, there is evidence of splenic marginal zone lymphoma in association with a splenic lymphangioma

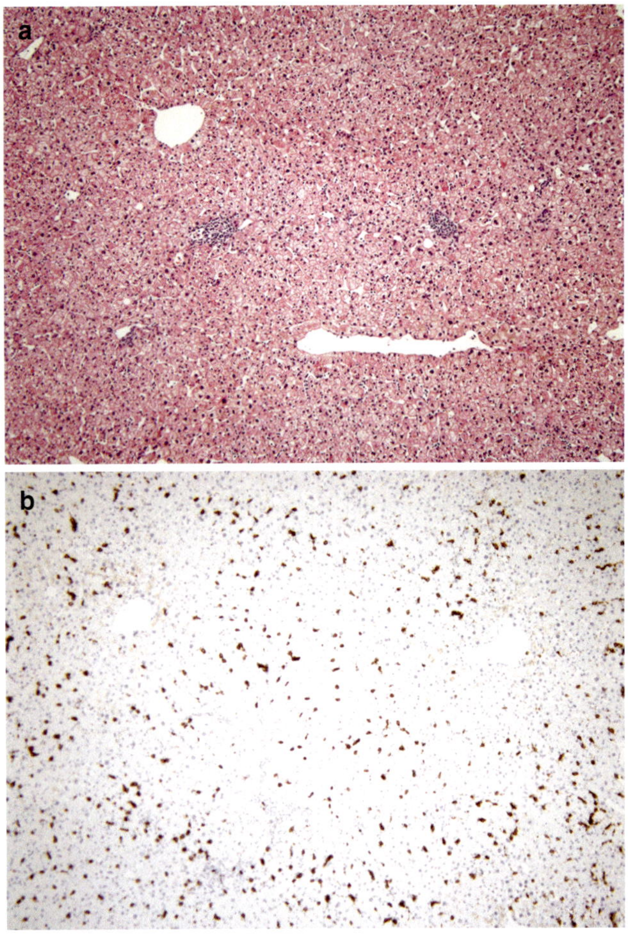

Fig. 3.17 Splenic marginal zone lymphoma, liver. In this liver biopsy from a patient with splenic marginal zone lymphoma, there is subtle involvement (**a**) in cords and portal triads, which is best appreciated by CD20 staining (**b**). (*Images courtesy of* L.J. Medeiros, Houston, TX.)

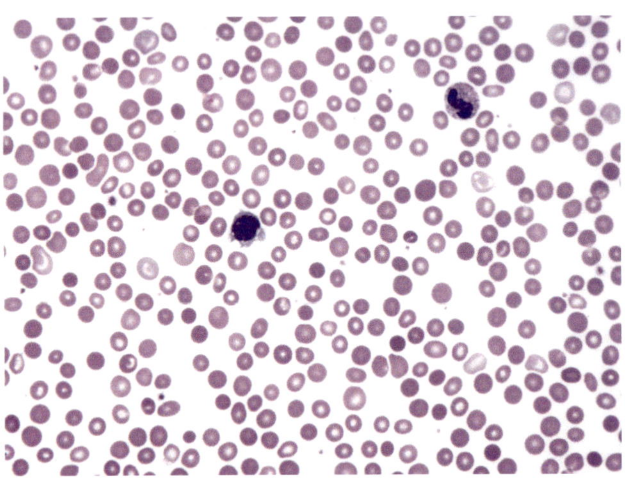

Fig. 3.18 Splenic marginal zone lymphoma, peripheral blood. Individual abnormal lymphocytes are often seen in the peripheral blood in splenic marginal zone lymphoma. These will often, but not always, have cytoplasmic projections (villous lymphocytes). This finding is not entirely specific to splenic marginal zone lymphoma

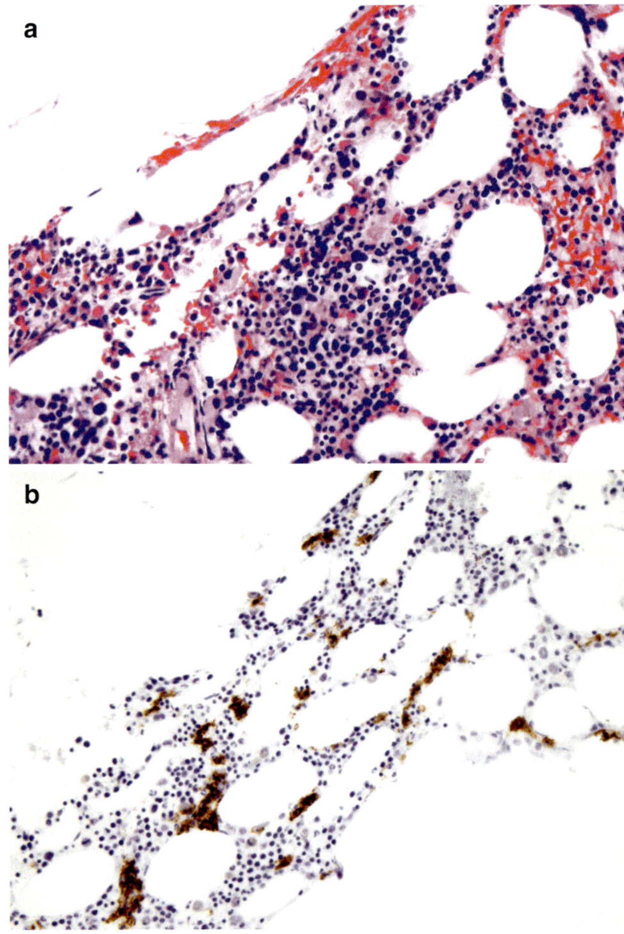

Fig. 3.19 Splenic marginal zone lymphoma, bone marrow. This bone marrow sample has subtle involvement by splenic marginal zone lymphoma (**a**) in an intrasinusoidal pattern. It can be more easily appreciated with CD20 staining, highlighting the linear arrangement of the abnormal lymphocytes (**b**)

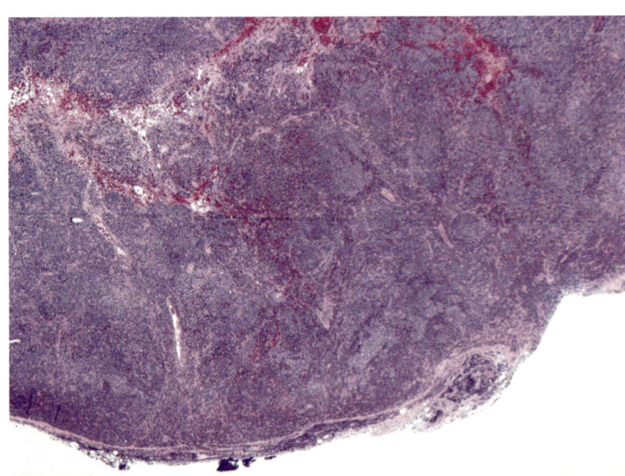

Fig. 3.20 Splenic marginal zone lymphoma, neck lymph node. Although not common, splenic marginal zone can involve distant lymph nodes, as in this case. This circumstance is quite difficult to distinguish from non-splenic marginal zone lymphomas (*eg*, extranodal marginal zone lymphoma) with secondary involvement of the spleen

3.2 Chronic Lymphocytic Leukemia

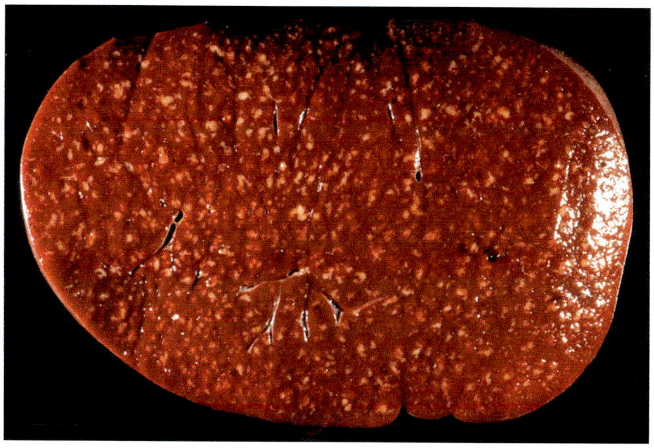

Fig. 3.21 Chronic lymphocytic leukemia (CLL). Gross image of a spleen enlarged by CLL. This spleen shows a seemingly normal pattern of white and red pulp. However, this spleen is massively enlarged (> 1 kg), with a relatively proportional expansion of red and white pulp. There are increases in abnormal lymphocytes within both the white and red pulp. (*Image courtesy of* A. Sohani, Boston, MA.)

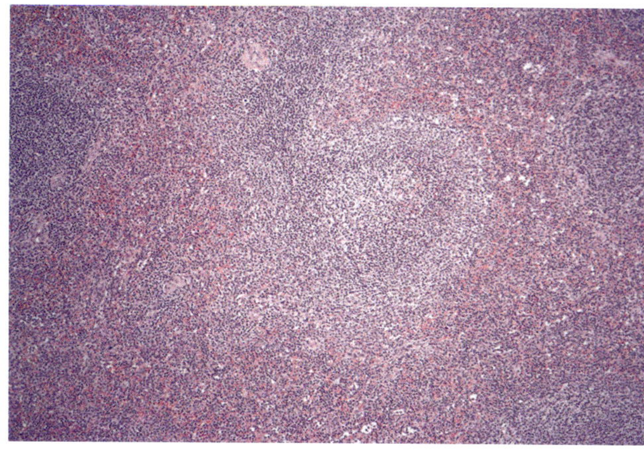

Fig. 3.23 Persistent polyclonal B-cell lymphocytosis. Spleen sections of persistent polyclonal B-cell lymphocytosis. This benign disorder is associated with increased circulating polyclonal B-cells. As is seen in this section, the spleen shows both enlargement of the white pulp nodules and increases in red pulp lymphocytes. This could easily be confused with splenic involvement by a B-cell lymphoma, especially CLL. Those affected are usually women smokers with an HLA-DR7 phenotype

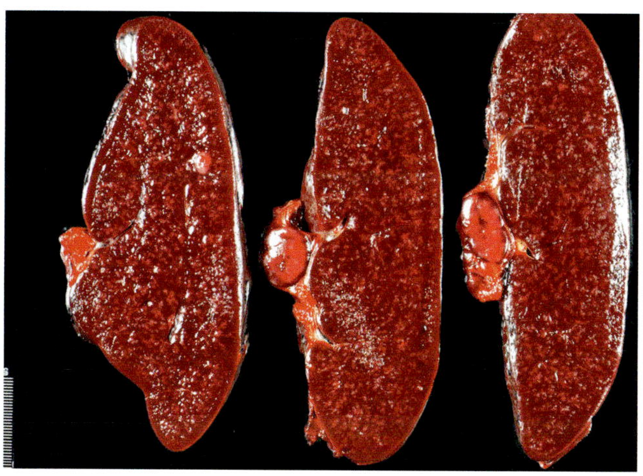

Fig. 3.22 Chronic lymphocytic leukemia. Sections of spleen with a diagnosis of CLL and refractory immune thrombocytopenic purpura. Both would contribute to splenic enlargement. In this case, there is an overall increase in white pulp structures accentuating the miliary appearance, which is seen in many B-cell lymphoma predominantly involving splenic white pulp. (*Image courtesy of* Z. Chakhachiro, Houston, TX.)

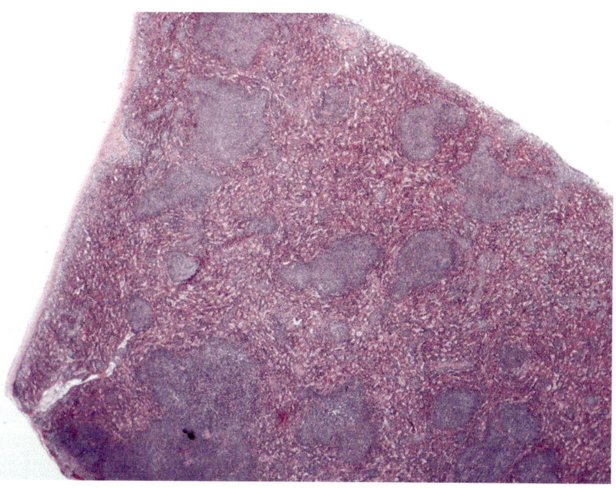

Fig. 3.24 Chronic lymphocytic leukemia. Low-magnification image of spleen with involvement by CLL. In this typical case, there is an overall increase in the number and size of white pulp nodules

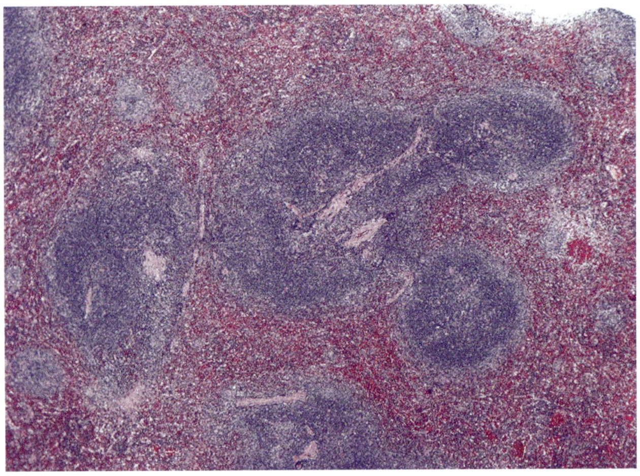

Fig. 3.25 Chronic lymphocytic leukemia. In this case, white pulp nodules are enlarged, as well as extension of lymphoid cells along arterioles, replacing the normal PALS region. In this case, there is a pale rim around the lymphoid nodules, raising the differential of residual marginal zone cells or marginal zone differentiation within the CLL cells

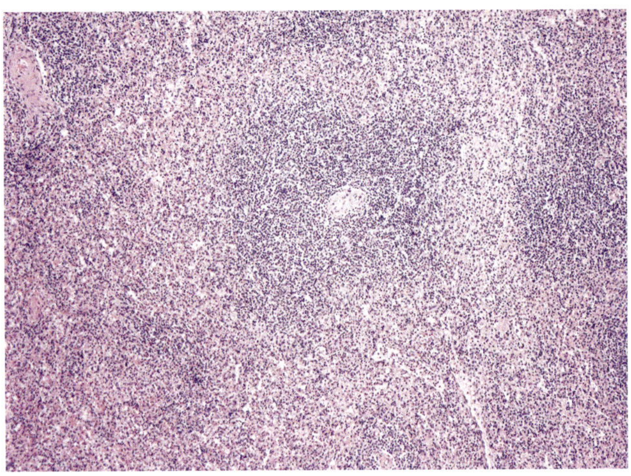

Fig. 3.26 Chronic lymphocytic leukemia. Early involvement of spleen with CLL. There is only minimal distortion of white pulp structures and an increase in red pulp small lymphocytes. Splenic findings of early CLL and "monoclonal B cell lymphocytosis" would be difficult to differentiate without clinical information

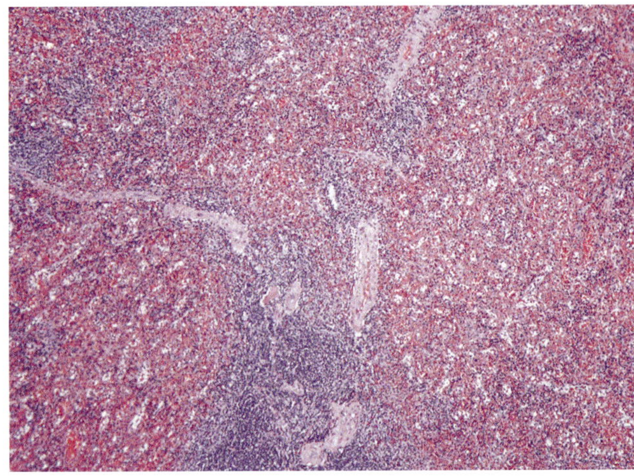

Fig. 3.27 Chronic lymphocytic leukemia. Extension of abnormal lymphocytes of CLL along arterioles, replacing the normal collections of T cells in these areas

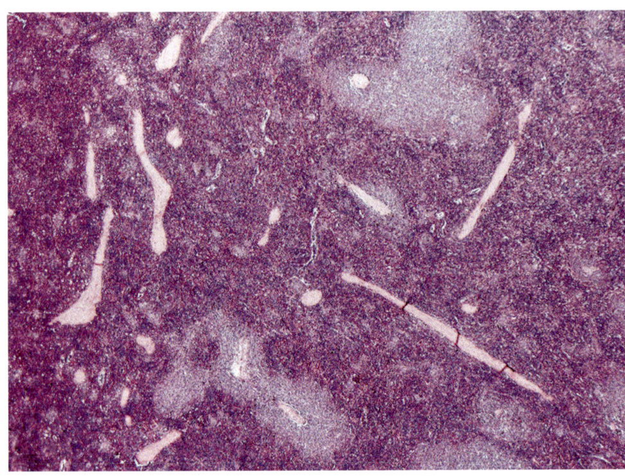

Fig. 3.28 Chronic lymphocytic leukemia. In this example of CLL, there is both involvement of white pulp (usual) and extensive involvement of red pulp, imparting an overall deep-blue appearance

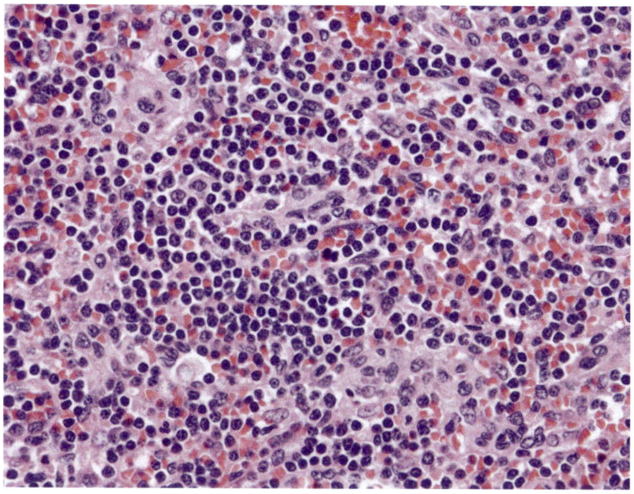

Fig. 3.29 Chronic lymphocytic leukemia. High magnification of splenic red pulp in CLL. There is an increase in small lymphocytes within cords and sinuses, as well as some nodular aggregates surrounding a small vessel (*center*), likely representing a PALS region

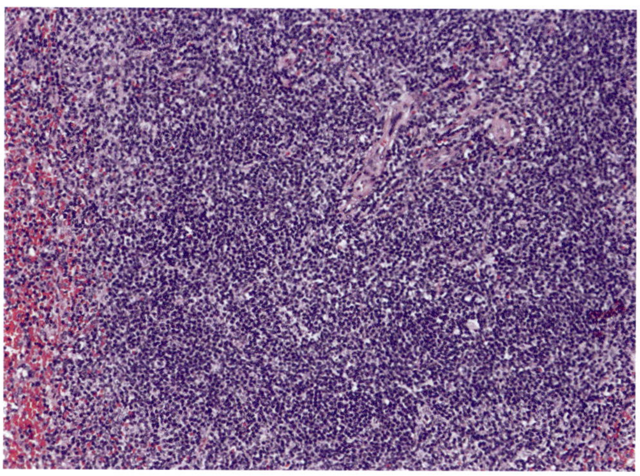

Fig. 3.30 Chronic lymphocytic leukemia. Intermediate magnification of white pulp nodule in CLL. The lymphocytes are mostly homogeneous with small nuclei, dense chromatin, and scant cytoplasm

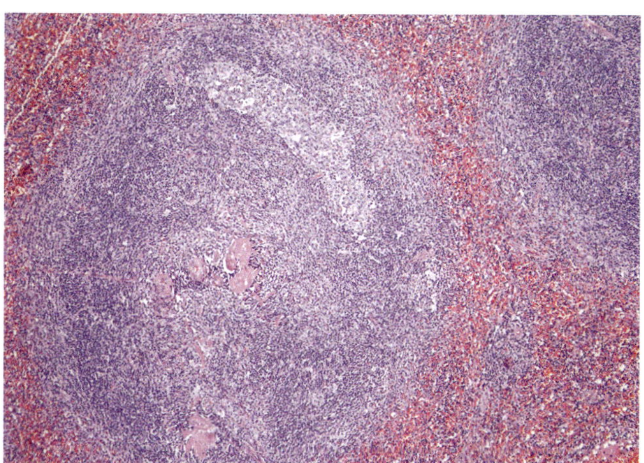

Fig. 3.31 Chronic lymphocytic leukemia. In this example of CLL, there is an expanded white pulp nodule present. In the upper portion, there is evidence of a residual germinal center (arc-shaped area), surrounded by the lymphocytes of CLL. A preserved marginal zone is also present

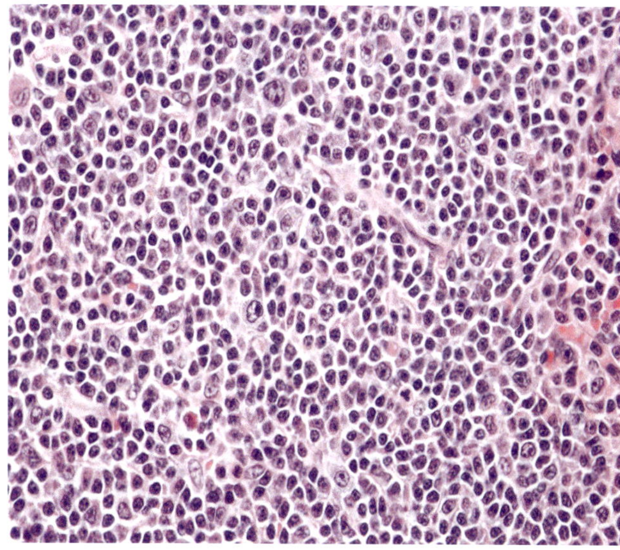

Fig. 3.32 Chronic lymphocytic leukemia. High magnification of CLL in spleen. In this case, the biphasic nature of CLL is highlighted. There are many small cells present. There are also large lymphocytes with fairly open chromatin, prominent nucleoli, and increased amounts of cytoplasm. These are prolymphocytes or para-immunoblasts

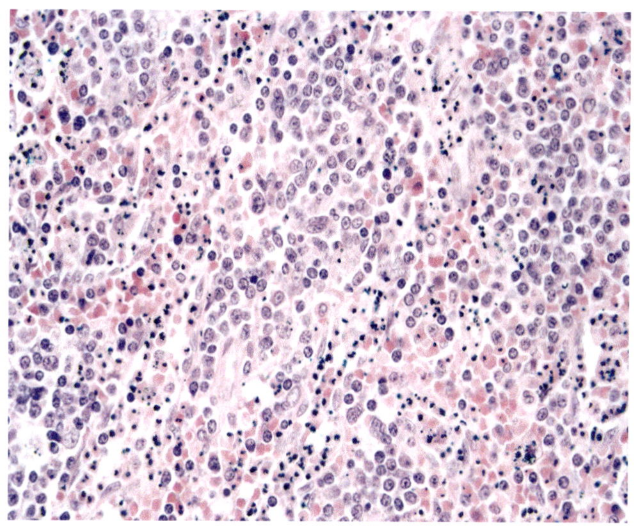

Fig. 3.33 Chronic lymphocytic leukemia with malaria. A rare case of concurrent CLL and *Plasmodium falciparum* malaria. The dark material is "malarial pigment"

3.3 Lymphoplasmacytic Lymphoma

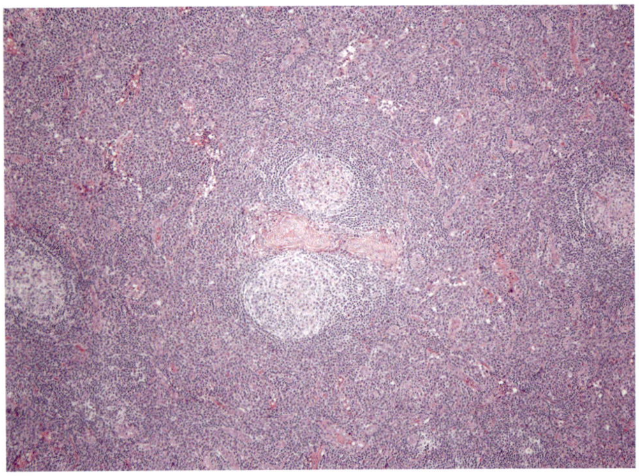

Fig. 3.34 Lymphoplasmacytic lymphoma. Low magnification of spleen with involvement by LPL. Spleens with LPL are rarely seen, likely due to rare significant involvement of spleen and little need for splenectomy as a diagnostic specimen

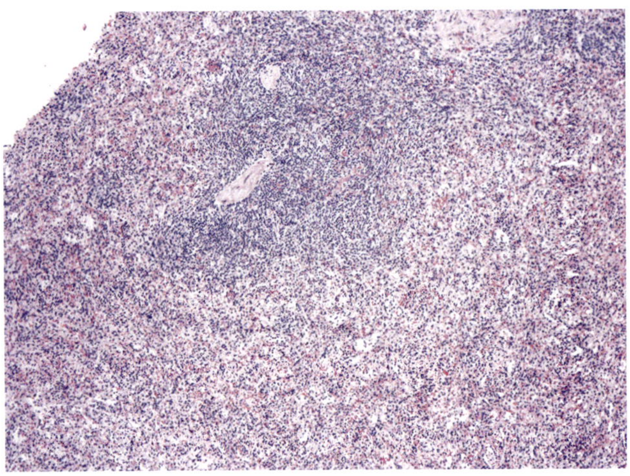

Fig. 3.35 Lymphoplasmacytic lymphoma. Another low magnification of LPL in the spleen. In this case, there is evidence of an increase in lymphocytes in the red pulp, with apparently normal white pulp

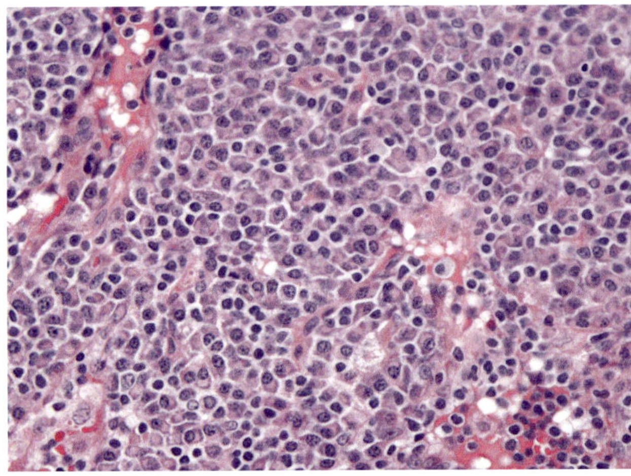

Fig. 3.36 Lymphoplasmacytic lymphoma. High magnification of LPL involving spleen. This case shows extensive plasma cell differentiation. The abnormal cells are predominantly present in the red pulp and areas present in cords and sinuses. Atypical plasma cells (large size, binucleated forms) are seen. No Dutcher bodies (intranuclear immunoglobulin inclusions) are seen, but they are often present in LPL

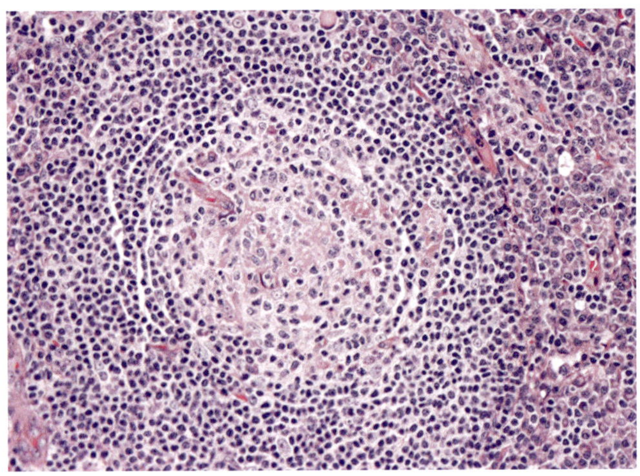

Fig. 3.37 Lymphoplasmacytic lymphoma. High magnification of a germinal center in splenic white pulp. There is no apparent involvement by LPL in white pulp, although increased plasma cells and abnormal lymphocytes can be seen in the adjacent red pulp

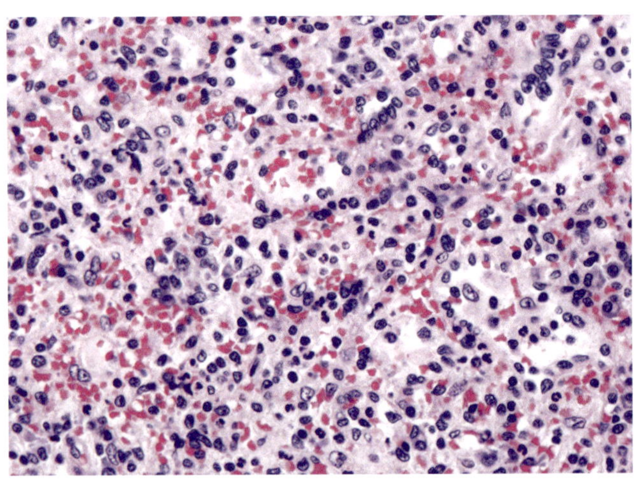

Fig. 3.38 Lymphoplasmacytic lymphoma. In this high magnification of LPL in the spleen, there are increased large and small lymphocytes and only occasional plasma cells are seen in the red pulp

3.4 Follicular Lymphoma

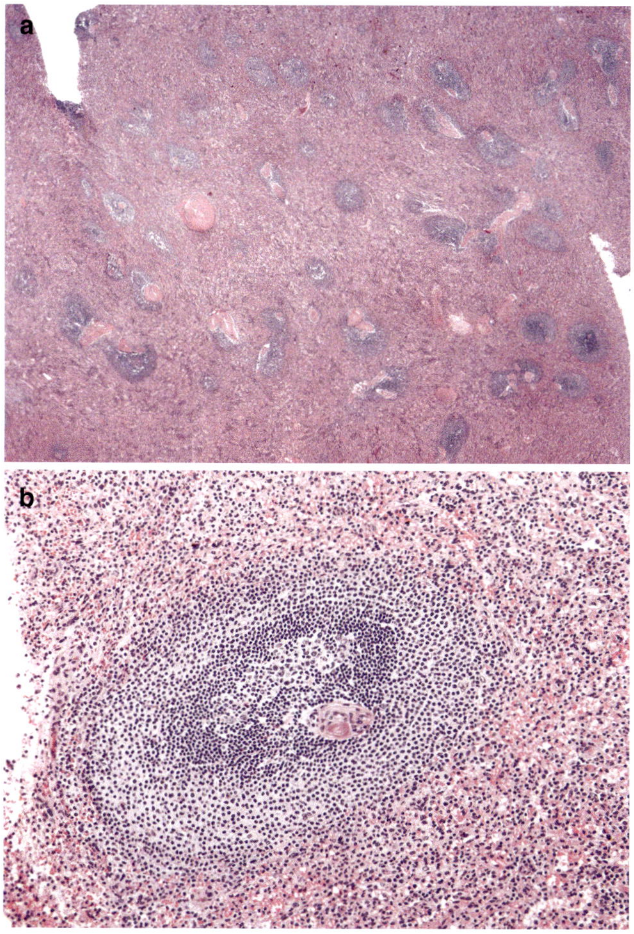

Fig. 3.39 Follicular lymphoma, in situ. Microscopic image of follicular lymphoma in situ (FLIS). FLIS is typified by localized involvement of germinal centers by neoplastic cells with both the phenotype and genetics of follicular lymphoma. However, in contrast to follicular lymphoma, only a small subset of FLIS go on to develop systemic follicular lymphoma. **a**, Low magnification; **b**, intermediate magnification. (*Image courtesy of* J. Cooke, Pittsburgh, PA.)

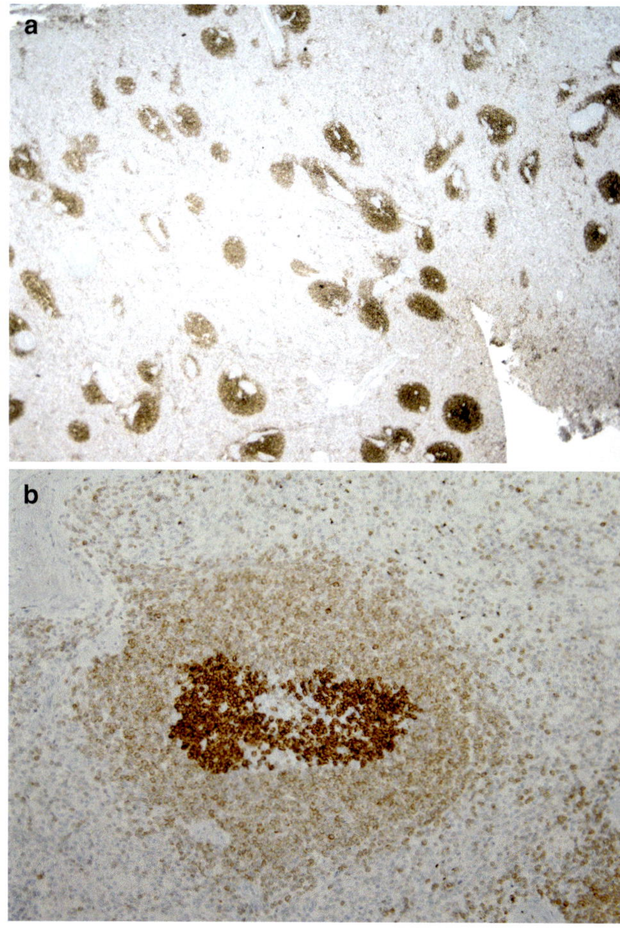

Fig. 3.40 Follicular lymphoma, in situ, Bcl-2 stain. Bcl-2 staining of FLIS. In contrast to normal follicles, which are negative, the abnormal cells of FLIS are strongly positive for Bcl-2. These abnormal cells are often strongly positive in comparison to usual expression seen in mantle and marginal zones of spleen. **a**, Low magnification; **b**, intermediate magnification. (*Image courtesy of* J. Cooke, Pittsburgh, PA.)

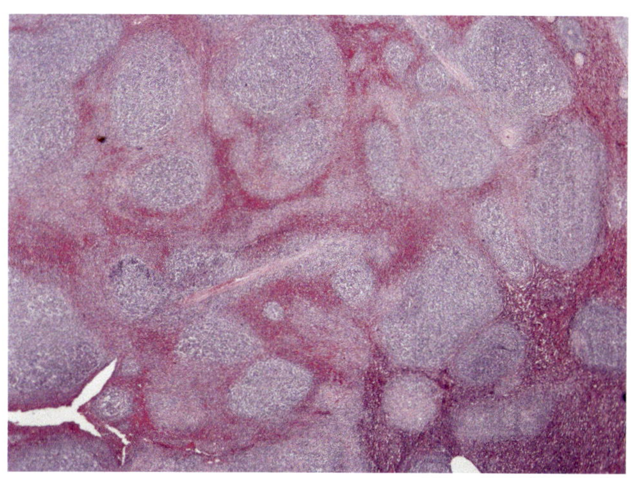

Fig. 3.41 Follicular lymphoma. Low-magnification image of follicular lymphoma in the spleen. There is a massive expansion of the white pulp in a nodular pattern. Note that there are some residual PALS regions, which appear partly infiltrated or expanded by the abnormal lymphocytes. Red pulp structures are preserved but crowded by the neoplastic white pulp elements

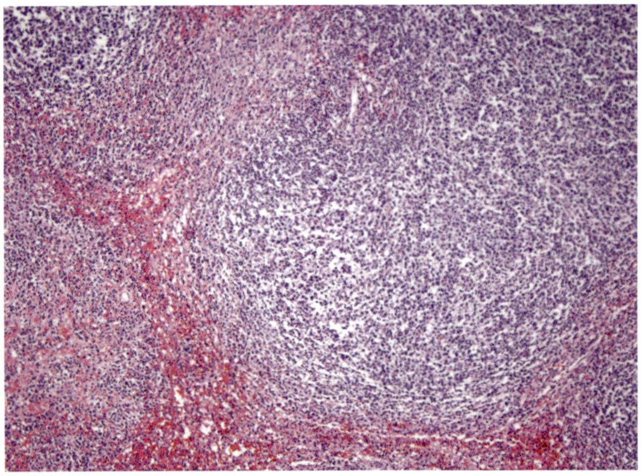

Fig. 3.42 Follicular lymphoma. High magnification of low-grade follicular lymphoma in the spleen. In this case, the abnormal follicular structure is composed of predominantly small irregular lymphocytes (centrocytes) with irregular nuclei and scant cytoplasm. There are only rare admixed larger transformed lymphocytes and mitotic figures are quite rare

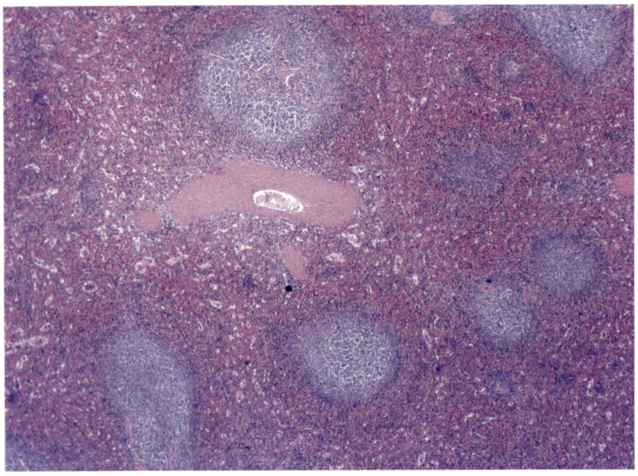

Fig. 3.43 Follicular lymphoma. Low magnification of follicular lymphoma in the spleen. At this magnification, most B-cell lymphomas that predominantly involve splenic white pulp cannot be reliably distinguished from one another

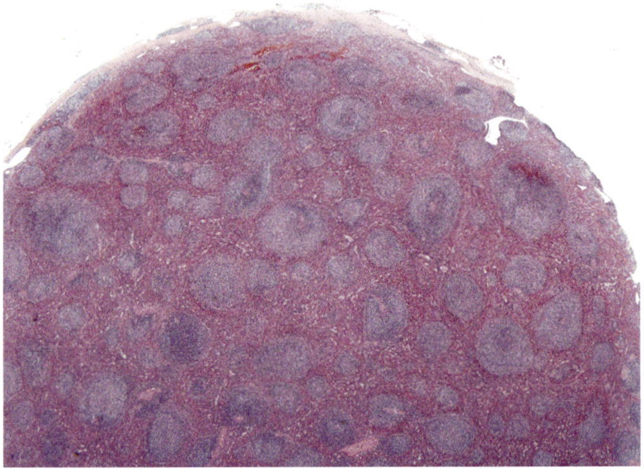

Fig. 3.44 Follicular lymphoma. Low magnification of follicular lymphoma involving an accessory spleen. Note the overall increase in number of white pulp structures/follicles

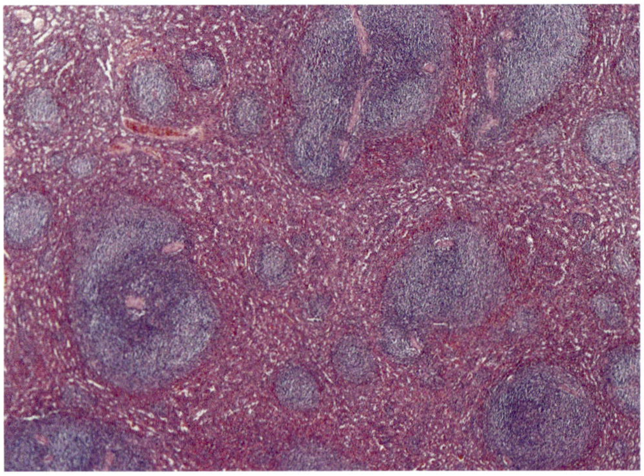

Fig. 3.45 Follicular lymphoma. In this example of a low-grade follicular lymphoma, the splenic white pulp nodules are variable in size but have a relatively uniform appearance of their cellular components

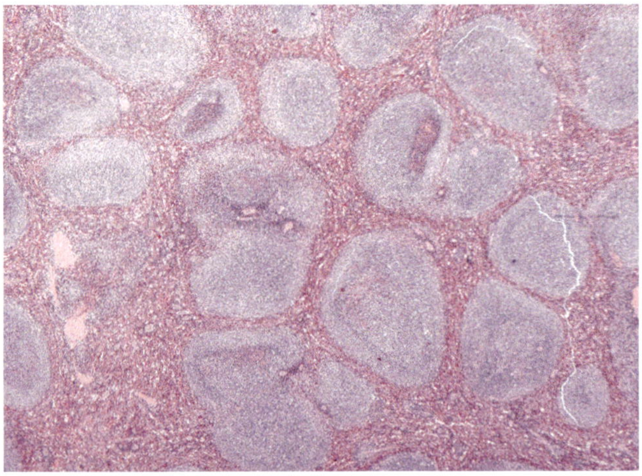

Fig. 3.46 Follicular lymphoma. Another example of low-grade follicular lymphoma in spleen. In this case, there are marked increased numbers of white pulp nodules, making the ratio of red to white pulp decrease dramatically. These follicular structures maintain their marginal zones

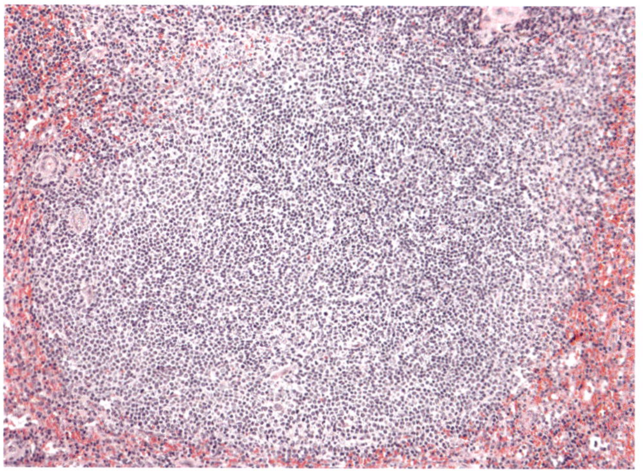

Fig. 3.47 Follicular lymphoma. Intermediate magnification of low-grade follicular lymphoma. Note the homogeneous appearance of the lymphocytes within the white pulp

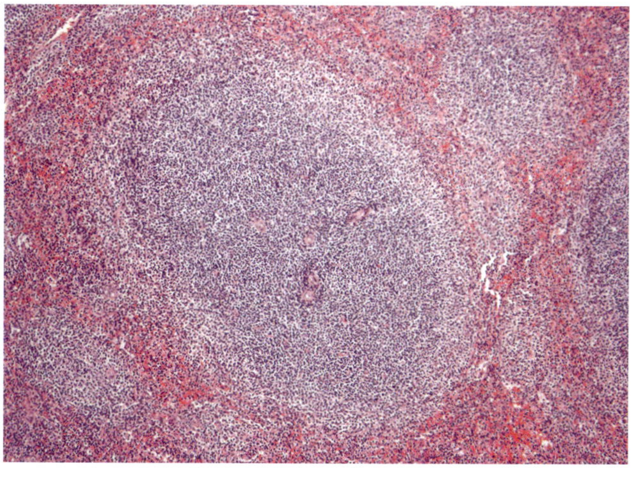

Fig. 3.48 Follicular lymphoma. An example of an abnormal follicle of low-grade follicular lymphoma, with preservation of the marginal zone

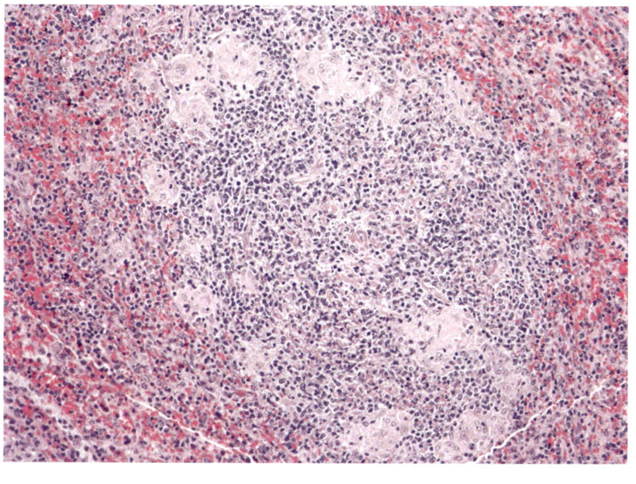

Fig. 3.49 Follicular lymphoma with granuloma. Granulomas can be seen in association with all types of B-cell lymphoma. In this case, small epithelioid granulomas are seen at the periphery of a white pulp nodule of follicular lymphoma

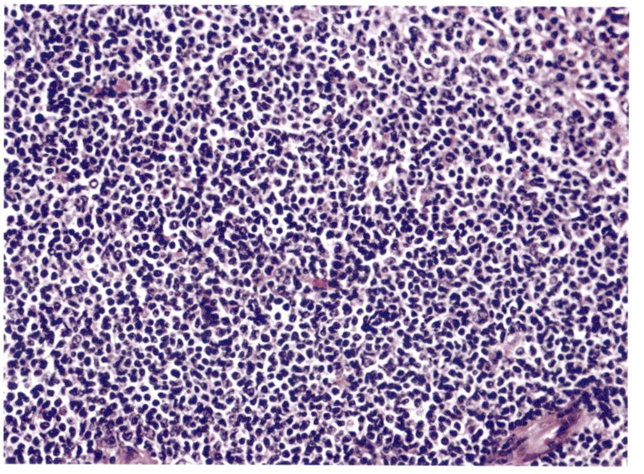

Fig. 3.50 Follicular lymphoma. Cytologic composition of follicular lymphoma grade 1. Note that the abnormal lymphocytes are predominantly small with irregular nuclei. Large cells are less than 5 per high-power field (HPF), although this number should be correlated with the composition of several follicles and averaged

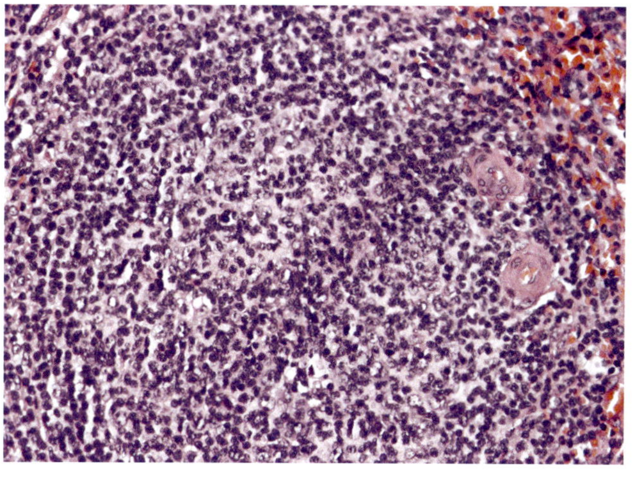

Fig. 3.51 Follicular lymphoma. Cytologic composition of follicular lymphoma grade 2. Note that the abnormal lymphocytes are predominantly small with irregular nuclei but there are some large transformed cells present (5–15 per HPF)

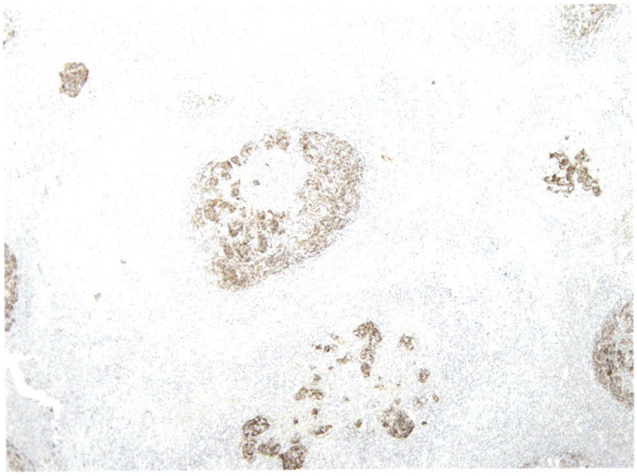

Fig. 3.52 Follicular lymphoma, CD21 stain. CD21 stains follicular dendritic cell (FDC) networks. In this case, the abnormal FDC networks of follicular lymphoma are highlighted by the stain

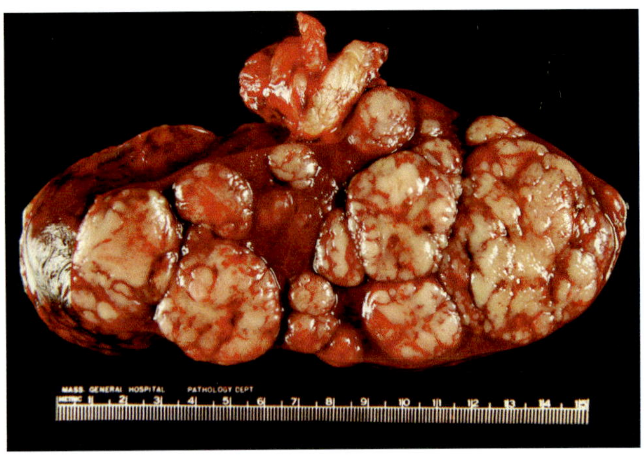

Fig. 3.53 Follicular lymphoma. Gross image of spleen with massive involvement by follicular lymphoma grade 3. The large fleshy nodules expand and replace the normal splenic parenchyma. They are a bizarre caricature of the normal nodular aggregates seen in sections of normal spleen. (*Image courtesy of* A. Sohani, Boston, MA.)

3.5 Mantle Cell Lymphoma

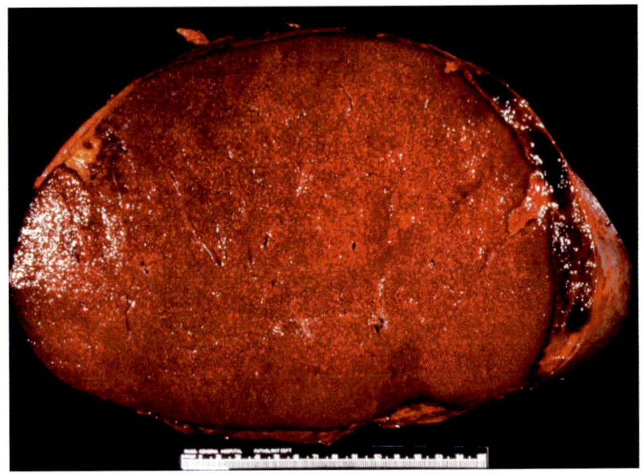

Fig. 3.54 Mantle cell lymphoma. Gross image of a spleen that is massively enlarged by involvement of mantle cell lymphoma (MCL). In this case, the spleen weighs 4 kg. There is both a massive increase in the number of white pulp follicles (small pale tan nodules) with a concurrent increase in the intervening red pulp. This would suggest an expansion of both red pulp and white pulp by the abnormal lymphocytes. (*Image courtesy of* A. Sohani, Boston, MA.)

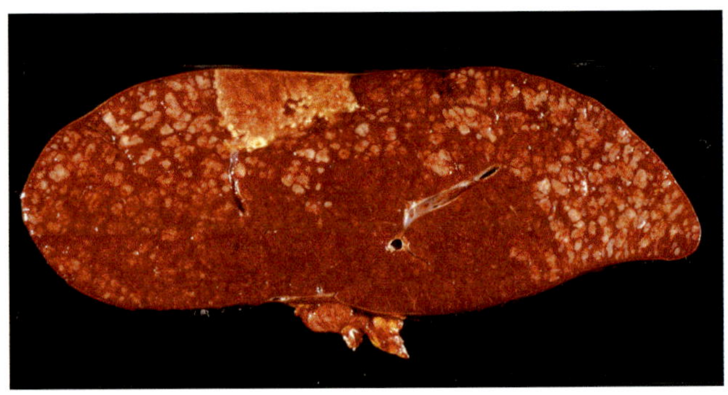

Fig. 3.55 Mantle cell lymphoma. Gross image of mantle cell lymphoma involving the spleen. In this case, there is evidence of infarction (upper, middle, wedge-shaped tan area). Infarction usually leads to pain and may necessitate removal of an enlarged spleen. In contrast to the previous example, this case has relatively larger and more irregular white pulp nodules with less expansion of the red pulp. (*Image courtesy of* A. Sohani, Boston, MA.)

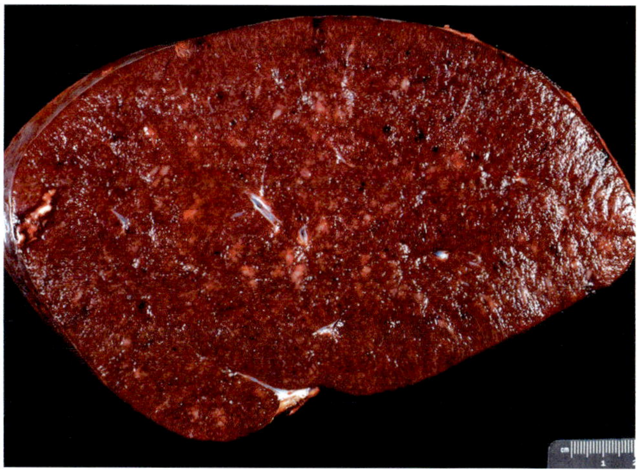

Fig. 3.56 Mantle cell lymphoma. A massively enlarged spleen with extensive involvement by mantle cell lymphoma. The miliary pattern typical of white pulp lymphomas is subtle in this case. (*Image courtesy of* W. Greaves, Houston, TX.)

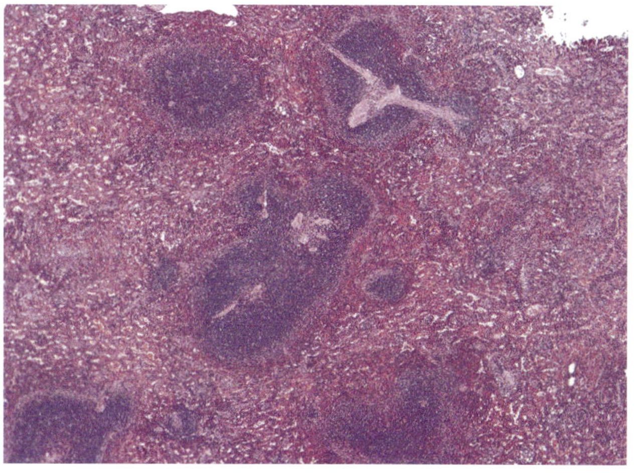

Fig. 3.58 Mantle cell lymphoma. In this low-magnification image, there is an overall increase in number and size of white pulp structures. In this example of mantle cell lymphoma, there are also some small nodules of lymphocytes within the red pulp. These features can be seen in other small B-cell lymphomas in the spleen

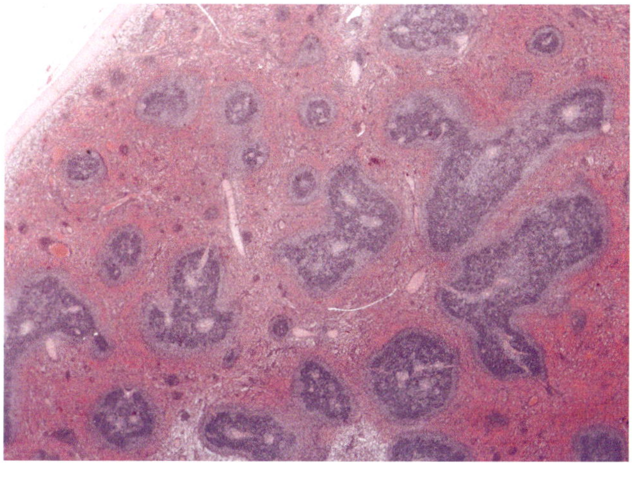

Fig. 3.57 Mantle cell lymphoma. Mantle cell lymphoma most often has a white pulp distribution in the spleen. In this case, there is preservation of marginal zone or "monocytoid" differentiation of the mantle cell lymphoma. Distinction of small B-cell lymphomas of spleen usually requires immunohistochemical staining

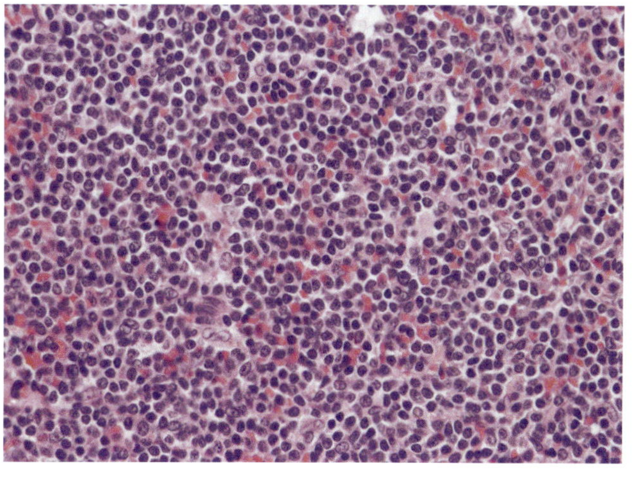

Fig. 3.59 Mantle cell lymphoma. High magnification of mantle cell lymphoma in the spleen. The individual cells of mantle cell lymphoma are round to slightly irregular, with mature chromatin and scant cytoplasm; larger transformed lymphocytes are quite rare

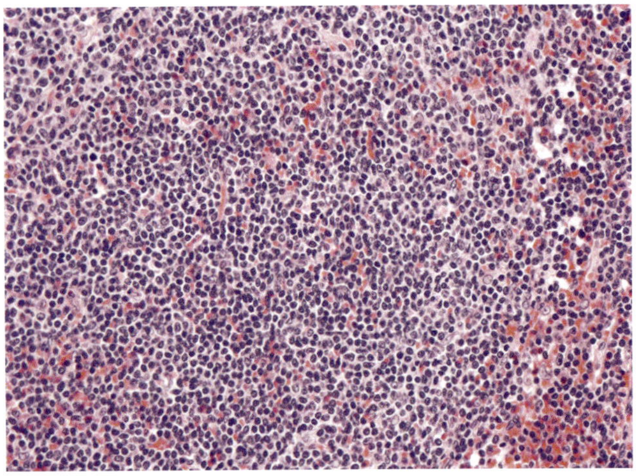

Fig. 3.60 Mantle cell lymphoma. Another example of the higher magnification features of mantle cell lymphoma in the spleen. This white pulp nodule is composed of a fairly monomorphic population of small lymphocytes. The presence of cyclin D1 staining would confirm the diagnosis in most cases

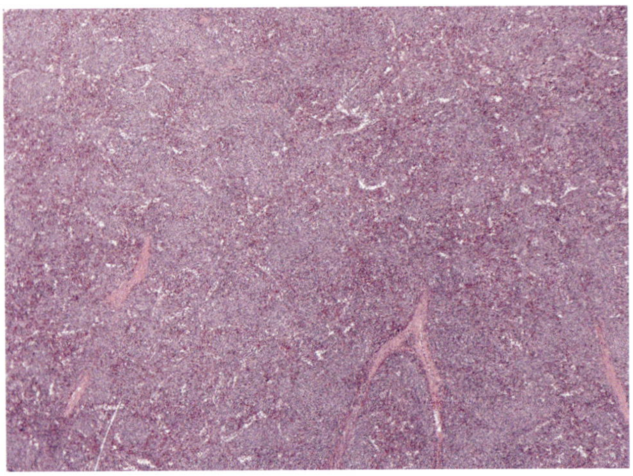

Fig. 3.61 Mantle cell lymphoma, blastoid variant. In this low-magnification image, the spleen is diffusely involved by lymphoma. In this case, the lymphoma cells have a more pale appearance than is typical. Blastoid variant of mantle cell lymphoma is associated with an aggressive clinical course

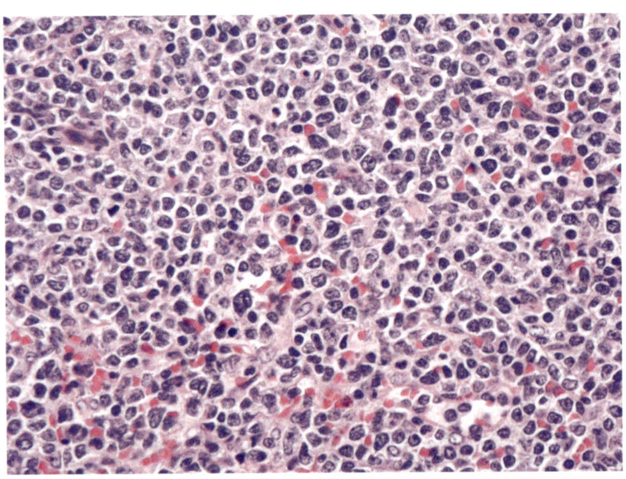

Fig. 3.62 Mantle cell lymphoma, blastoid variant. High magnification of mantle cell lymphoma, blastoid variant, pleomorphic type. There are two types of blastoid mantle cell lymphomas: pleomorphic type and blastic type. In this case, the cells are large in size with irregular pleomorphic nuclei

3.6 Plasma Cell Myeloma

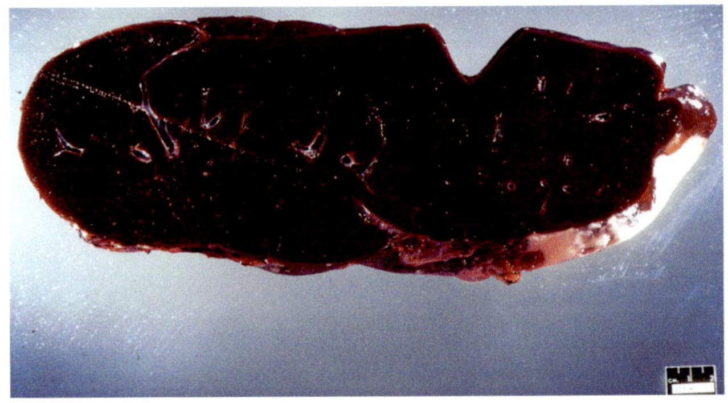

Fig. 3.63 Plasma cell myeloma. Gross image of spleen involvement by plasma cell myeloma. In this case, there is a diffuse red pulp pattern, imparting a "beefy" appearance. Myeloma only rarely involves the spleen diffusely, as in this case. (*Image courtesy of* D. Farhi, Atlanta, GA.)

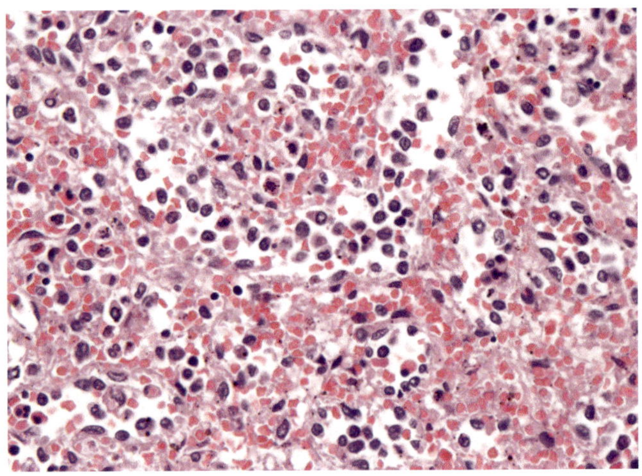

Fig. 3.64 Plasma cell myeloma. High magnification of splenic involvement by plasma cell myeloma. Only occasional plasma cells are seen, including some with Dutcher bodies. Many of the cells present are histiocytes or are extramedullary hematopoiesis

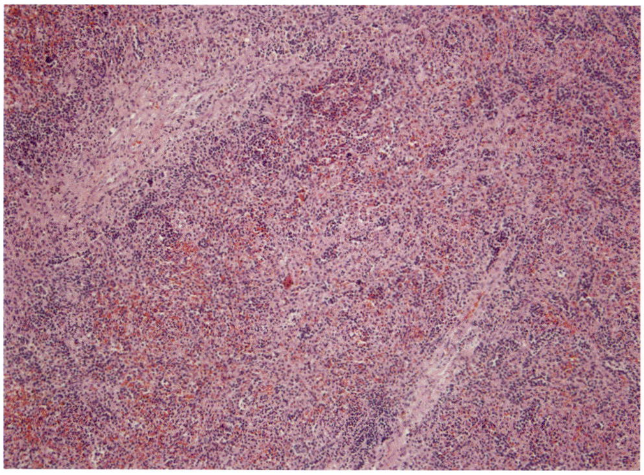

Fig. 3.65 Plasma cell myeloma (PCM), extramedullary hematopoiesis. Intermediate magnification of spleen in patient with plasma cell myeloma. There is no histologic evidence of PCM in this spleen. Rather, the spleen is enlarged due to a massive increase in extramedullary hematopoiesis. This is likely a result of neoplastic plasma cells crowing out normal marrow elements

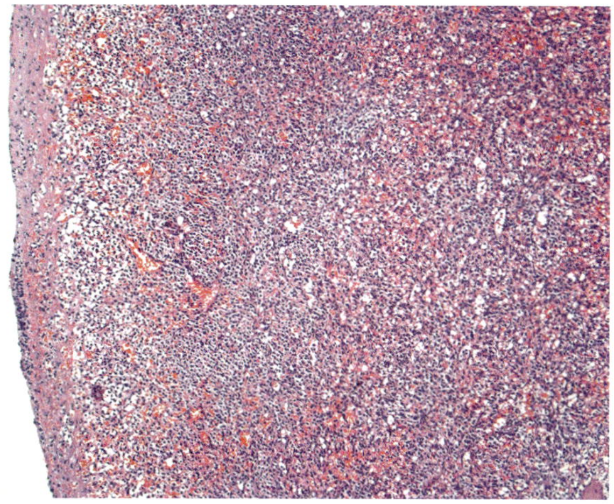

Fig. 3.67 Hairy cell leukemia. In this low-magnification image of spleen, there is diffuse involvement by a proliferation of small lymphocytes; many of these have increased pale cytoplasm. There are no white pulp structures present, as this is diffuse red pulp involvement. There is some infiltration of the capsule. The weakened capsule would be more susceptible to rupture with even minor trauma

3.7 Hairy Cell Leukemia

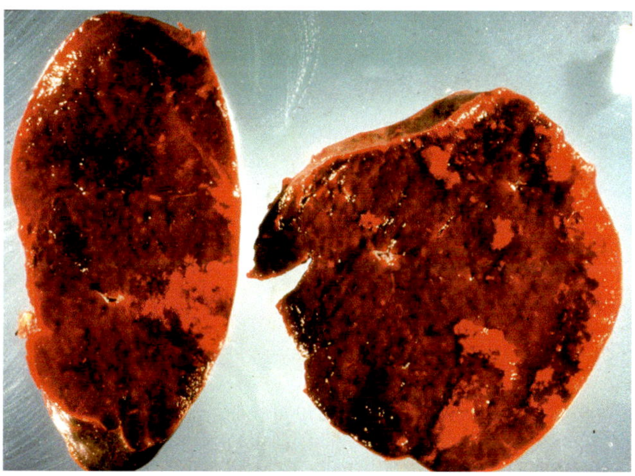

Fig. 3.66 Hairy cell leukemia. Gross image of hairy cell leukemia in the spleen. There is an expansion of the red pulp, imparting a "beefy" appearance typical of red pulp disease. In addition, there are small nodules of blood, corresponding to the "blood lakes" seen by histology. In this case, there are areas of infarction present, which have a paler and uniform appearance compared to the remaining spleen. (*Image courtesy of* D. Farhi, Atlanta, GA.)

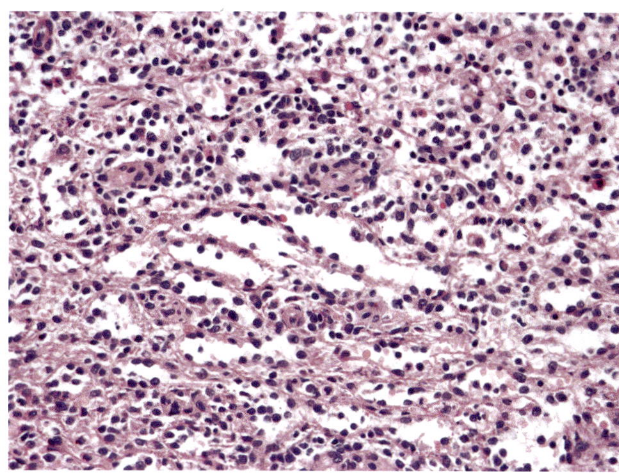

Fig. 3.68 Hairy cell leukemia. In this example, there is an expansion of red pulp structures and a reduction in size of white pulp structures

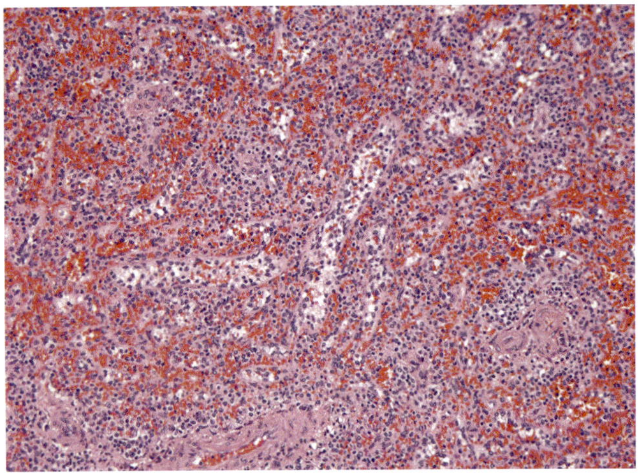

Fig. 3.69 Hairy cell leukemia. Intermediate magnification of hairy cell leukemia. The abnormal lymphocytes are present in the red pulp, expanding the cords and are present within sinuses

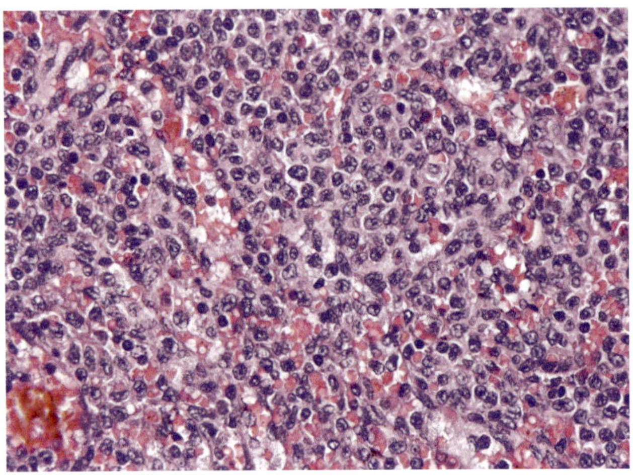

Fig. 3.70 Hairy cell leukemia. The cytologic appearance of hairy cell leukemia is generally round nuclei of small to intermediate size. These have marginated or coarsely granular chromatin and moderate amounts of pink cytoplasm. The cytoplasmic projections (hairs) are not easily seen on hematoxylin and eosin (H&E) stain sections

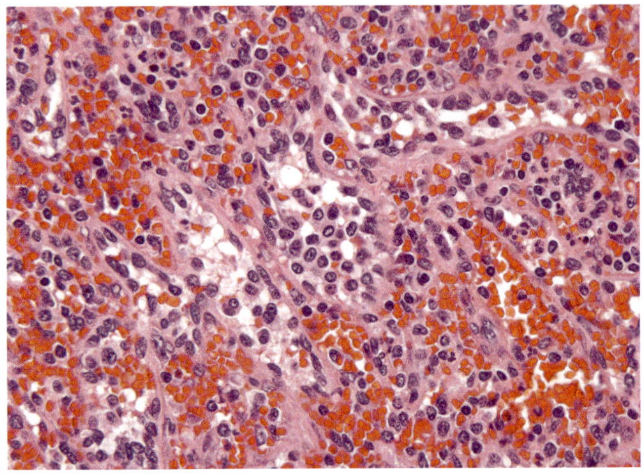

Fig. 3.71 Hairy cell leukemia. Another high magnification image of hairy cell leukemia. In this case, the cytoplasmic borders are more irregular with more vesicular chromatin. This may be fixation artifact

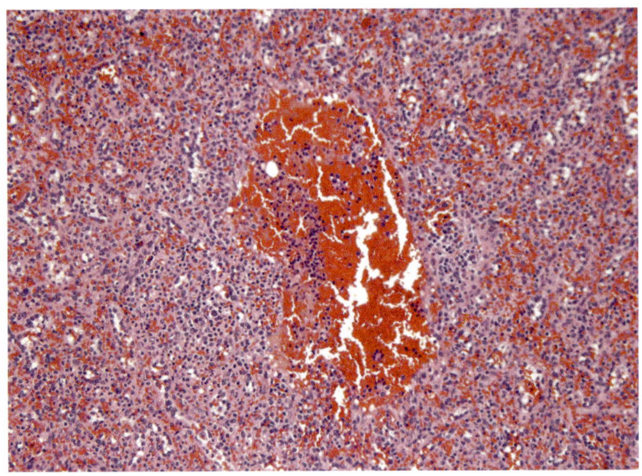

Fig. 3.72 Hairy cell leukemia. A "blood lake" in hairy cell leukemia. Blood lakes are small areas of blood that are not in vascular spaces but in spaces lined by abnormal lymphocytes. Although usual in hairy cell leukemia, they are not pathognomonic for this diagnosis

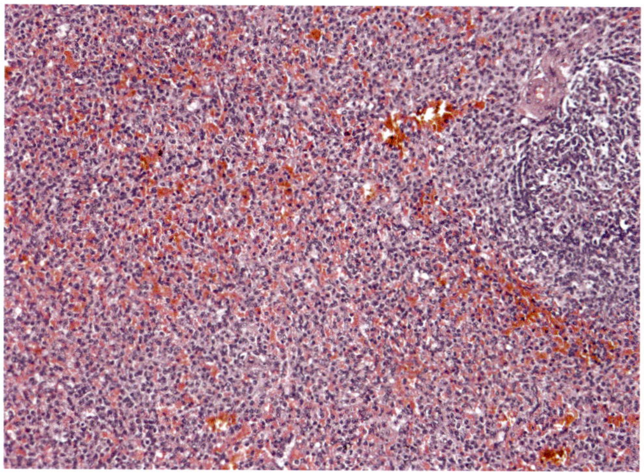

Fig. 3.73 Hairy cell leukemia. In this example of hairy cell leukemia, the abnormal lymphocytes have greatly expanded the red pulp, but there is preservation of normal-appearing white pulp structures (follicle, marginal zone; *middle right*)

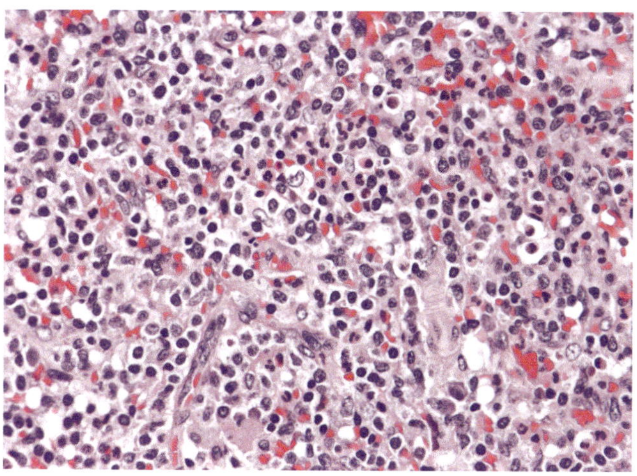

Fig. 3.74 Hairy cell leukemia. In this example of hairy cell leukemia, the cytoplasm of the abnormal lymphocytes is clear. The appearance of round nuclei with ample clear cytoplasm has traditionally been referred to as a "fried egg" appearance

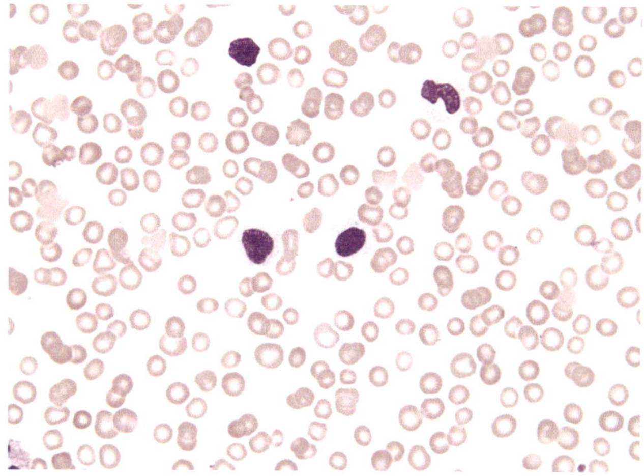

Fig. 3.75 Hairy cell leukemia, peripheral blood. A peripheral blood smear with two abnormal hairy cell leukemia lymphocytes (*center*) and two granulocytes (*upper* and *left*). The lymphocyte nuclei are round to ovoid, with ample pale cytoplasm, and subtle cytoplasmic projections ("hairs")

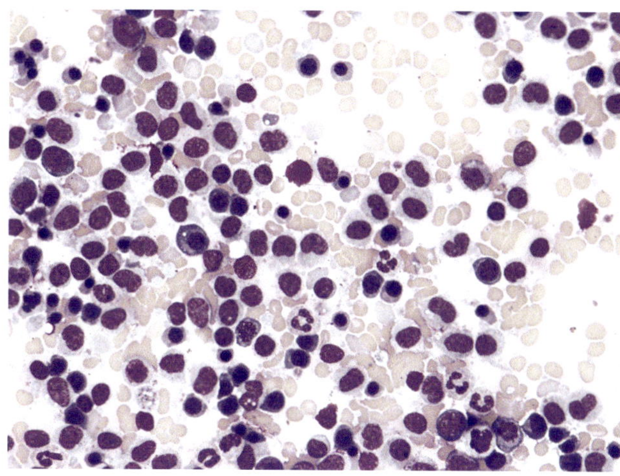

Fig. 3.76 Hairy cell leukemia, bone marrow aspirate. Not commonly seen in aspirate smears due to marrow fibrosis, this is an example of hairy cell leukemia in a bone marrow smear. The nuclei are round to ovoid with condensed chromatin and moderate amounts of pale blue-pink cytoplasm

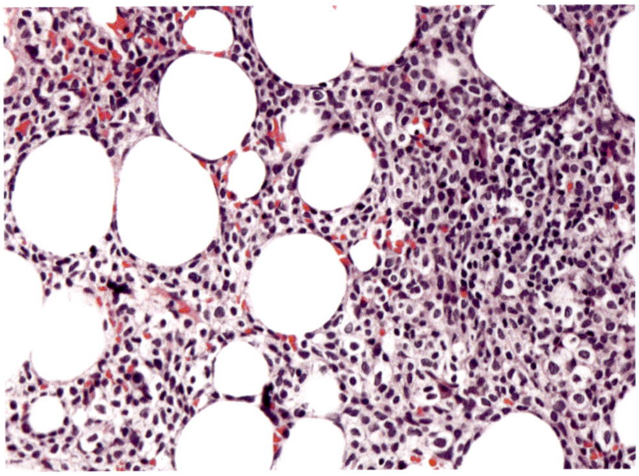

Fig. 3.77 Hairy cell leukemia, bone marrow core. Most of the lower portion of the image is composed of hairy cell leukemia. The abnormal lymphocytes have dense blue chromatin and ample pale/clear cytoplasm, imparting a "fried egg" appearance

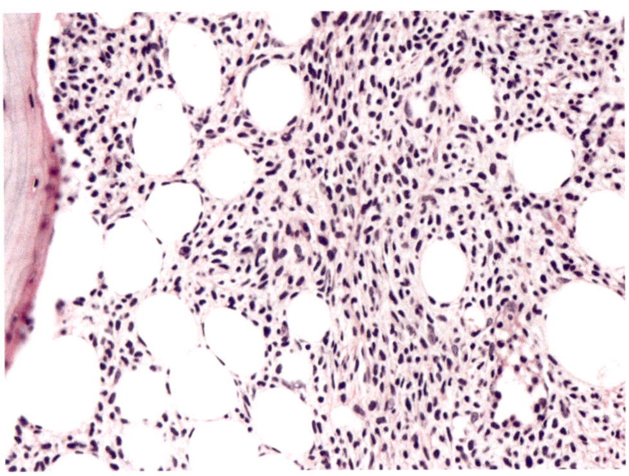

Fig. 3.78 Hairy cell leukemia, bone marrow core. Marrow involvement by hairy cell leukemia, in this case a spindled variant. The spindle shape of the cells may be due to distortion of the lymphocytes by the severe stromal fibrosis present

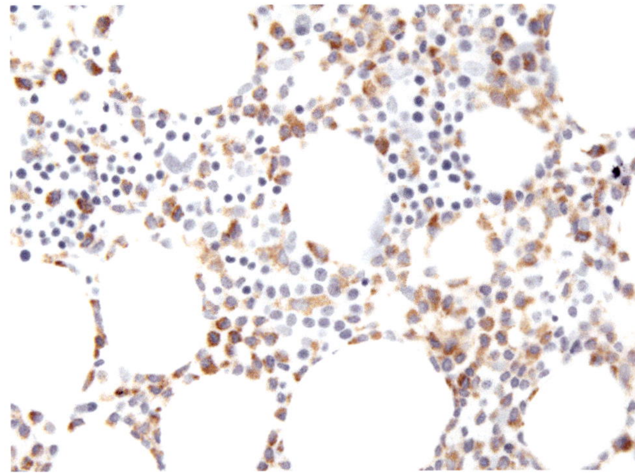

Fig. 3.79 Hairy cell leukemia, bone marrow, TRAP stain (an immunohistochemical stain for tartrate-resistant acid phosphatase). The cytoplasmic staining has a granular quality in hairy cell leukemia

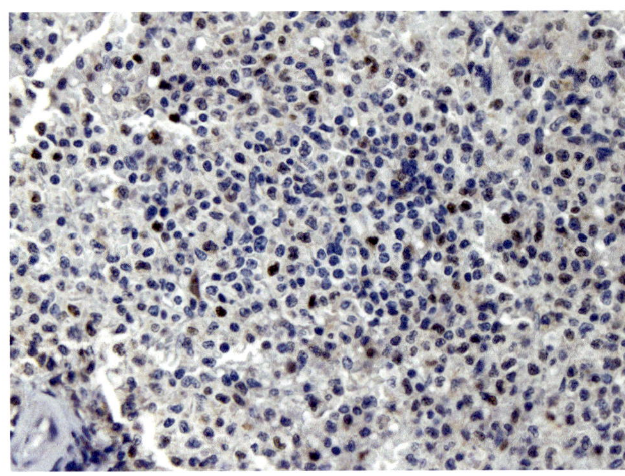

Fig. 3.80 Hairy cell leukemia, cyclin D1 stain. Hairy cell leukemia is positive for cyclin D1 but often is more pale and patchy than that seen in mantle cell lymphoma

3.8 Hairy Cell Leukemia-Variant

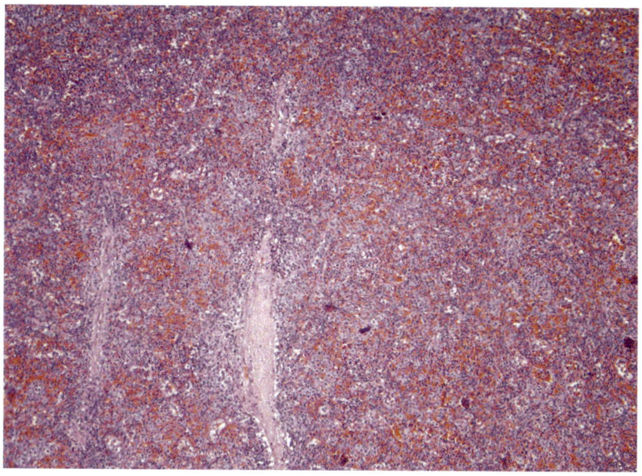

Fig. 3.81 Hairy cell leukemia-variant (HCL-V). Low magnification of HCL-V. In contrast to typical hairy cell leukemia, HCL-V has either immunophenotypic variation from HCL (absence of CD25, annexin A1 or TRAP) or has a prolymphocytoid appearance of individual cells

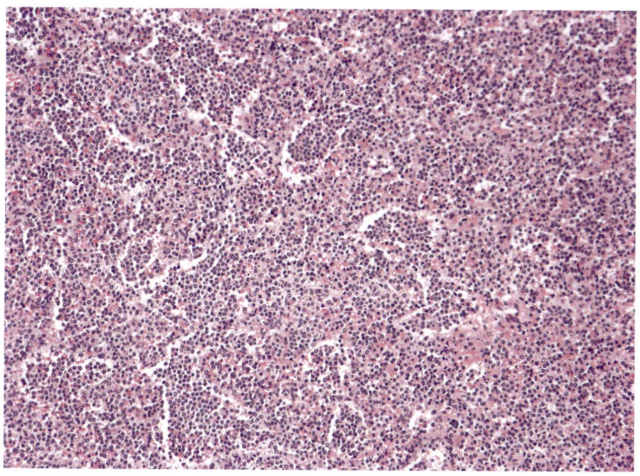

Fig. 3.82 Hairy cell leukemia-variant. Diffuse red pulp involvement, as seen in this image, is typical of HCL-V

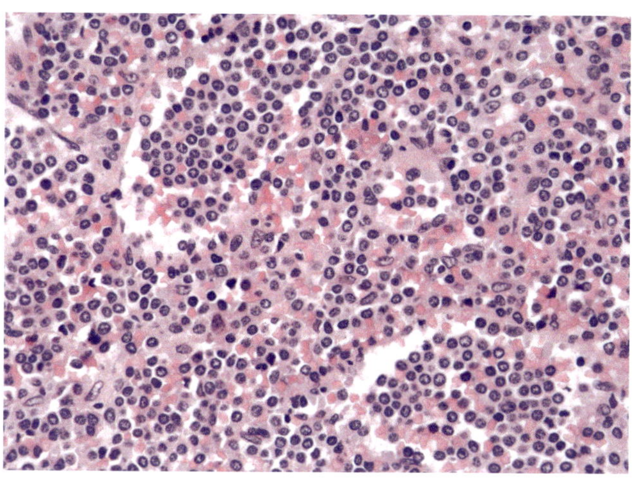

Fig. 3.83 Hairy cell leukemia-variant. In the spleen, the prominent nucleoli, seen in the peripheral blood, are not always seen in HCL-V

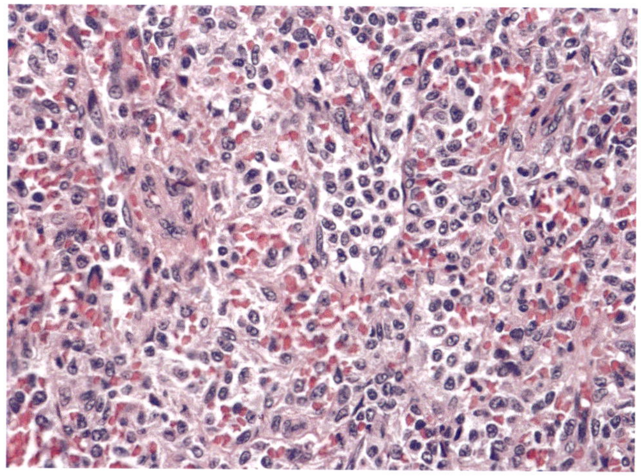

Fig. 3.84 Hairy cell leukemia-variant. In this example of HCL-V, the lymphocytes are small to intermediate in size and have irregular nuclei. As in typical cases, there are moderate amounts of pink, granular cytoplasm

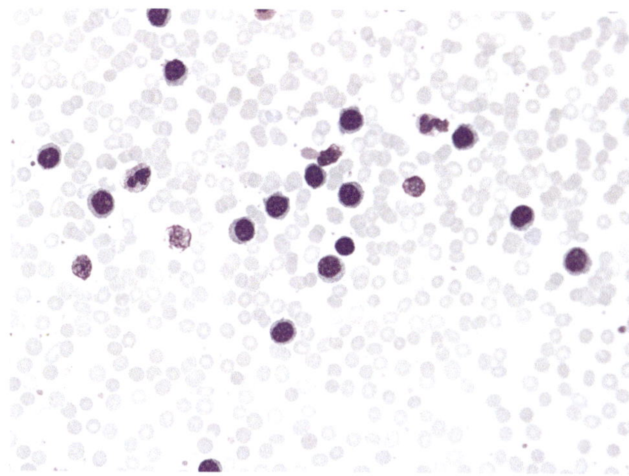

Fig. 3.85 Hairy cell leukemia-variant, peripheral blood. Cell of HCL-V. In this case, some small nucleoli are present. However, the designation as HCL-V was based on a lack of CD25 staining, with other markers typical for hairy cell leukemia

3.9 Splenic Diffuse Red Pulp Small B-Cell Lymphoma

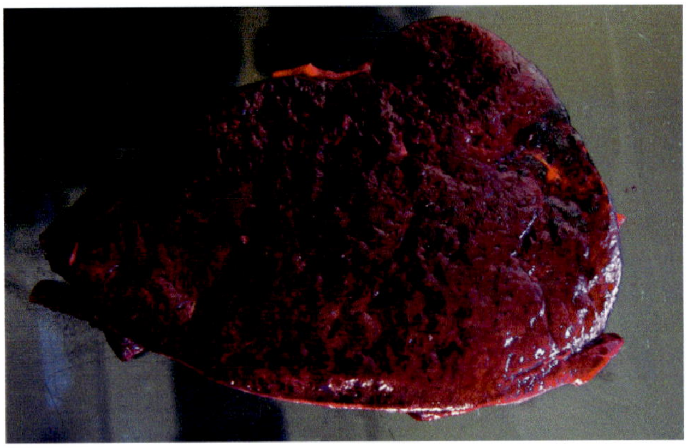

Fig. 3.86 Splenic diffuse red-pulp small B-cell lymphoma. Gross image of spleen involved by splenic diffuse red-pulp small B-cell lymphoma. This lymphoma has exclusively red pulp involvement and it imparts a grossly red, beefy appearance. There is little or no appearance of the small pale or tan flecks representing white pulp, which are seen in normal spleens. (*Image courtesy of* A. Sohani, Boston, MA.)

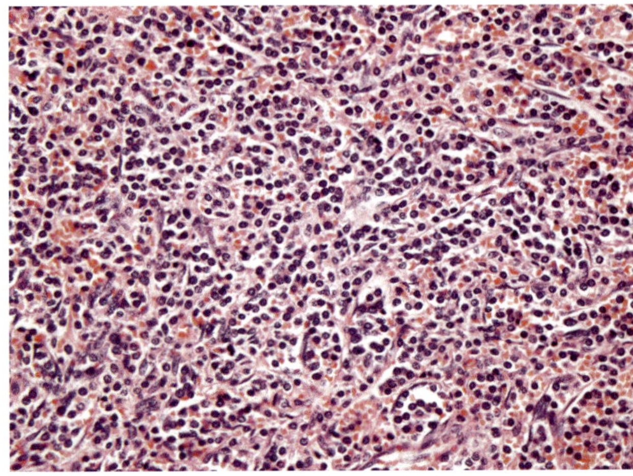

Fig. 3.88 Splenic diffuse red-pulp small B-cell lymphoma. High magnification of splenic diffuse red-pulp small B-cell lymphoma. The cells are seen within splenic cords and sinuses. The cells are round to slightly irregular with small to moderate amounts of pink cytoplasm

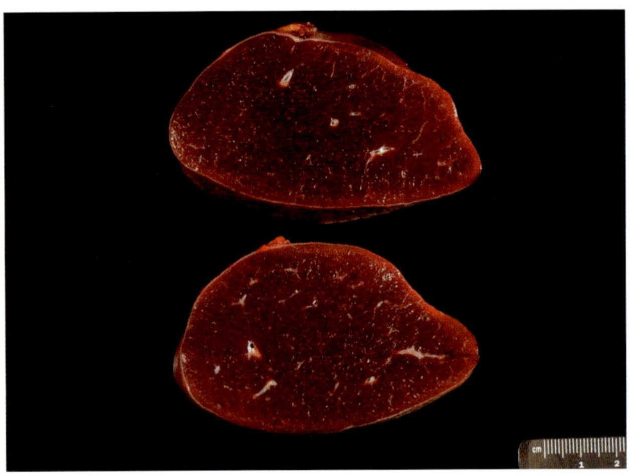

Fig. 3.87 Splenic diffuse red-pulp small B-cell lymphoma. Diffuse red pulp involvement is present in this enlarged spleen after chemotherapy. A similar appearance could be seen in HCL-V. (*Image courtesy of* W. Greaves, Houston, TX)

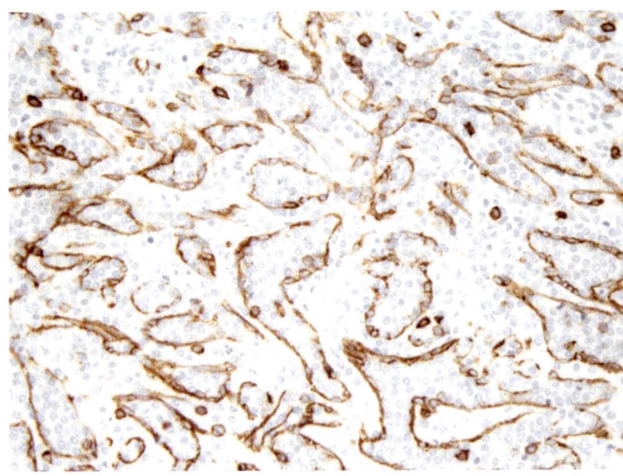

Fig. 3.89 Splenic diffuse red-pulp small B-cell lymphoma, CD8 stain. This highlights the expansion of the splenic cords and the distribution of the abnormal lymphocytes within splenic sinuses

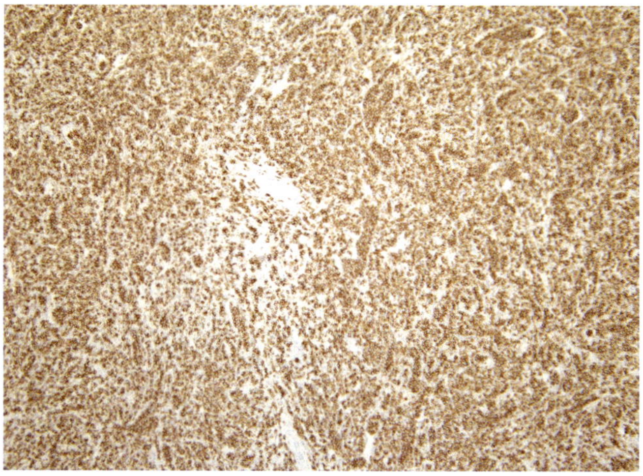

Fig. 3.90 Splenic diffuse red-pulp small B-cell lymphoma, CD20 stain. CD20 stain in splenic diffuse red-pulp small B-cell lymphoma. The diffuse distribution of B cells is characteristic of this lymphoma

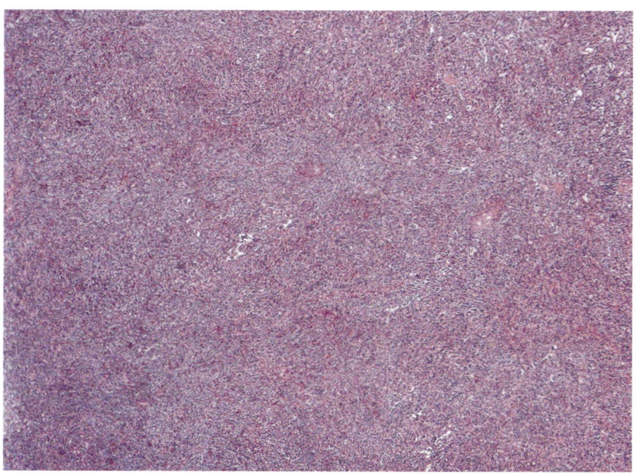

Fig. 3.91 Splenic diffuse red-pulp small B-cell lymphoma. Low-magnification microscopic image of splenic diffuse red-pulp small B-cell lymphoma. No white pulp structures are apparent at this magnification. The entire splenic architecture is overtaken by a proliferation of small lymphocytes

3.10 Prolymphocytic Leukemia

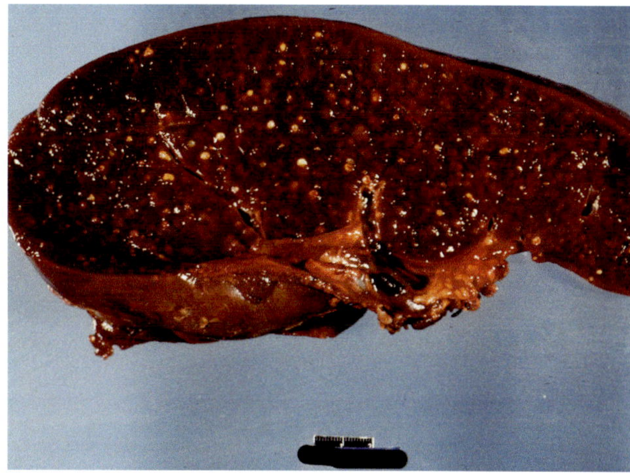

Fig. 3.92 B-cell prolymphocytic leukemia. Gross image of B-cell prolymphocytic leukemia (B-PLL) in spleen. This spleen is massively enlarged with prominent and enlarged white pulp structures. B-PLL needs to be distinguished from other B-cell disorders of the spleen, especially mantle cell lymphoma with a prolymphocytoid appearance. (*Image courtesy of* D. Farhi, Atlanta, GA.)

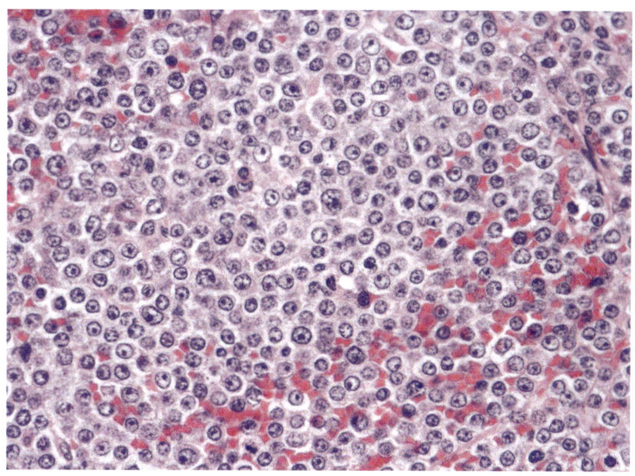

Fig. 3.93 B-cell prolymphocytic leukemia. High magnification of B-cell prolymphocytic leukemia in spleen. Involvement is usually in the red pulp but often completely effaces the normal splenic architecture. The cells are large in size, with round nuclei, prominent nucleoli, and small to moderate amounts of pale cytoplasm

3.11 Lymphoblastic Leukemia/Lymphoma

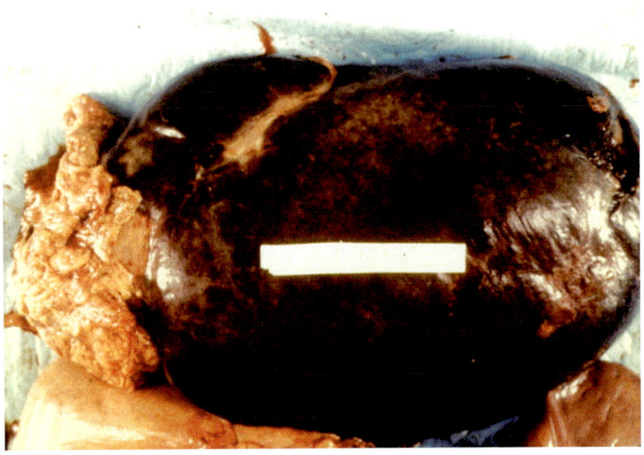

Fig. 3.94 Lymphoblastic leukemia/lymphoma. Gross image of spleen involved by lymphoblastic lymphoma/leukemia. No specific findings are apparent in the unsectioned spleen expect for a significant increase in size. The measuring tape is approximately 7 cm, suggesting markedly enlarged spleen

3.12 Hepatosplenic T-cell Lymphoma

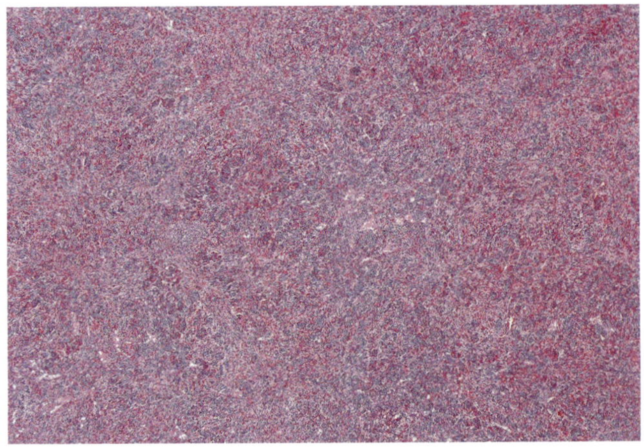

Fig. 3.95 Hepatosplenic T-cell lymphoma. Low-magnification image of hepatosplenic T-cell lymphoma involving spleen. Note that white pulp structures are not apparent, and there is a massive increase in lymphoid cells in the red pulp. Numerous abnormal lymphocytes are in the splenic cords and sinuses

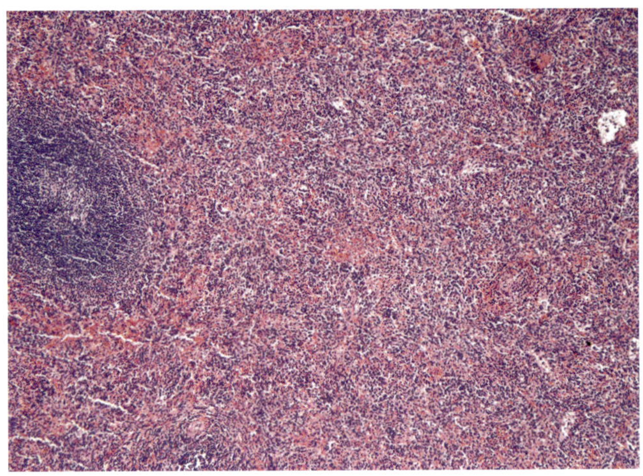

Fig. 3.96 Hepatosplenic T-cell lymphoma. In this example, lymphocytes in both cords and sinuses are small-to-intermediate in size, with condensed chromatin, round to slightly irregular nuclei, and only small amounts of cytoplasm. Note the large numbers of cells within the splenic sinuses

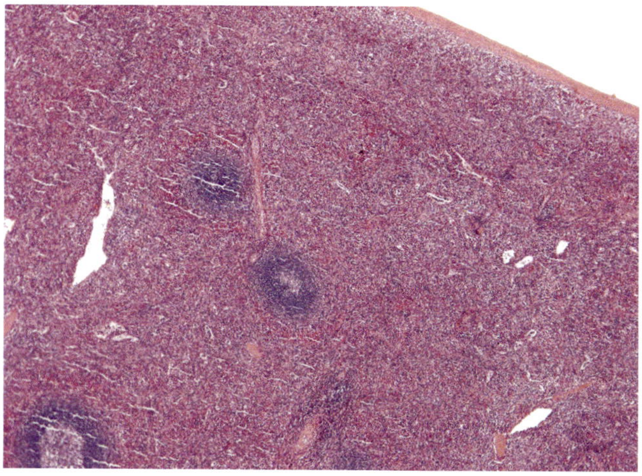

Fig. 3.97 Hepatosplenic T-cell lymphoma. Red pulp structures are expanded by small lymphocytes, with apparent sparing of white pulp structures in this example of hepatosplenic T-cell lymphoma

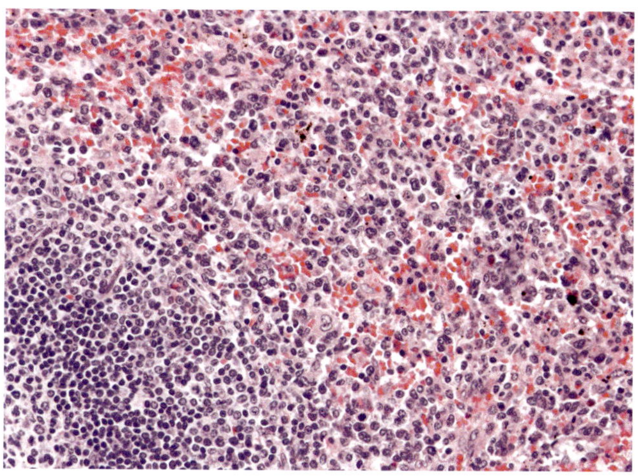

Fig. 3.98 Hepatosplenic T-cell lymphoma. Hepatosplenic T-cell lymphoma with preserved white pulp structure (*lower left*), and increased abnormal lymphocytes in the red pulp. The lymphocytes are slightly polymorphous in appearance. Cells are round to slightly irregular, with mature chromatin and small amounts of cytoplasm

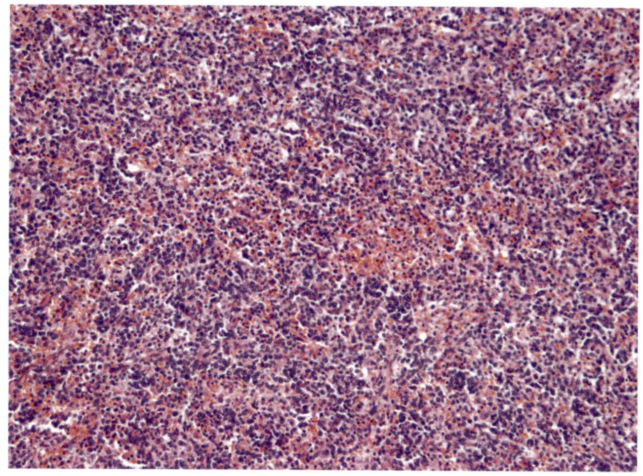

Fig. 3.99 Hepatosplenic T-cell lymphoma. In this example of hepatosplenic T-cell lymphoma, the abnormal lymphocytes are intermediate in size, with round to ovoid nuclei

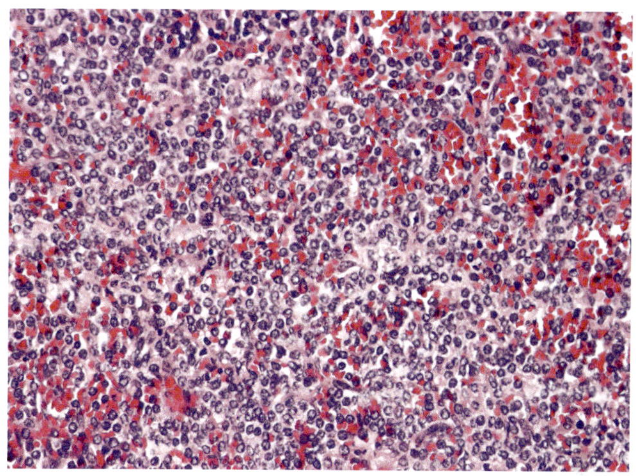

Fig. 3.100 Hepatosplenic T-cell lymphoma. In this example of hepatosplenic T-cell lymphoma, the cells are more intermediate to large in size, with vesicular chromatin. Most cases of hepatosplenic T-cell lymphoma express gamma/delta T-cell receptors

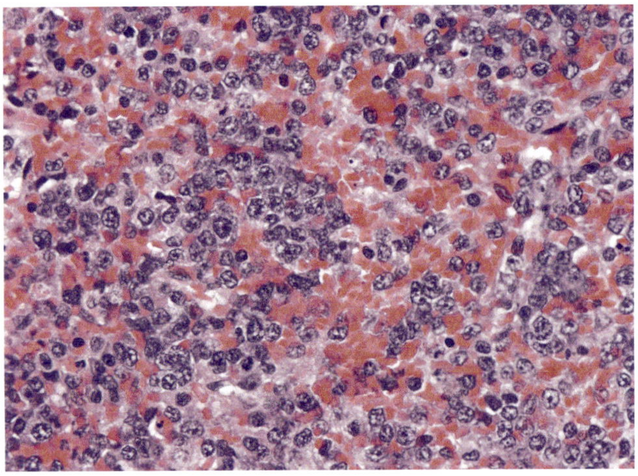

Fig. 3.101 Hepatosplenic T-cell lymphoma. Not all cases of hepatosplenic T-cell lymphoma are composed of small or intermediate cells. In this case, the lymphoma cells are large in size with irregular nuclei and occasional prominent nucleoli

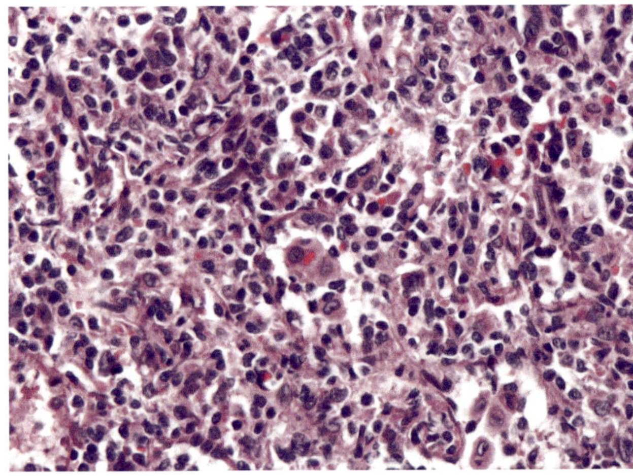

Fig. 3.103 Hepatosplenic T-cell lymphoma, hemophagocytosis. This example of hepatosplenic T-cell lymphoma shows predominantly small lymphocytes. In addition, this case has hemophagocytosis (*center*). Histiocytes ingest cellular components (often erythroid elements). Hemophagocytosis can be seen in association with all types of lymphoma, but is relatively more frequent with hepatosplenic T-cell lymphoma

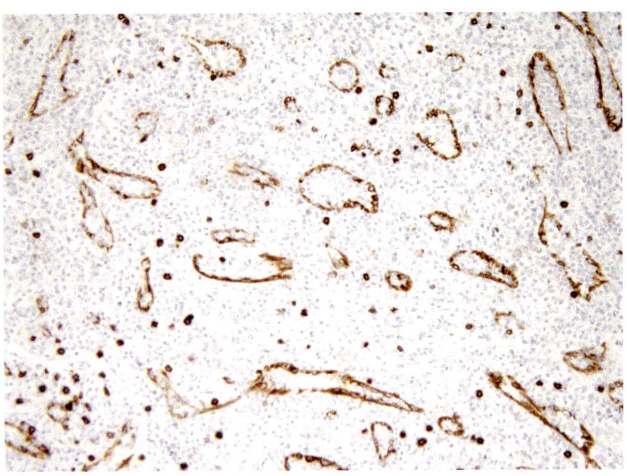

Fig. 3.102 Hepatosplenic T-cell lymphoma, CD8 stain. Most cases of hepatosplenic T-cell lymphoma are negative for both CD4 and CD8. In this CD8 stain, the red pulp is expanded by lymphocytes, but only rare, residual normal cytotoxic T cells are positive for CD8. Splenic sinuses (spread apart by the neoplastic infiltrate) also express CD8

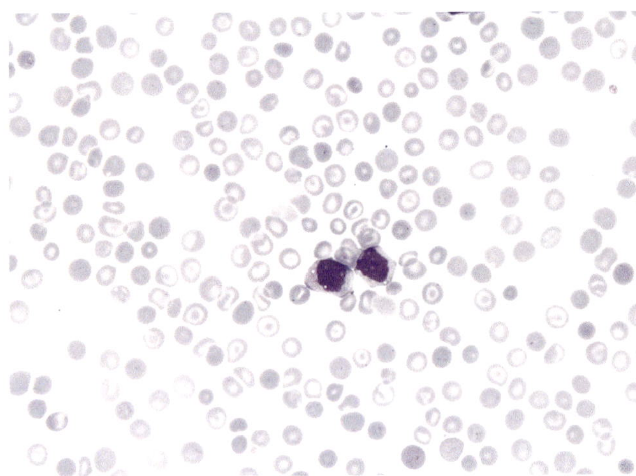

Fig. 3.104 Hepatosplenic T-cell lymphoma, peripheral blood. Peripheral blood smear of hepatosplenic T-cell lymphoma. The abnormal lymphocytes (*center*) have intermediate to large nuclei with irregular nuclear contours and moderate amounts of cytoplasm

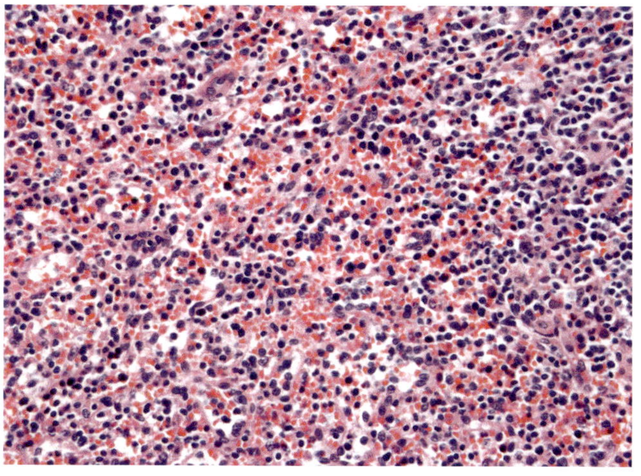

Fig. 3.105 Hepatosplenic T-cell lymphoma, bone marrow. Bone marrow involvement by hepatosplenic T-cell lymphoma. The abnormal lymphocytes are apparent, and are larger in size that granulocytes and erythroid precursors. In some cases of hepatosplenic T-cell lymphoma, bone marrow infiltration can be subtle and is only apparent by immunohistochemical staining. Involvement is often in bone marrow sinuses

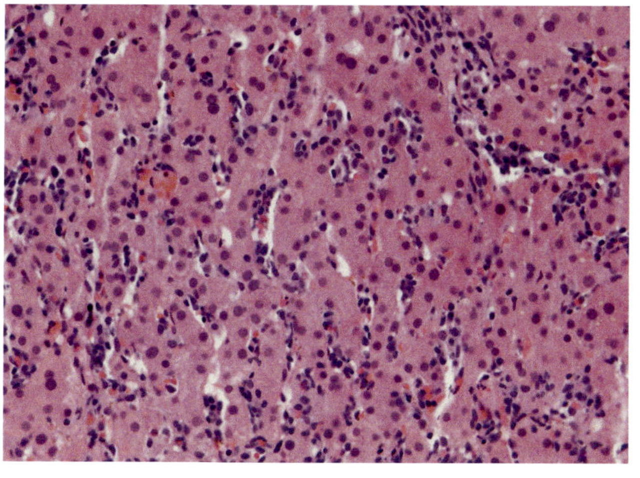

Fig. 3.106 Hepatosplenic T-cell lymphoma, liver. Liver biopsy with involvement by hepatosplenic T-cell lymphoma. The abnormal lymphocytes are predominantly present within the hepatic sinuses. (*Image courtesy of* L.J. Medeiros, Houston, TX.)

3.13 T-cell Large Granular Lymphocytic Leukemia

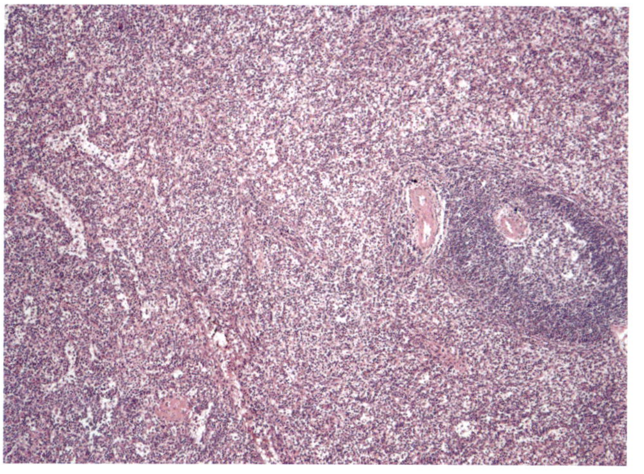

Fig. 3.107 T-cell large granular lymphocytic leukemia. Low-magnification image of spleen with involvement by T-cell large granular lymphocytic leukemia. At this magnification, there are no significant abnormalities except a mild to moderate increase in small nucleated cells (lymphocytes) within the red pulp. Sinuses are maintained as are white pulp structures

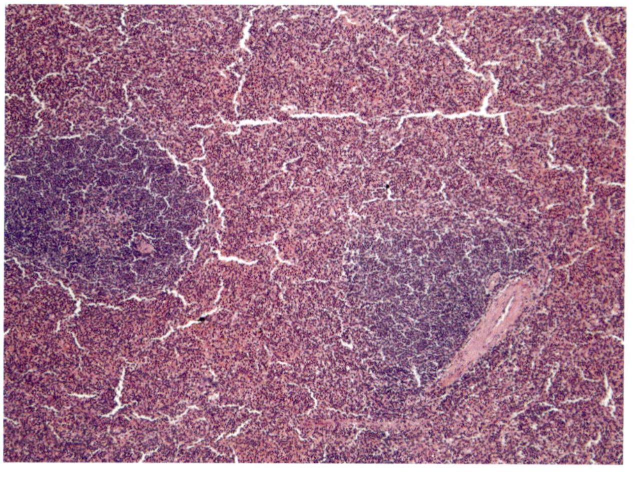

Fig. 3.108 T-cell large granular lymphocytic leukemia. Another example of T-cell large granular lymphocytic leukemia at low magnification. In this case, there is a notable increase in small lymphocytes within the red pulp. T-cell large granular lymphocytic leukemia is often identified by flow cytometry of peripheral blood. Splenectomy is only rarely performed for diagnostic purposes

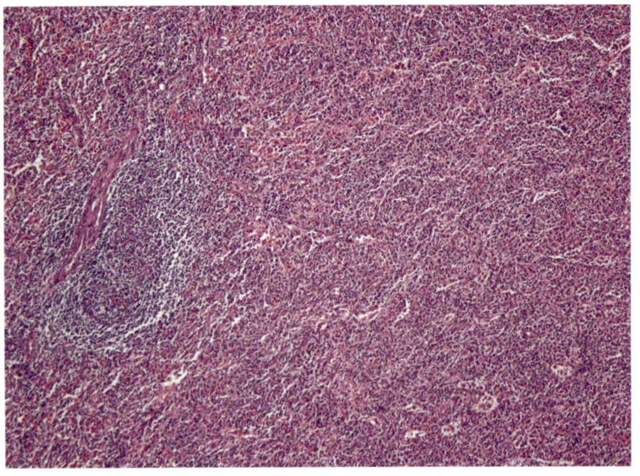

Fig. 3.109 T-cell large granular lymphocytic leukemia. In this example of T-cell large granular lymphocytic leukemia, there are many lymphocytes in the red pulp and a blurring of the borders of the white pulp. In more advanced disease, T-cell large granular lymphocytic leukemia involves the PALS regions of the spleen

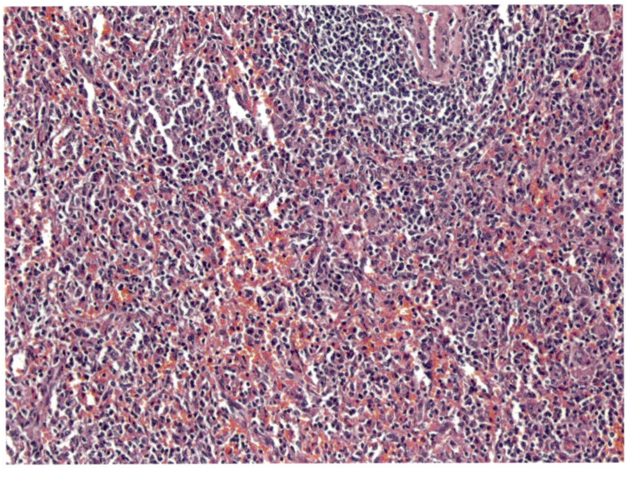

Fig. 3.110 T-cell large granular lymphocytic leukemia. An increase in lymphocytes of T-cell large granular lymphocytic leukemia in the red pulp of the spleen. At this magnification, the involvement of both red pulp cords and sinuses is noted. The typical immunophenotype for T-cell large granular lymphocytic leukemia is expression of CD3 (surface), CD8 (a cytotoxic T-cell marker), and aberrant expression of CD57 and CD16. Most cases also have dim expression or loss of CD5

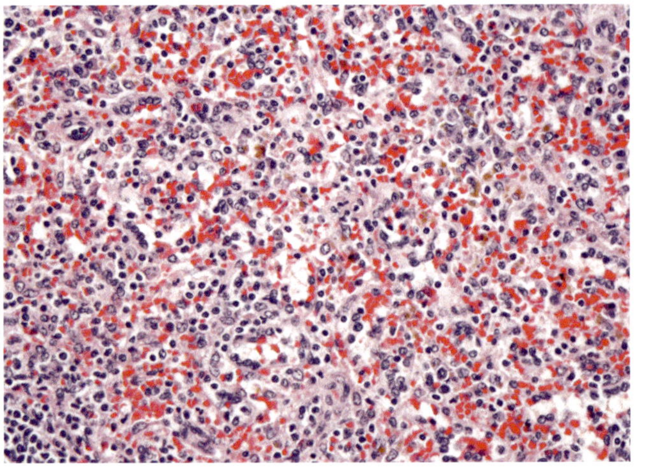

Fig. 3.111 T-cell large granular lymphocytic leukemia. High magnification of T-cell large granular lymphocytic leukemia in splenic red pulp. The lymphocytes mostly have small nuclei, with some minimal nuclear irregularity. The cells have minimal cytoplasm or small amounts of pink cytoplasm

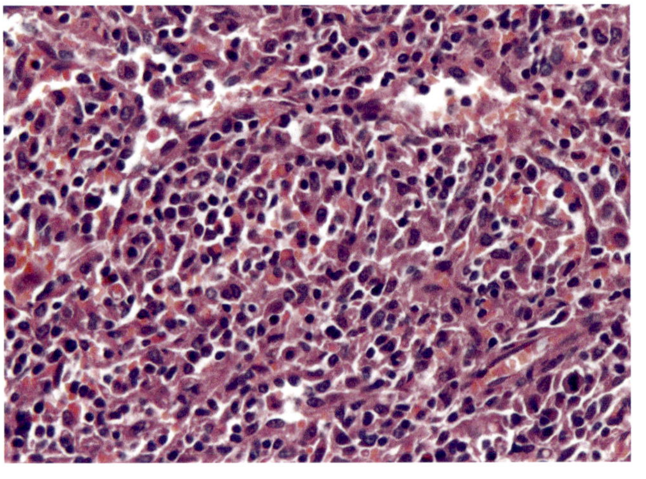

Fig. 3.112 T-cell large granular lymphocytic leukemia. Another example of high magnification in T-cell large granular lymphocytic leukemia. In this case, the abnormal lymphocytes show more cellular pleomorphism than in the previous illustrated case

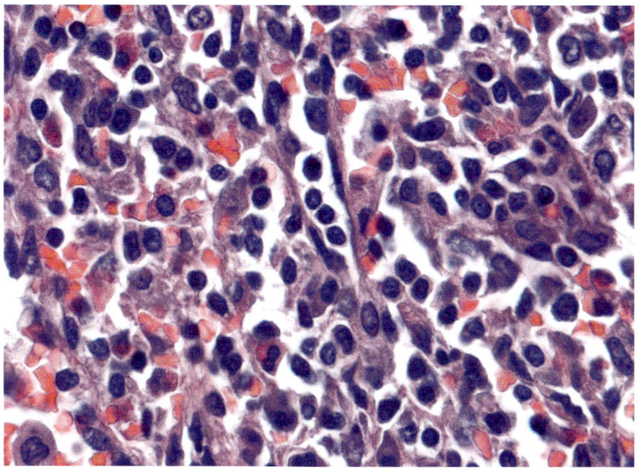

Fig. 3.113 T-cell large granular lymphocytic leukemia. Another high magnification image of T-cell large granular lymphocytic leukemia. In this case, a granulocyte is present (*middle, lower*), which suggests the size (mostly small) of the abnormal lymphoid cells

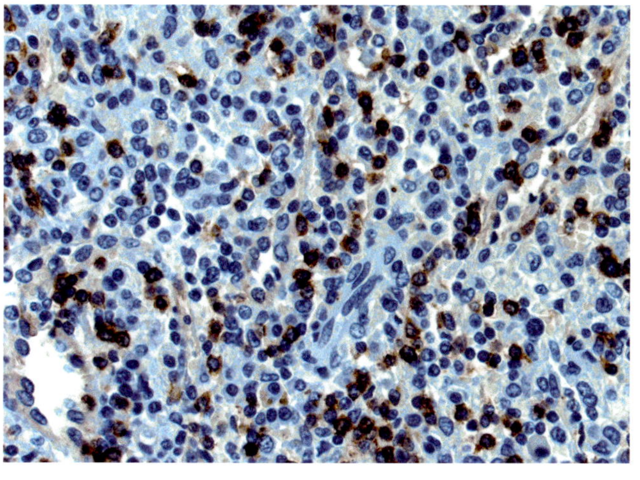

Fig. 3.114 T-cell large granular lymphocytic leukemia, CD8 stain. Immunohistochemical staining for T-cell large granular lymphocytic leukemia with CD8, a cytotoxic T-cell marker. The abnormal T-cells are present mostly within sinuses but also are present within cords

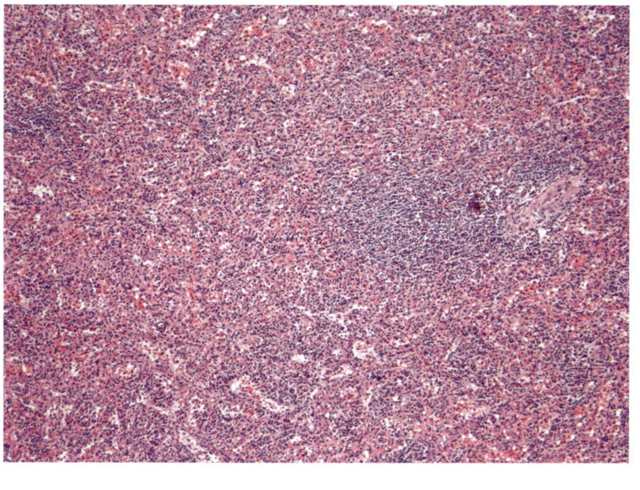

Fig. 3.115 Natural killer (NK)-cell large granular lymphocytic leukemia. Low-magnification image of a chronic leukemia, composed of NK cells, involving the spleen. This is termed *chronic lymphoproliferative disorder of NK cells* in the 2008 World Health Organization classification

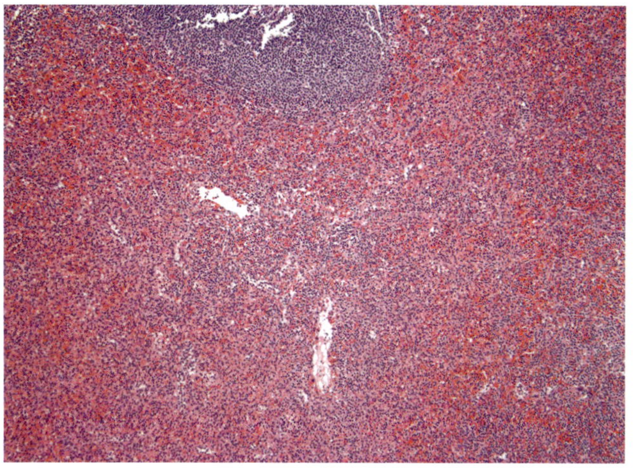

Fig. 3.116 NK-cell large granular lymphocytic leukemia. Another low magnification of a chronic leukemia of NK cells in the spleen. This has a comparable histologic appearance to T-cell large granular lymphocytic leukemia, but differs in immunophenotype. NK cells will lack expression of surface CD3 and some other pan T-cell antigens. In most cases, they will express NK-associated marker CD56, and often CD57 and CD16

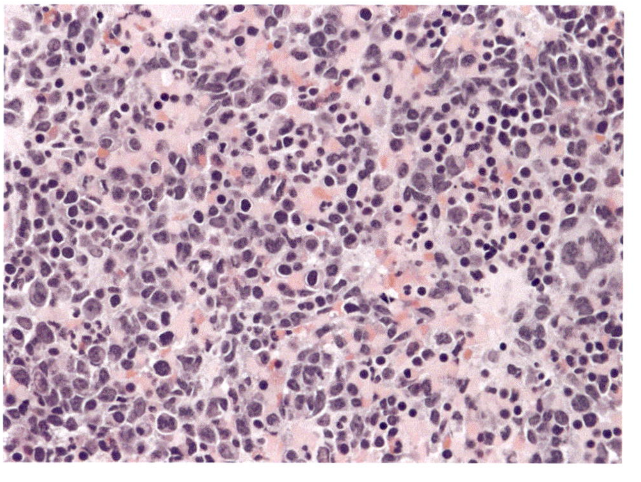

Fig. 3.117 NK-cell large granular lymphocytic leukemia. High magnification of chronic lymphoproliferative disorder of NK cells. In morphology, this is quite similar to T-cell large granular lymphocytic leukemia, with small lymphocytes involving the red pulp

3.14 Other T-cell Lymphomas

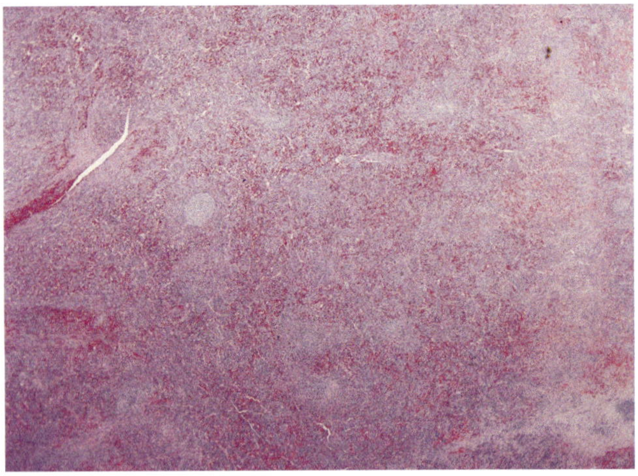

Fig. 3.118 Angioimmunoblastic T-cell lymphoma. Low magnification of spleen with involvement by angioimmunoblastic T-cell lymphoma. In this example, there are some preserved follicles present. In addition, there is expansion of the PALS region, with blurring of the borders between red and white pulp. The red pulp also shows infiltration by large numbers of lymphoid cells

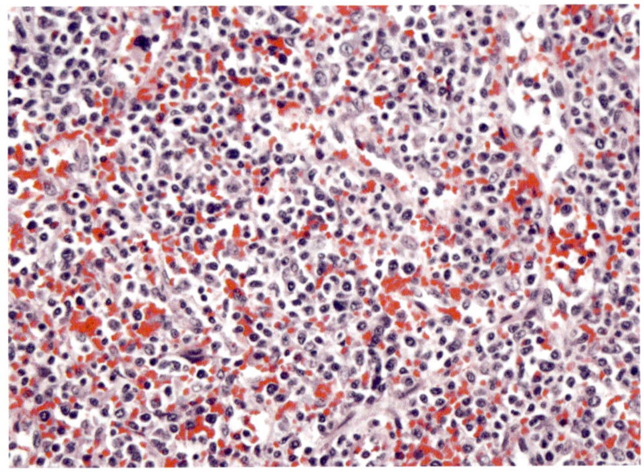

Fig. 3.120 Angioimmunoblastic T-cell lymphoma. The polymorphous cellular nature of angioimmunoblastic T-cell lymphoma in spleen. In addition to many lymphocytes, there are also admixed plasma cells. In many cases, extramedullary hematopoiesis may also be seen

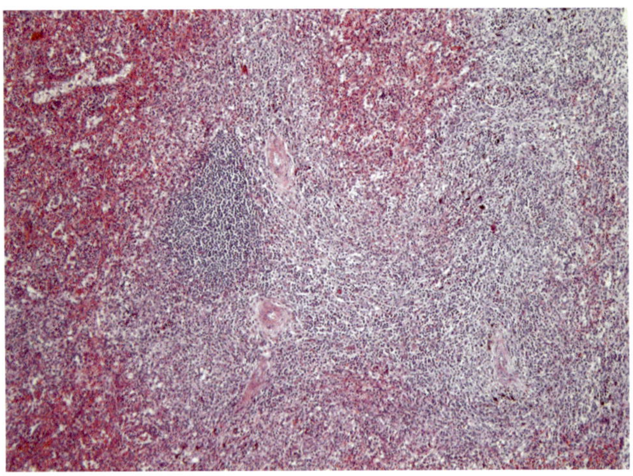

Fig. 3.119 Angioimmunoblastic T-cell lymphoma. In this example of angioimmunoblastic T-cell lymphoma in spleen, the red pulp is mostly uninvolved. However, the PALS region is expanded by atypical lymphoid cells with a polymorphous appearance, which are pale in comparison to the cells of the white pulp follicle (primary follicle)

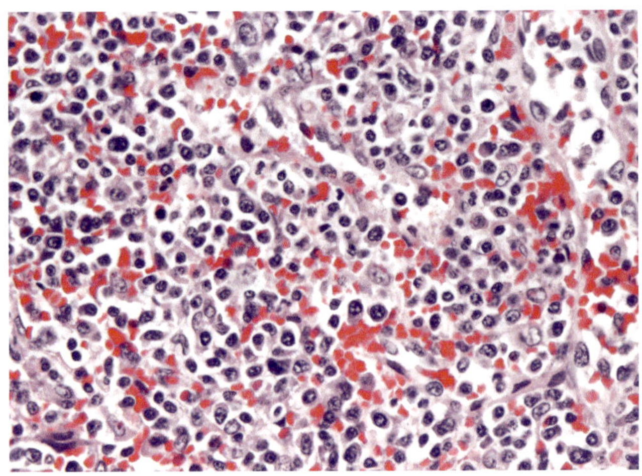

Fig. 3.121 Angioimmunoblastic T-cell lymphoma. High magnification of angioimmunoblastic T-cell lymphoma in spleen. There are many large- and intermediate-sized lymphocytes present. Plasma cells are prominent. The largest cells present (ovoid nuclei, marginated chromatin) are stromal elements from cords and/or sinuses

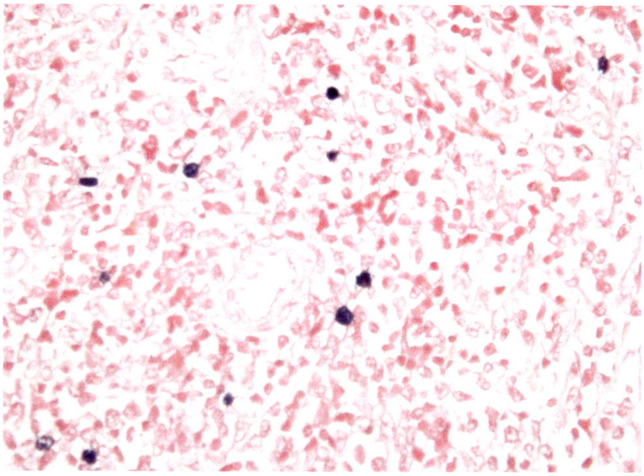

Fig. 3.122 Angioimmunoblastic T-cell lymphoma, Epstein-Barr virus (EBV) in situ stain. In situ staining for EBV in angioimmunoblastic T-cell lymphoma. Scattered cells are typically positive for EBV, but these are not specific for this diagnosis as several T-cell lymphoma types may have EBV-positive cells

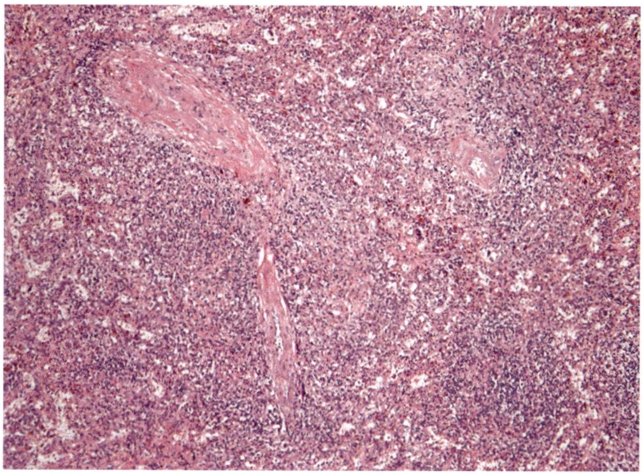

Fig. 3.123 Enteropathy-type T-cell lymphoma. Enteropathy-type T-cell lymphoma is most often diagnosed in the small intestine or other areas of the gastrointestinal tract. Although it has some characteristic findings that could support a diagnosis in spleen, splenic involvement is very rare and would be seen as a site of diffuse systemic involvement. In this example, the involvement is mostly confined to the PALS region of the white pulp

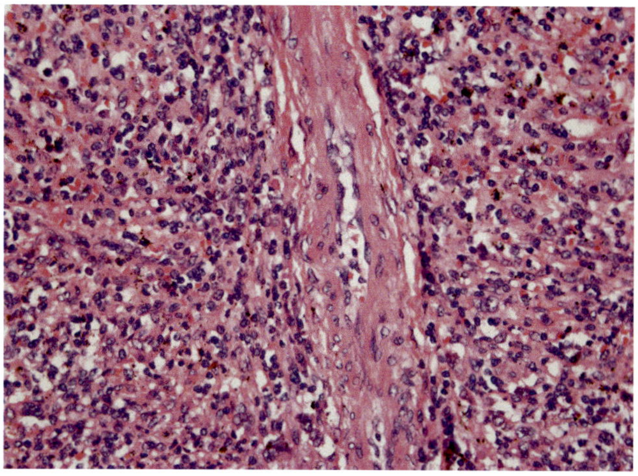

Fig. 3.124 Enteropathy-type T-cell lymphoma. High magnification of enteropathy-type T-cell lymphoma in the spleen. The abnormal lymphocytes are small to intermediate in size with moderate cellular pleomorphism and coarse chromatin

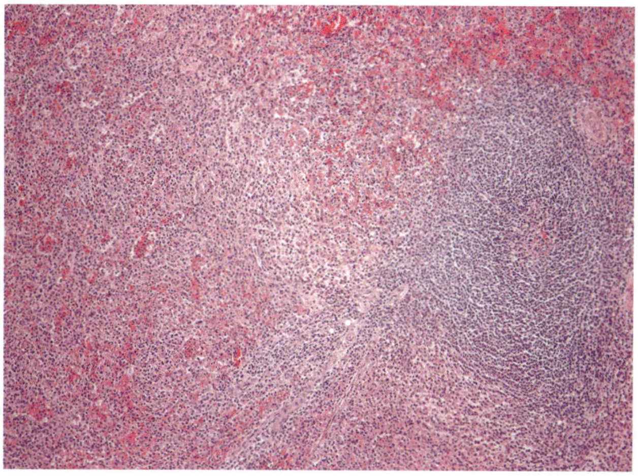

Fig. 3.125 Peripheral T-cell lymphoma. Low magnification of spleen with involvement by peripheral T-cell lymphoma. In this case, the abnormal lymphocytes are present in splenic red pulp. They are seen within the cords and sinuses

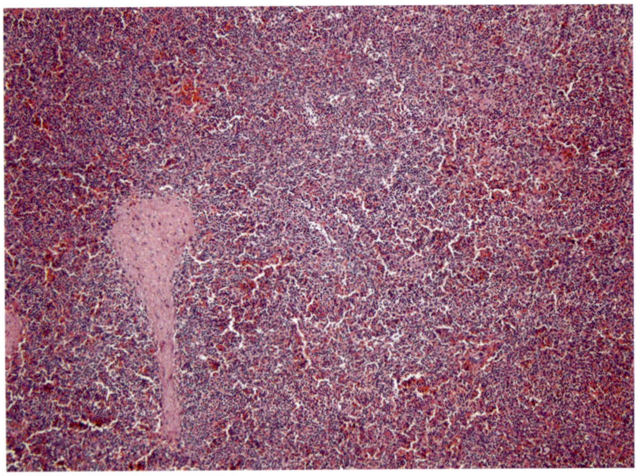

Fig. 3.126 Peripheral T-cell lymphoma. In this example of peripheral T-cell lymphoma in spleen, there is an increase in lymphocytes (*eg*, lymphoma cells) in both the red pulp and in the PALS region. The border between the red and white pulp is indistinct due to the infiltrate of abnormal lymphocytes. The diagnosis of peripheral T-cell lymphoma is used when the features of a more specific type of T-cell lymphoma cannot be applied to a particular case

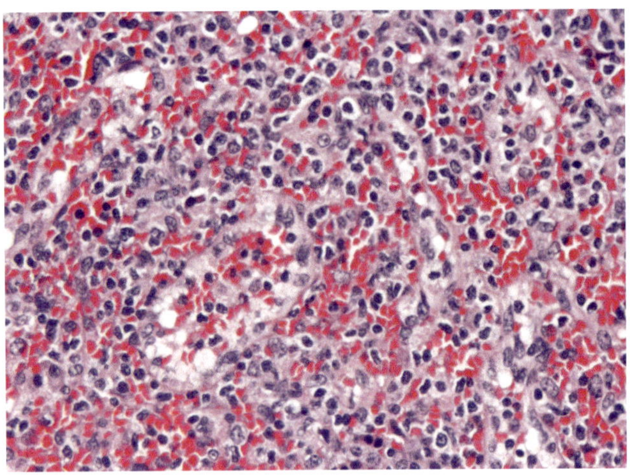

Fig. 3.127 Peripheral T-cell lymphoma. High magnification of peripheral T-cell lymphoma with red pulp involvement. Abnormal lymphocytes are within the splenic cords and some are in sinuses. They are small to intermediate in size with irregular nuclei

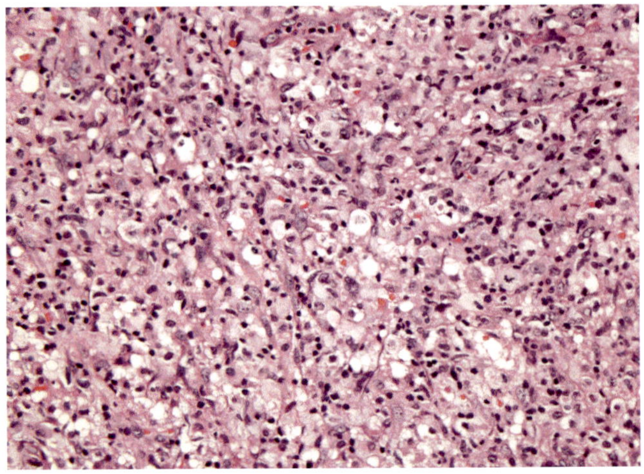

Fig. 3.128 Peripheral T-cell lymphoma. A highly pleomorphic example of peripheral T-cell lymphoma. The abnormal lymphocytes have large amounts of clear or pale cytoplasm. Some have large, Hodgkin-like nuclei

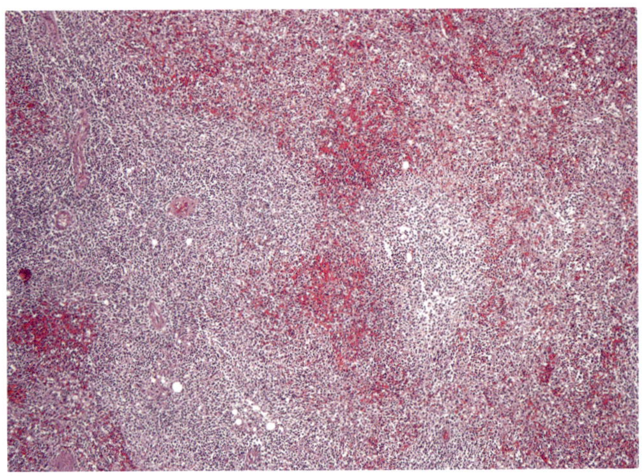

Fig. 3.129 Mycosis fungoides. This low magnification of spleen shows involvement by mycosis fungoides. Although typically an indolent T-cell lymphoma confined to skin, in some cases, diffuse systemic involvement may be present. The spleen shows massive expansion of the PALS regions by abnormal lymphocytes

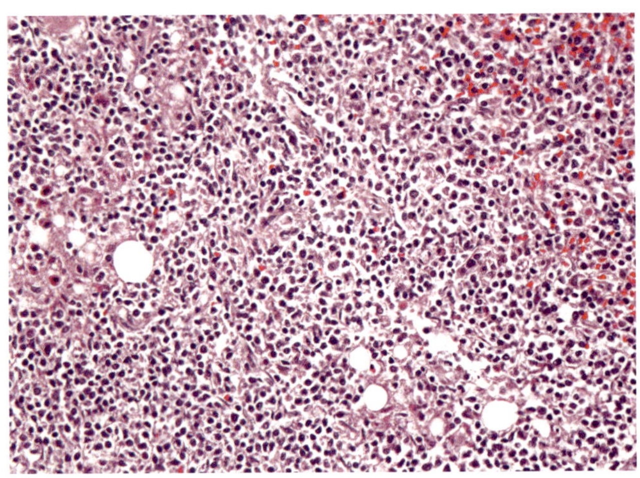

Fig. 3.130 Mycosis fungoides. High magnification of mycosis fungoides involving spleen. The abnormal lymphocytes are small in size with slightly irregular nuclei and scant cytoplasm. This image shows an area of expanded white pulp (PALS region)

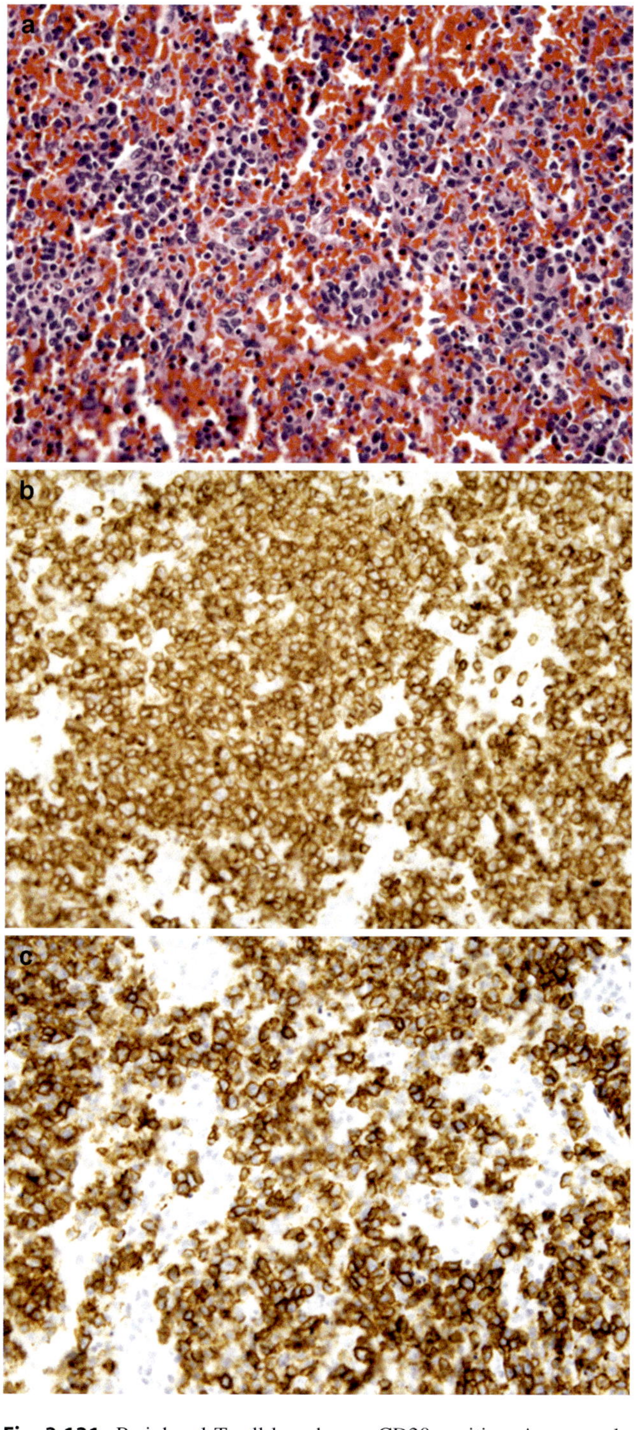

Fig. 3.131 Peripheral T-cell lymphoma, CD20 positive. An example of peripheral T-cell lymphoma with an extremely rare immunophenotype (**a**). This T-cell lymphoma involving spleen showed expression of CD3 (**b**) in addition to CD20 (**c**). The lymphoma cells showed other pan T-cell antigen expression but lacked any other B-cell markers

3.15 Diffuse Large B-cell Lymphomas

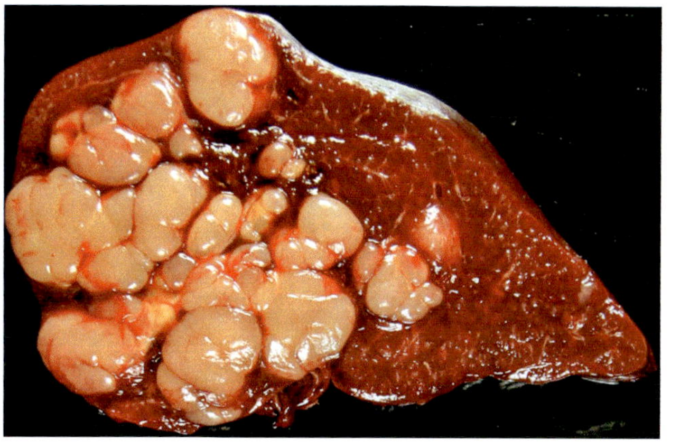

Fig. 3.132 Diffuse large B-cell lymphoma. Gross image of diffuse large B-cell lymphoma (DLBCL) in spleen. The pale, fleshy cut surface of the nodules are typical of the gross appearance of DLBCL in any site. (*Image courtesy of* A. Sohani, Boston, MA.)

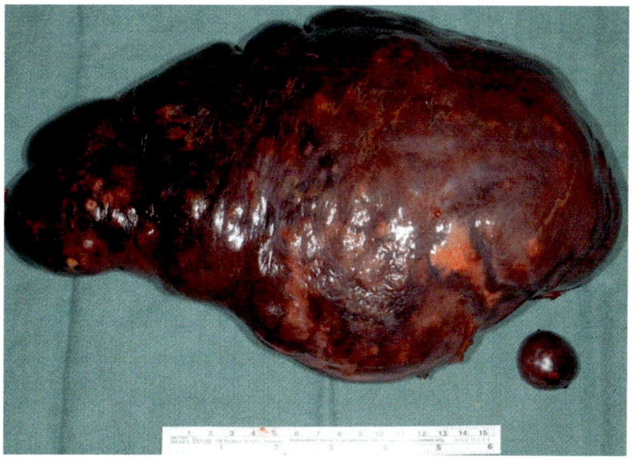

Fig. 3.133 Diffuse large B-cell lymphoma. An example of an unsectioned spleen and accessory spleen with massive involvement by DLBCL. The abnormal pale lumps of tumor stretch and abut the capsule. Some areas of capsular thinning are apparent. Tumors such as this can be associated with splenic rupture. (*Image courtesy of* S. Konoplev, Houston, TX.)

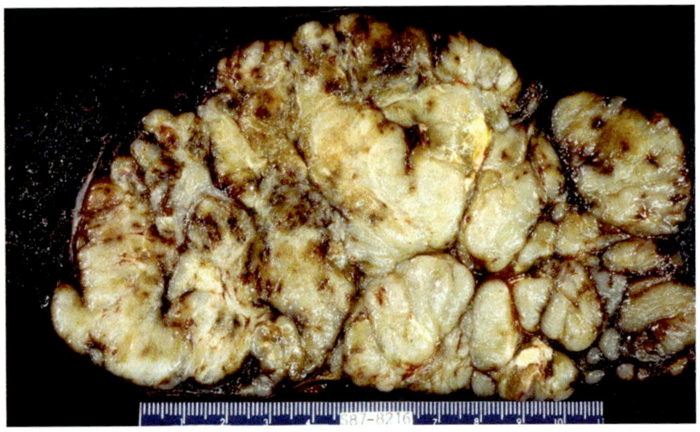

Fig. 3.134 Diffuse large B-cell lymphoma. Gross image of diffuse large B-cell lymphoma. Compressed normal spleen parenchyma (*upper left*) is present. The lymphoma is yellow with areas of hemorrhage. It is also present in large macro-nodules. Some foci of necrosis may be present in the central portions of some nodules. (*Image courtesy of* D. Farhi, Atlanta, GA.)

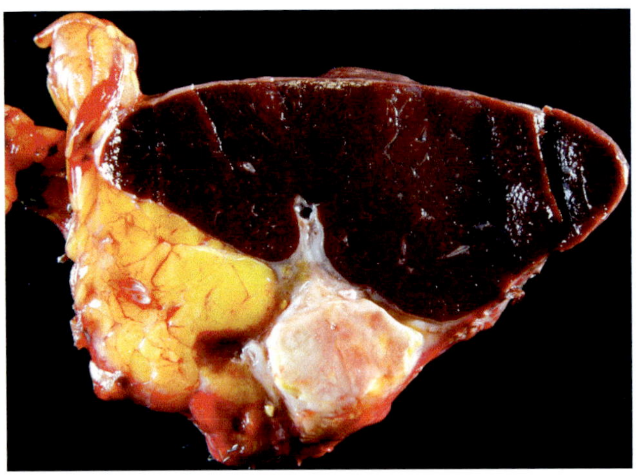

Fig. 3.135 Diffuse large B-cell lymphoma. Gross image of spleen (*upper*) and diffuse large B-cell lymphoma with massive involvement of splenic hilar lymph node. Note the uniform pink, fleshy appearance of lymphoma compared to the deep red of normal splenic parenchyma. (*Image courtesy of* A. Sohani, Boston, MA.)

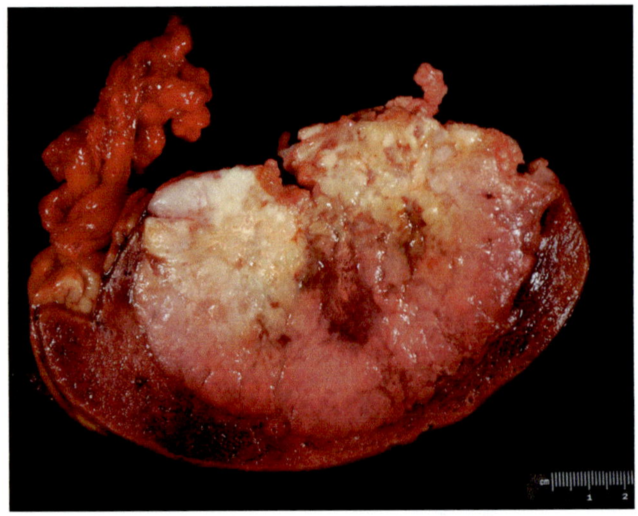

Fig. 3.136 Diffuse large B-cell lymphoma. Gross image of diffuse large B-cell lymphoma of spleen. Areas of necrosis are present that impart a chalky, pale appearance to the otherwise fleshy pink of the lymphoma. Compressed unaffected spleen is seen at the periphery. (*Image courtesy of* W. Greaves, Houston, TX.)

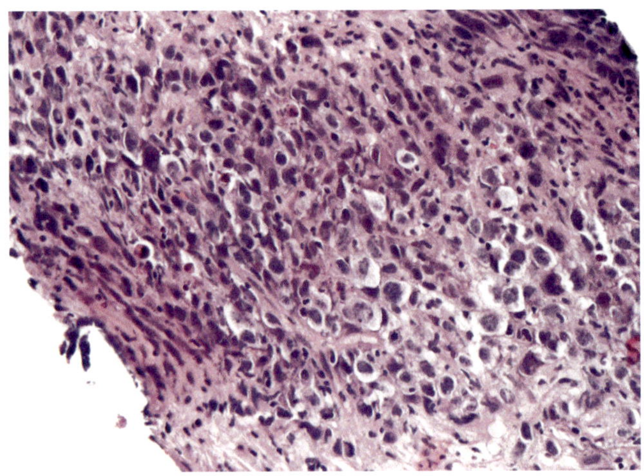

Fig. 3.137 Diffuse large B-cell lymphoma. Splenic needle biopsy showing diffuse large B-cell lymphoma. Needle biopsies are often non-diagnostic for many splenic disorders, however, in some cases, such as diffuse large B-cell lymphoma, a needle biopsy may be informative

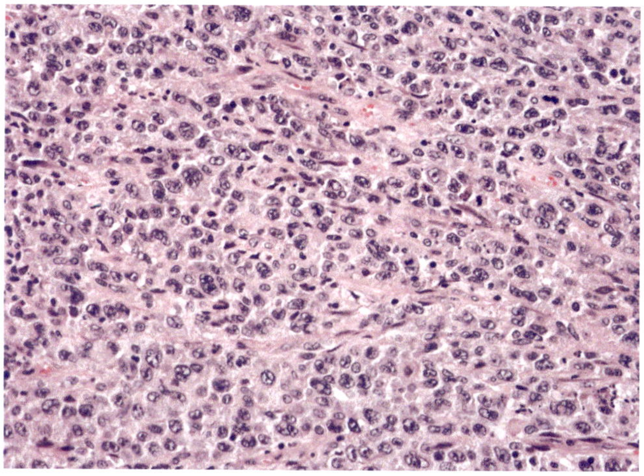

Fig. 3.138 Diffuse large B-cell lymphoma. An image of a diffuse large B-cell lymphoma. While the morphologic appearance is not specific, this case represents a large cell transformation in a patient with a history of nodular lymphocyte-predominant Hodgkin lymphoma (NLPHL)

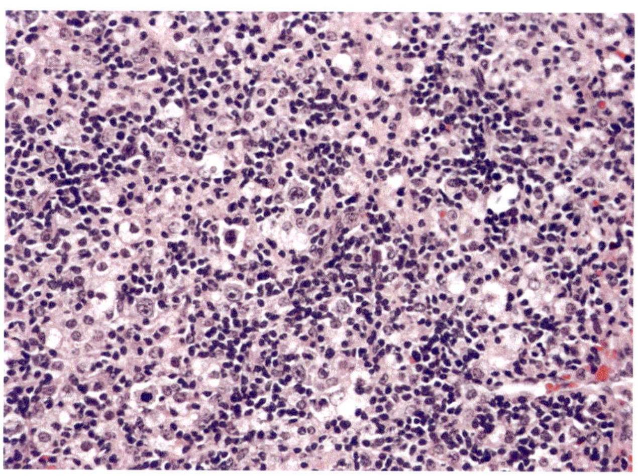

Fig. 3.139 Nodular lymphocyte-predominant Hodgkin lymphoma: T-cell/histiocyte-rich large B-cell lymphoma. This patient has a history of NLPHL. Scattered, very large, abnormal lymphocytes are present. This could represent NLPHL, but it is difficult to distinguish from T-cell/histiocyte-rich large B-cell lymphoma in this case

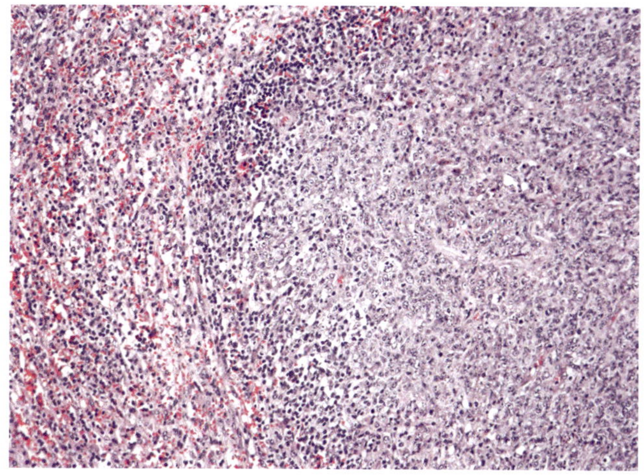

Fig. 3.140 Diffuse large B-cell lymphoma. Example of a nodular aggregate of diffuse large B-cell lymphoma adjacent to uninvolved red pulp. Different patterns of involvement can be seen in the spleen including nodular (*eg*, white pulp), red pulp, or diffuse

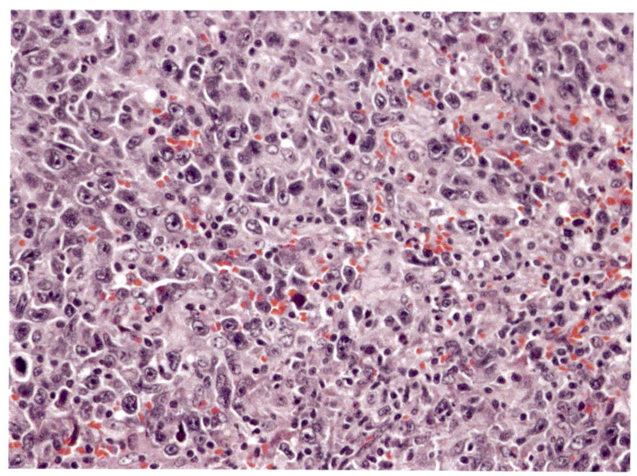

Fig. 3.141 Diffuse large B-cell lymphoma. High magnification of diffuse large B-cell lymphoma of spleen. Individual tumor cells are large in size with irregular nuclear contours, vesicular chromatin, and prominent nucleoli

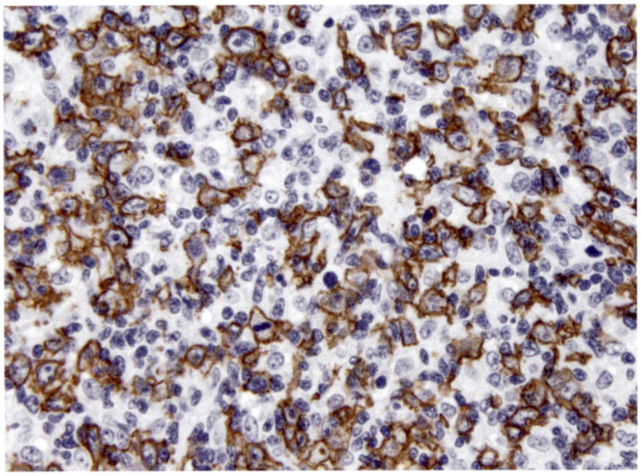

Fig. 3.142 Diffuse large B-cell lymphoma, CD20 stain. CD20 stains of diffuse large B-cell lymphoma involving red pulp. Individual large lymphoma cells are outlined by CD20 immunohistochemical stain. These lymphoma cells are within cords and sinuses, with many interspersed small lymphocytes (likely T cells) and macrophages

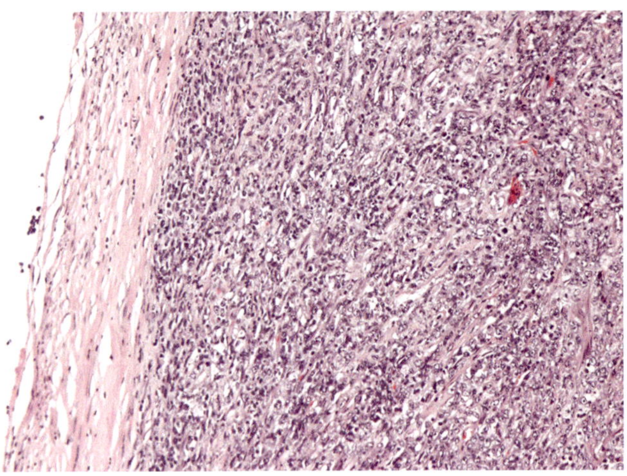

Fig. 3.143 Diffuse large B-cell lymphoma. Diffuse large B-cell lymphoma adjacent to splenic capsule. The splenic capsule is not typically infiltrated by tumor cells in lymphoma. The capsule would likely rupture if this were to happen

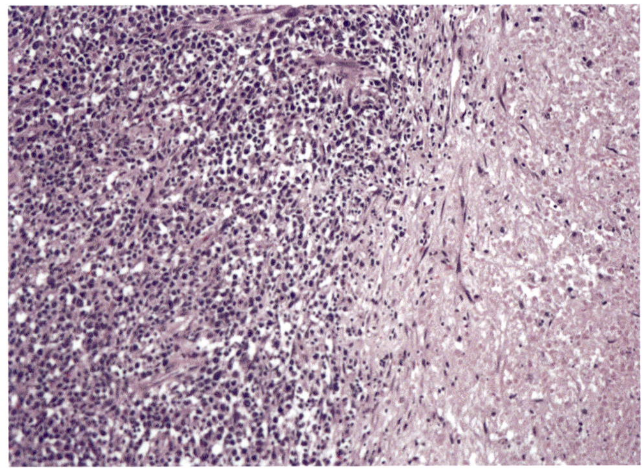

Fig. 3.144 Diffuse large B-cell lymphoma. Diffuse large B-cell lymphoma (*left*) adjacent to an area of coagulative necrosis (*right*). The lymphoma cells have a diffuse pattern. The areas of necrosis show cell ghosts. Likely there was compromise of the vascular supply leading to the area of necrosis

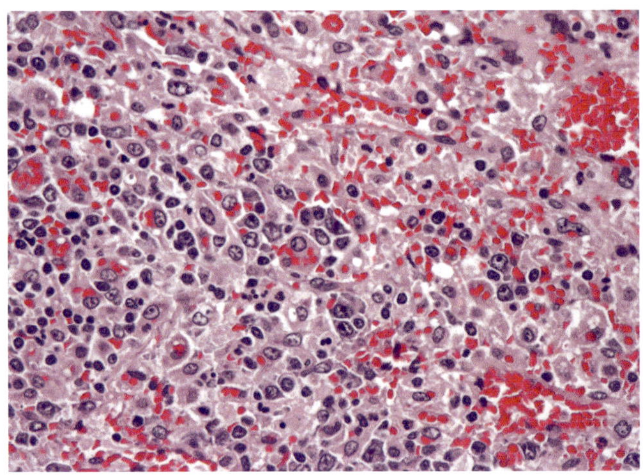

Fig. 3.145 Diffuse large B-cell lymphoma with hemophagocytosis. Although not a specific finding, this example of diffuse large B-cell lymphoma also has focal hemophagocytosis. Macrophages with ingested red blood cells and other cellular elements are seen

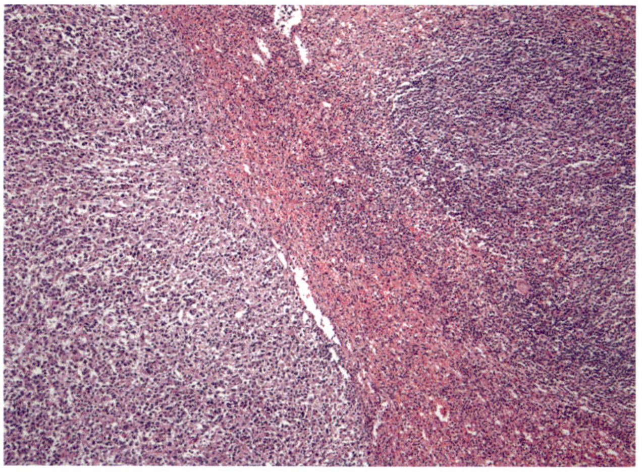

Fig. 3.146 Follicular lymphoma and diffuse large B-cell lymphoma. Composite lymphoma consists of follicular lymphoma of white pulp (*right*) and diffuse large B-cell lymphoma (*left*). Studies have shown that coincident lymphoma can share an origin with the low-grade component (*eg*, follicular lymphoma) or arise as de novo malignancies

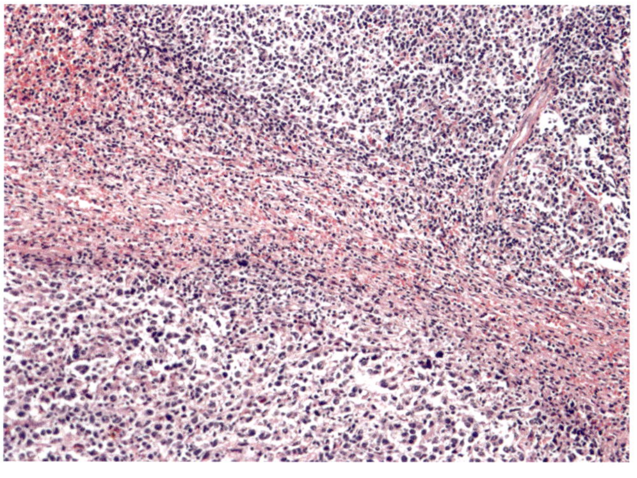

Fig. 3.147 Follicular lymphoma and diffuse large B-cell lymphoma. High magnification highlighting diffuse large B-cell lymphoma (*lower*) and coincident follicular lymphoma (*upper*)

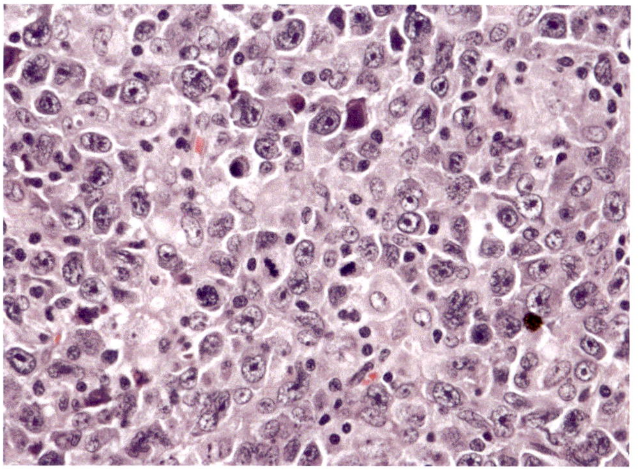

Fig. 3.148 Diffuse large B-cell lymphoma. High magnification of diffuse large B-cell lymphoma of spleen. Mitotic figures and apoptotic bodies are easy to identify in this case

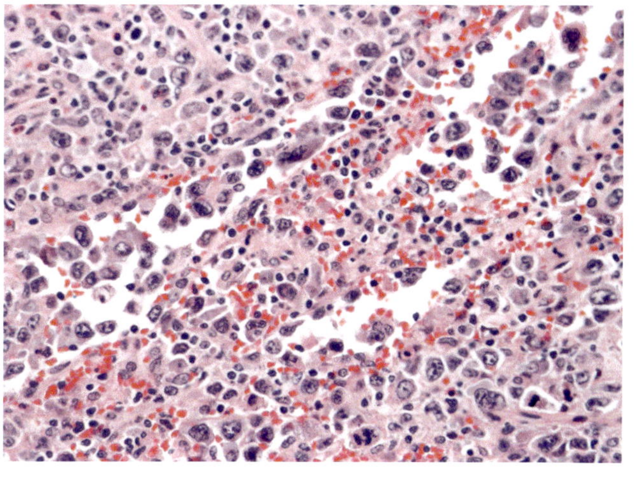

Fig. 3.149 Intravascular large B-cell lymphoma. In this example of large B-cell lymphoma of the spleen, there is a preferential distribution of the large abnormal lymphocytes within splenic sinuses. This can be referred to as intravascular large B-cell lymphoma

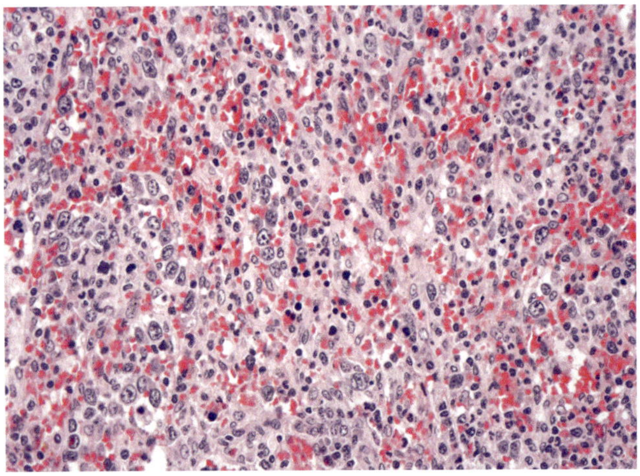

Fig. 3.150 Intravascular large B-cell lymphoma. Another example of intravascular large B-cell lymphoma with large cells mostly within splenic sinuses. In this case, there is some focal erythroid extramedullary hematopoiesis (small, dark, round nuclei)

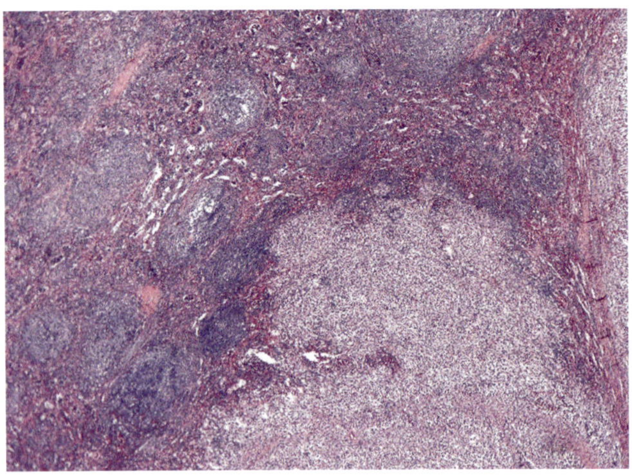

Fig. 3.151 Richter transformation. Low-magnification image of a composite lymphoma. In the upper portions of the image, there is expansion of white pulp structures by small dark blue lymphocytes. There is also an indication of increased small dark-blue lymphocytes in the red pulp. This is in contrast to the nodule or pale cells (*lower right*). This represents chronic lymphocytic leukemia (*upper*) with focal transformation to diffuse large B-cell lymphoma, also known as Richter syndrome

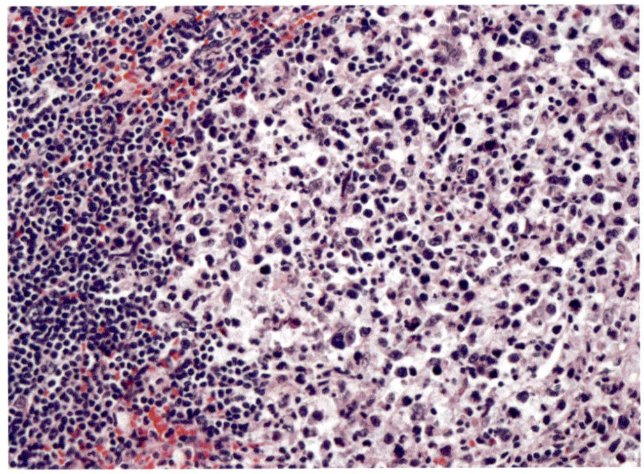

Fig. 3.152 Richter transformation. High magnification of CLL in spleen (*left*) and Richter syndrome (*right*). The Richter transformation is most traditionally seen as a transformation of CLL to large B-cell lymphoma

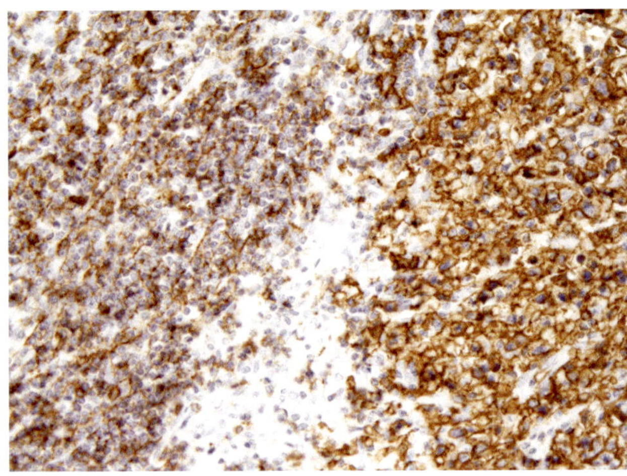

Fig. 3.153 Richter transformation, CD20 stain. CD20 staining in composite CLL with Richter syndrome. Note that the CLL (*left*) has weaker and less uniform staining for CD20 (a common finding in CLL) whereas the large B-cell transformation (*right*) has strong and uniform staining for CD20

3.16 Hodgkin Lymphoma

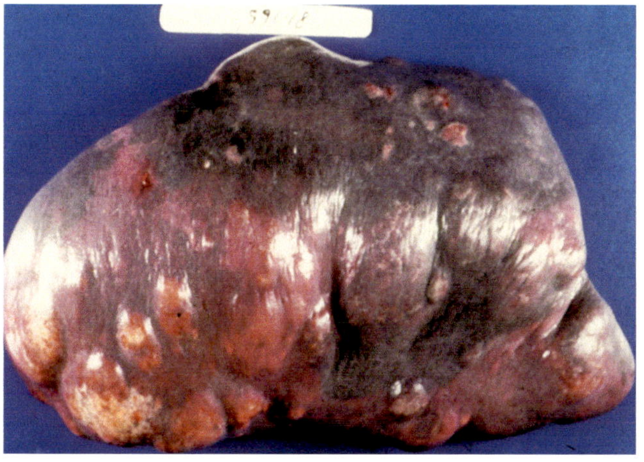

Fig. 3.154 Classical Hodgkin lymphoma (CHL). Gross image of an unsectioned spleen with massive involvement by Hodgkin lymphoma. Note that there are multiple nodules of varying sizes. Splenectomy specimens are rarely seen in association with Hodgkin lymphoma. In the past, splenectomy was a usual procedure for disease staging. However, this has been replaced by imaging studies. (*Image courtesy of* D. Farhi, Atlanta, GA.)

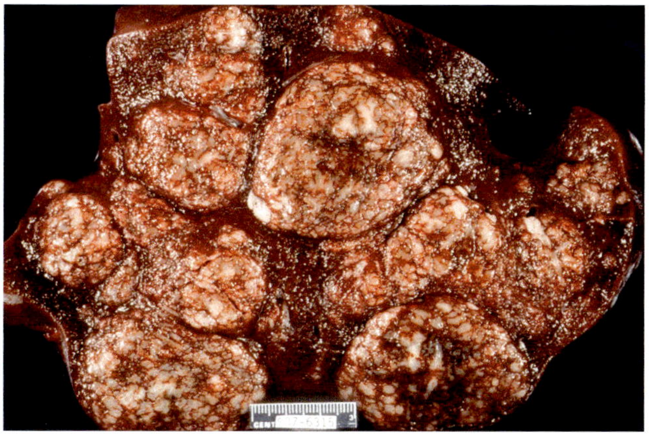

Fig. 3.155 Classical Hodgkin lymphoma. Gross image of a sectioned spleen with massive involvement by CHL. Although a firm tan nodule is typical, it can also be seen in a broad range of other lymphoid and nonlymphoid processes in the spleen. (*Image courtesy of* D. Farhi, Atlanta, GA.)

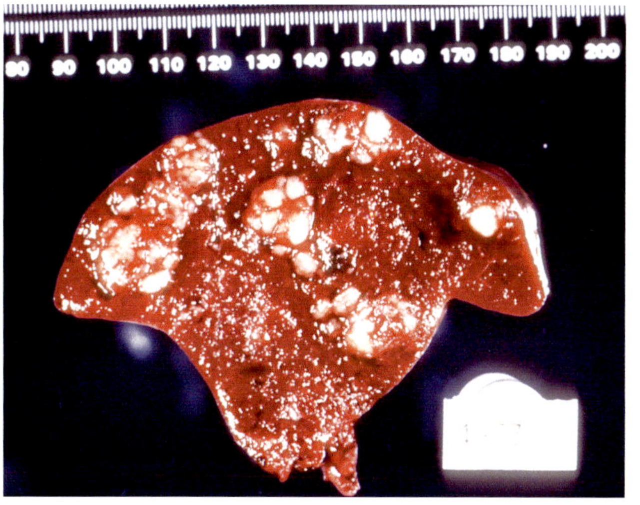

Fig. 3.156 Classical Hodgkin lymphoma. A more typical appearance of CHL in the spleen. This gross section of a spleen shows several small nodules (< 1–2 cm in diameter) scattered throughout the spleen. Each nodule is a site of Hodgkin lymphoma. Because of their small size, section of the spleen at intervals of 1–1.5 cm is recommended for thorough gross examination. (*Image courtesy of* D. Farhi, Atlanta, GA.)

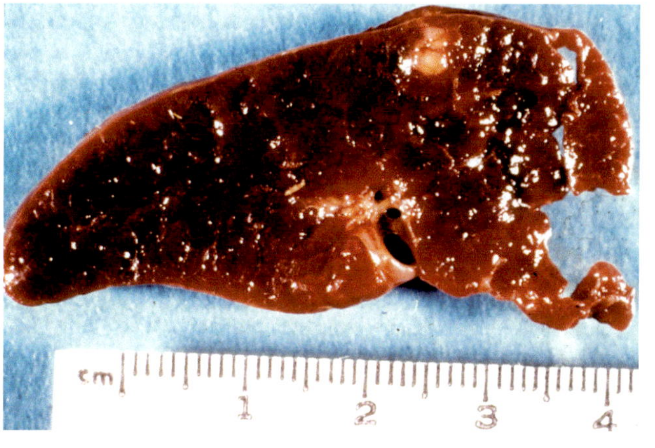

Fig. 3.157 Classical Hodgkin lymphoma. In this gross section of spleen, only a single nodule of Hodgkin lymphoma is present (*top, right*). Degrees of involvement can be quite variable as shown in the previous series of images. (*Image courtesy of* D. Farhi, Atlanta, GA.)

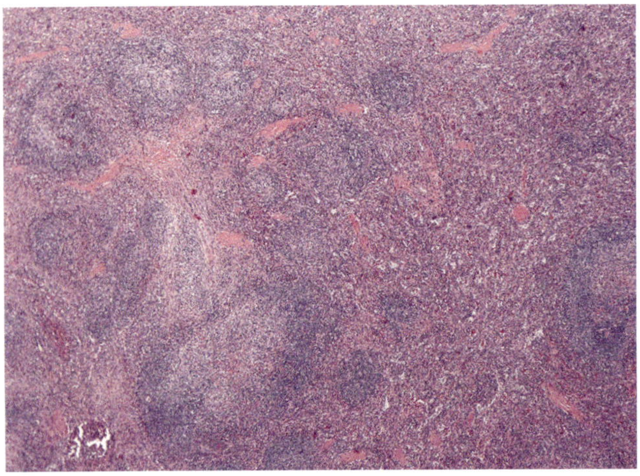

Fig. 3.158 Classical Hodgkin lymphoma. Low magnification of splenic involvement by CHL. In general, there is preservation of both red and white pulp structures; adjacent to the white pulp, however, are some hypocellular, pale-appearing areas. When present in small nodules, CHL is most often seen adjacent to white pulp or blood vessels

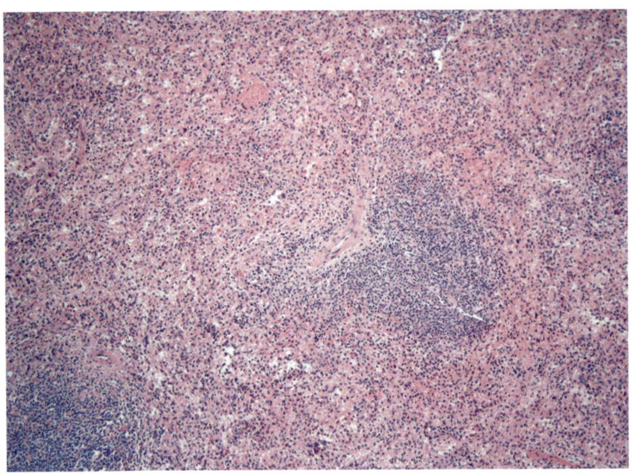

Fig. 3.159 Classical Hodgkin lymphoma. Intermediate magnification of CHL in spleen. In this example, there is an abnormal cellular composition in the central region of the image, with an increase in fibrosis, histiocytes, and eosinophils. In this background, careful inspection will reveal Hodgkin or Reed-Sternberg cells

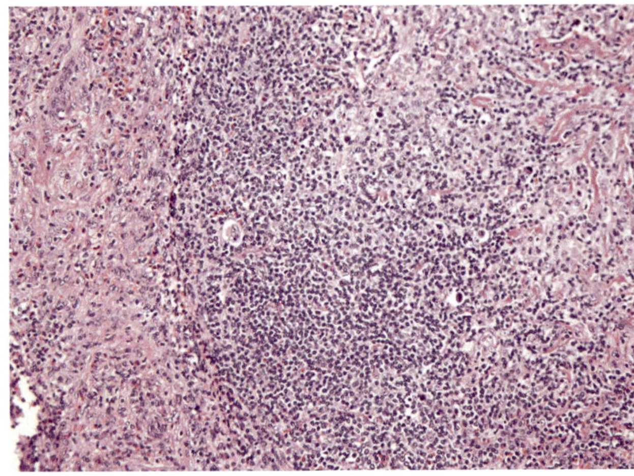

Fig. 3.160 Classical Hodgkin lymphoma. In this image, a single Hodgkin/Reed-Sternberg cell is seen (*center, left*). It is quite large in size and stands out from the background. To confirm the diagnosis, the typical immunophenotype (CD15+, CD30+, PAX-5 weak+, CD45-) should be present

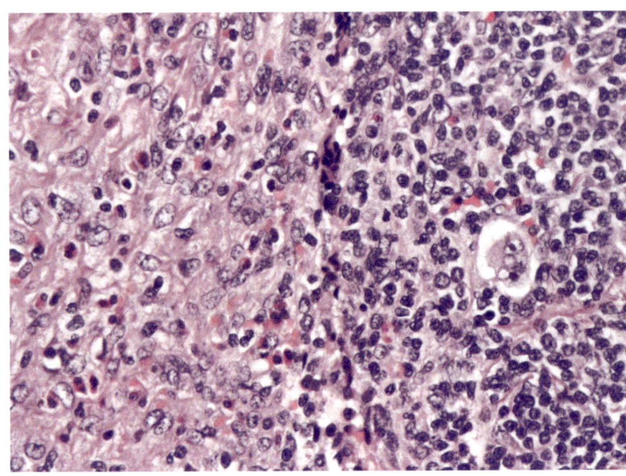

Fig. 3.161 Classical Hodgkin lymphoma. The same Hodgkin/Reed-Sternberg cell as above seen at high magnification. Note the prominent red macronucleoli. Also note the fibrosis and increased eosinophils (*left*)

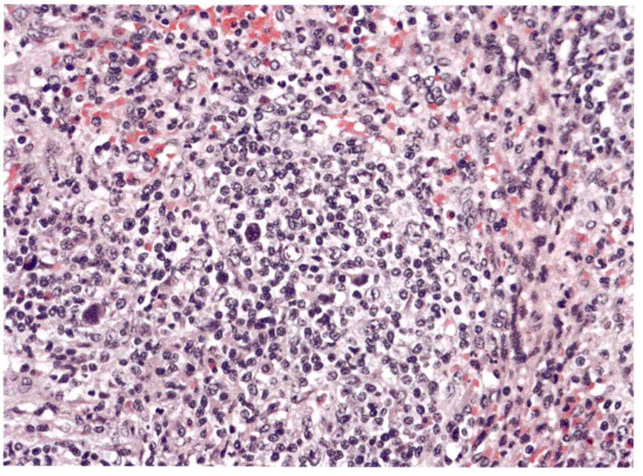

Fig. 3.162 Classical Hodgkin lymphoma. In this image, several Hodgkin cells are present, including "mummy cells" (darkened, shrunken nuclei in very large cells) and some large cells with single nuclei

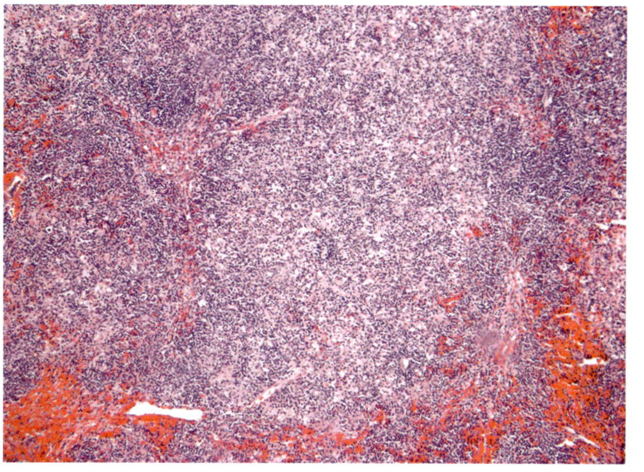

Fig. 3.163 Nodular lymphocyte-predominant Hodgkin lymphoma. NLPHL only rarely involves the spleen. When present, it has a histologic appearance comparable to CHL. The areas of involvement are paucicellular, with histiocytes and a polymorphous cellular infiltrate. Rare large abnormal cells (LP cells) are present. Distinguishing NLPHL from CHL is typically based on history and immunohistochemical evaluation

4 Myeloid and Related Disorders

Because of its microenvironment, the spleen will often have manifestations of bone marrow disease. Specifically, marrow-derived abnormalities of the myeloid series are apparent in the spleen. Proliferations of myeloid elements due to administration of cytokines, especially granulocyte colony-stimulating factor (G-CSF), are illustrated. This is an important mimic of other myeloid disorders in the spleen, including acute myeloid leukemia (AML). AML in the spleen is rare but can be seen both as a primary process and a transformation of chronic myeloid disorders. Splenic manifestations are most commonly seen in chronic myeloid disorders such as chronic myelogenous leukemia, primary myelofibrosis, polycythemia vera, and chronic myelomonocytic leukemia. Each of these has some unusual and occasionally characteristic findings in the spleen. Although more rare, myelodysplastic syndromes may also present with splenic pathology. Systemic mastocytosis often involves the spleen and the pathologic findings are illustrated. Histiocytic sarcoma, a rare and aggressive disorder, is illustrated. Both gross and histologic images are included. Where appropriate, immunohistochemical stains are also shown.

D.P. O'Malley, *Atlas of Spleen Pathology*, Atlas of Anatomic Pathology,
DOI 10.1007/978-1-4614-4672-9_4, © Springer Science+Business Media New York 2013

4.1 Extramedullary Hematopoiesis

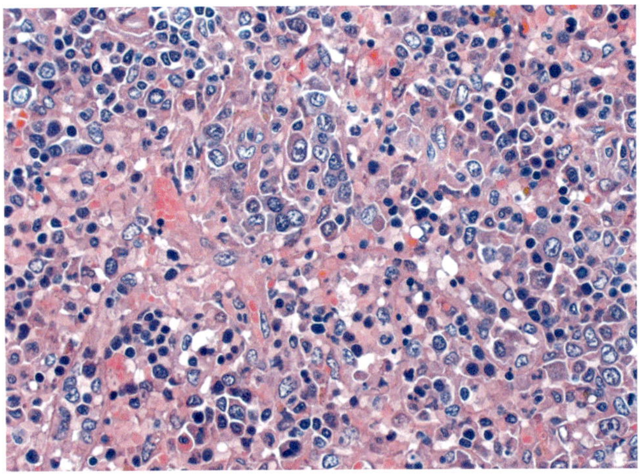

Fig. 4.1 Extramedullary hematopoiesis. In this example of autoimmune hemolytic anemia, many erythroid precursors appear in the red pulp of the spleen and there are also immature myeloid elements, with slightly irregular nuclei, coarse chromatin, and moderate amounts of pink granular cytoplasm

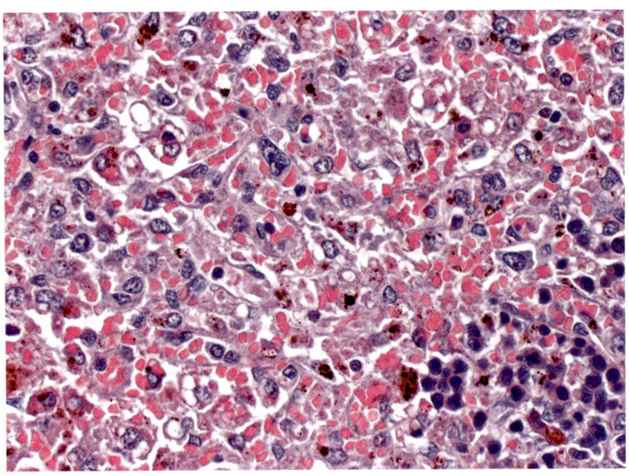

Fig. 4.2 Extramedullary hematopoiesis. In this spleen, the patient underwent autologous bone marrow transplant. Immature erythroid (*lower right*) and myeloid elements are seen throughout the splenic red pulp. Extensive hemosiderin deposition (brown pigment) is likely due to transfusions and excessive breakdown of red blood cells

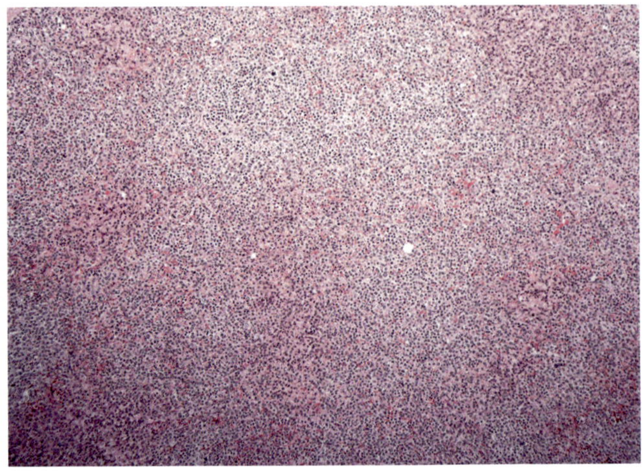

Fig. 4.3 Extramedullary hematopoiesis, G-CSF effect. Low magnification of a spleen in a patient treated with G-CSF. In some cases, this causes a massive proliferation of myeloid elements in the spleen. Rarely, as in this case, there is acute splenic rupture due to the enlargement

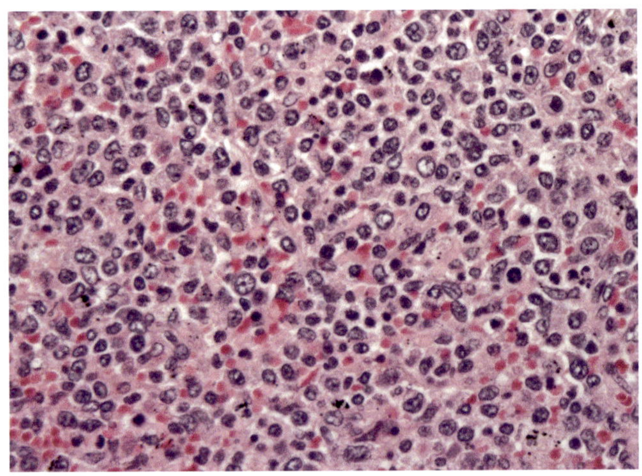

Fig. 4.4 Extramedullary hematopoiesis, G-CSF effect. High magnification of extramedullary proliferation of myeloid cells due to G-CSF therapy. Without appropriate history, this proliferation would be difficult to distinguish from AML/myeloid sarcoma

4.2 Myeloid Sarcoma/Acute Myeloid Leukemia

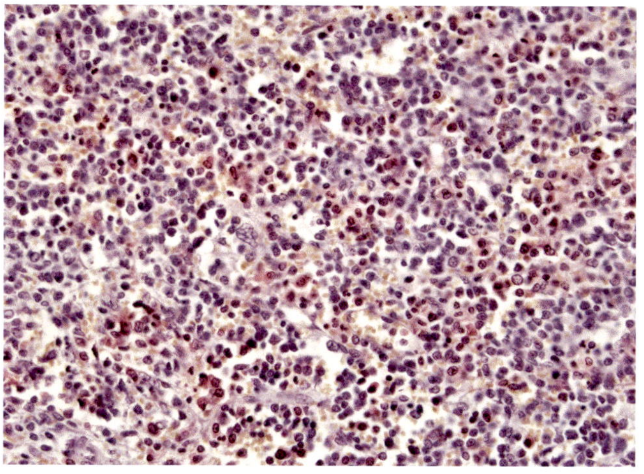

Fig. 4.5 Myeloid sarcoma, chloroacetate esterase stain. An example of the Leder stain (chloroacetate esterase [CAE]) in spleen. This patient has AML and some of the blasts cells are positive for CAE

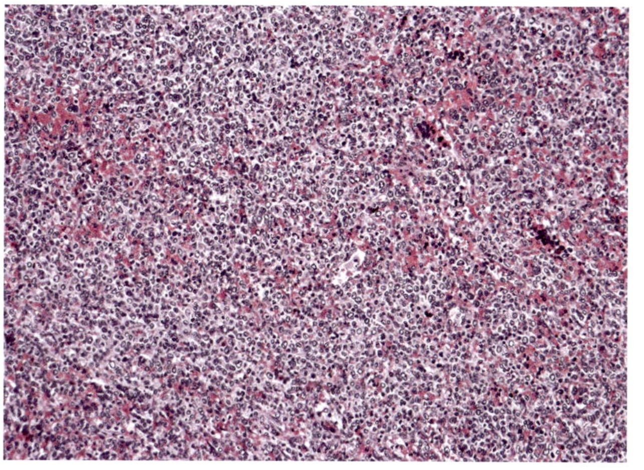

Fig. 4.7 Myeloid sarcoma/acute myeloid leukemia. Higher magnification showing a diffuse proliferation of immature myeloid elements within the red pulp structures. It is difficult to distinguish splenic sinuses from cords, although both are filled with myeloblasts. Some rare nucleated red blood cell precursors are also seen in this case (small, round, dark blue-black nuclei)

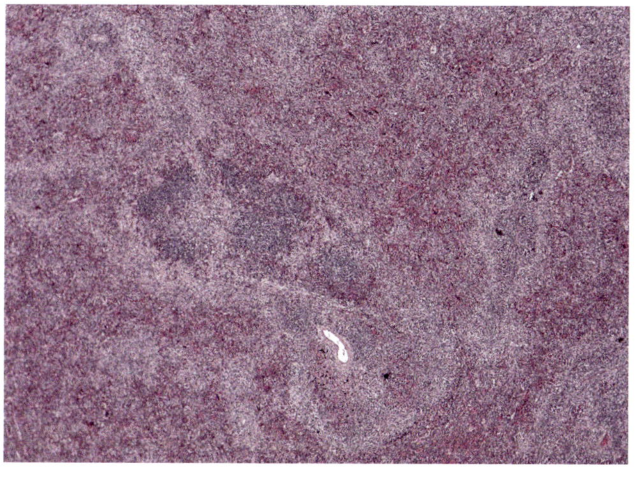

Fig. 4.6 Myeloid sarcoma/acute myeloid leukemia. Low magnification image of AML involving spleen. Note that although much of the involvement is in the red pulp, white pulp structures also appear abnormal

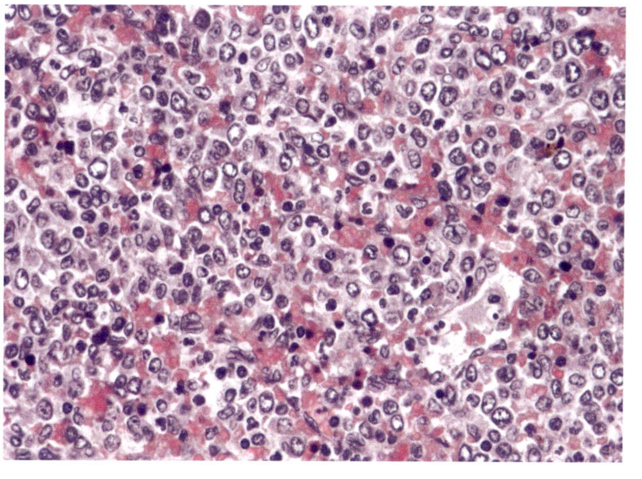

Fig. 4.8 Myeloid sarcoma/acute myeloid leukemia. High magnification of myeloblasts of AML in the splenic red pulp. Without appropriate immunophenotypic or clinical information, these cells would be difficult to distinguish from other malignancies, including lymphoma

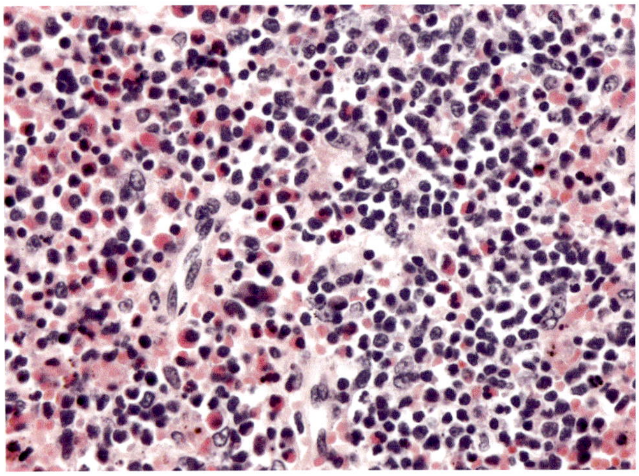

Fig. 4.9 Myeloid sarcoma/acute myeloid leukemia. AML in spleen with prominent eosinophilia. The presence of prominent eosinophilia in this case would raise the possibility of an AML associated with inversion of chromosome 16 [inv(16)], or possibly a transformation of chronic myelogenous leukemia (CML), which can have prominent splenic eosinophilia. The blasts are relatively small in this case

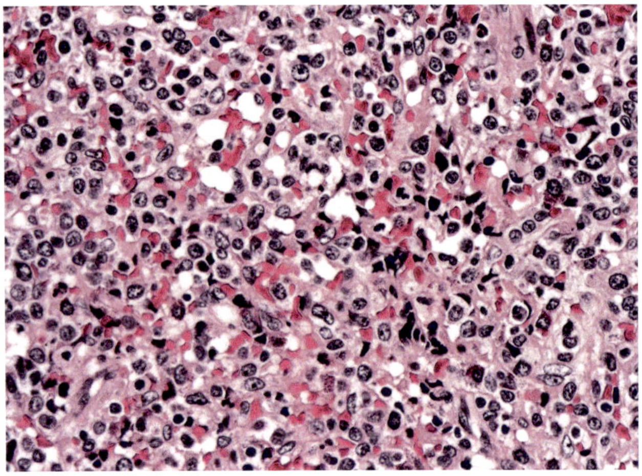

Fig. 4.10 Myeloid sarcoma/acute myeloid leukemia. An example of AML in the spleen. These blasts have slightly irregular nuclei with moderate amounts of pink, granular cytoplasm. Although not specific, these findings could be seen in AML with differentiation, acute promyelocytic leukemia, or AML with monocytic differentiation

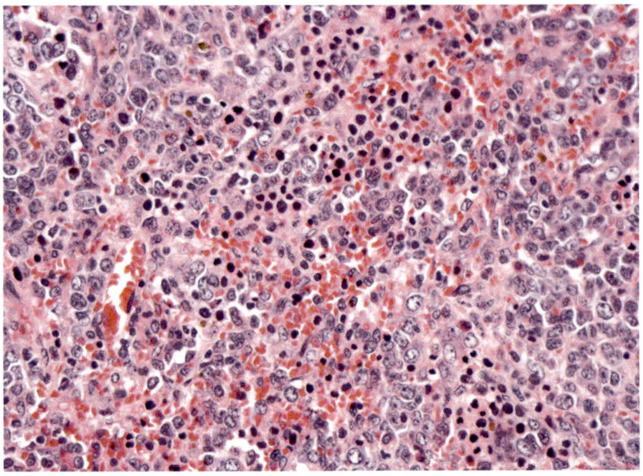

Fig. 4.11 Myeloid sarcoma/acute myeloid leukemia. This example of AML in the spleen is positive for a BCR/ABL translocation [t(9;22)], also known as the Philadelphia chromosome (Ph+). Based on morphology, it can be almost impossible to distinguish a *de novo* Ph+AML from a blast phase of CML. However, clinical history should help to differentiate the two entities

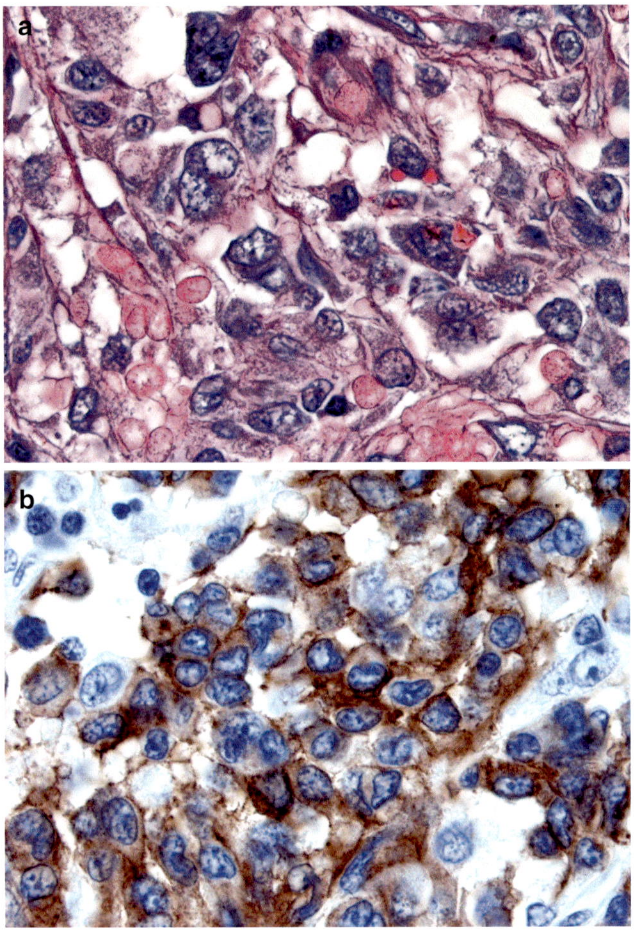

Fig. 4.12 Mast cell leukemia. High magnification of leukemic cells in the spleen (**a**). Mast cell leukemia is extremely rare and has no specific features to distinguish it from AML. This patient had a history of systemic mastocytosis, and besides positivity for CD117 (**b**), the patient had other evidence of mast cell differentiation in the leukemia cells

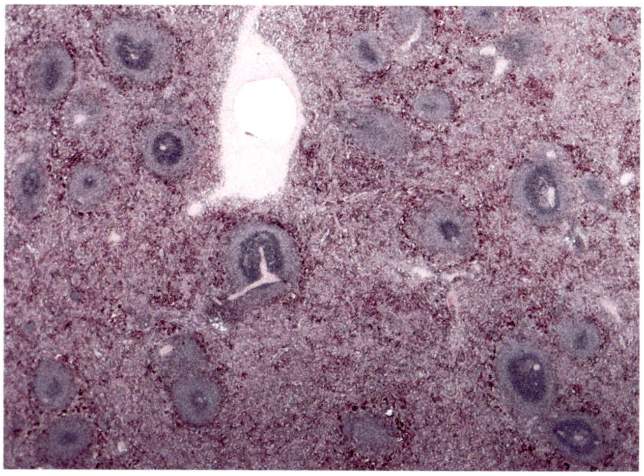

Fig. 4.13 Erythroleukemia. A low-magnification image of a spleen with involvement by erythroleukemia. In this case, the involvement is predominantly in the splenic red pulp, with preservation of white pulp structures. Note the prominent white pulp marginal zones

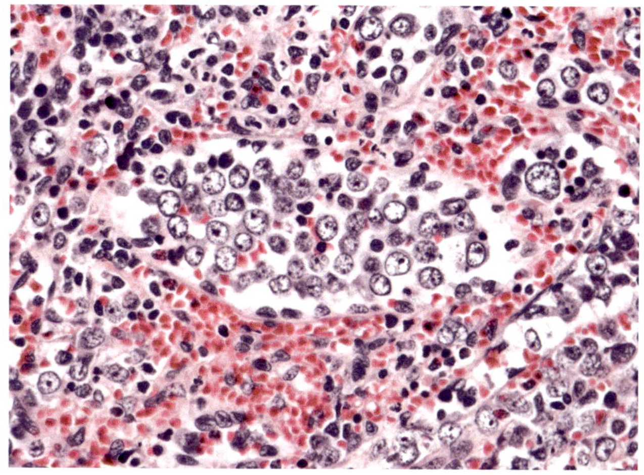

Fig. 4.15 Erythroleukemia. High magnification of erythroleukemia in spleen. In this example, the large blast cells have round nuclei, delicate marginated chromatin, and occasional prominent nucleoli. There appears to be a predilection for splenic sinusoidal distribution. It should be noted that erythroleukemia are not sufficiently different from other leukemias to be reliably distinguished by morphology alone. Immunophenotypic findings are critical in evaluating leukemia subtypes in the spleen

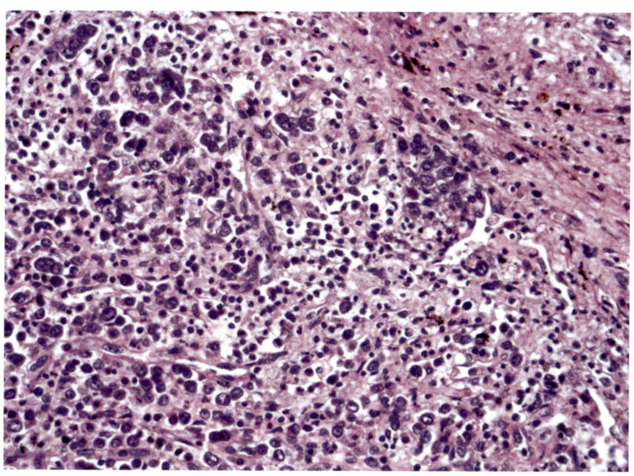

Fig. 4.14 Erythroleukemia. Intermediate magnification of splenic involvement by erythroleukemia. In this example, the immature elements have a biphasic appearance. Some immature cells have round nuclei, with dispersed chromatin, whereas others appear smaller with more condensed chromatin. This may represent erythroid and myeloid (*eg*, myeloblastic) differentiation

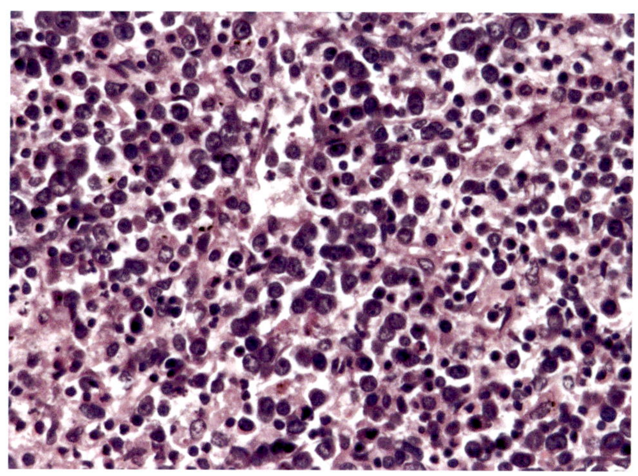

Fig. 4.16 Erythroleukemia. Another example of erythroleukemia involving the spleen. In this case, the erythroblasts appear relatively uniform in appearance. They have round nuclei and deeply colored chromatin and cytoplasm

4.3 Myeloproliferative Neoplasms

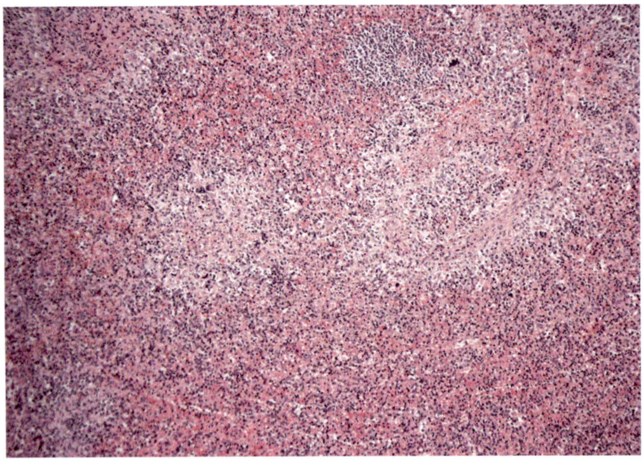

Fig. 4.17 Chronic myelogenous leukemia. Low-magnification image of CML in spleen. This case represents an accelerated phase of CML, with clusters and aggregates of immature myeloid elements that appear pale

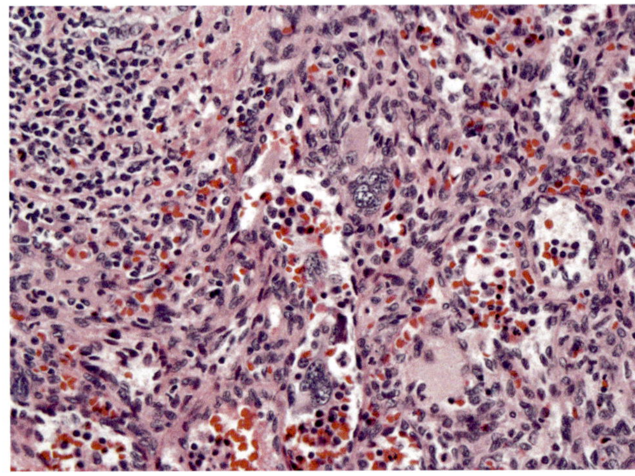

Fig. 4.19 Chronic myelogenous leukemia. High magnification showing neoplastic myeloid proliferation associated with CML. Within sinuses are erythroid precursors (small dark nuclei), and within and adjacent to sinuses are large atypical megakaryocytes. The marrow elements in spleen of CML often include trilineage elements

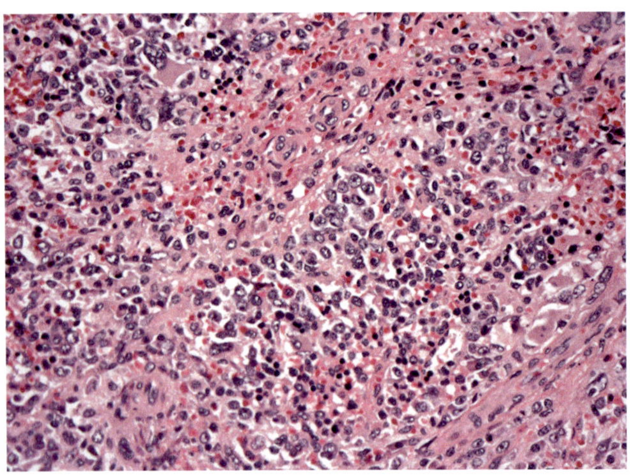

Fig. 4.18 Chronic myelogenous leukemia. Intermediate magnification of CML in accelerated phase. Note occasional erythroid precursors, rare megakaryocytes, and a predominance of immature myeloid elements

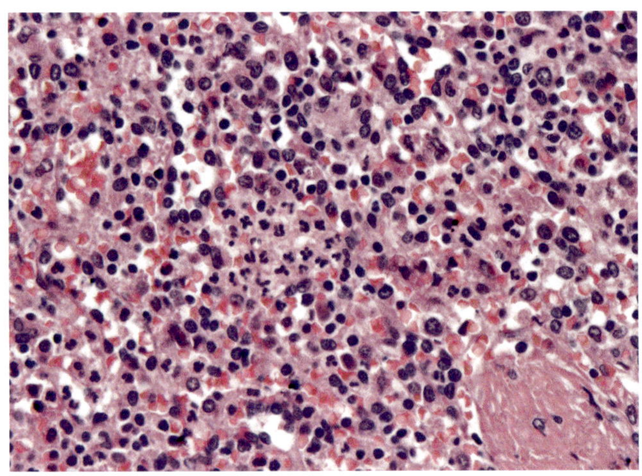

Fig. 4.20 Chronic myelogenous leukemia. In this example, red pulp and perivascular elements have increases in myeloid elements including mature forms (clusters of neutrophils) and more immature elements (round nuclei, granular pink cytoplasm)

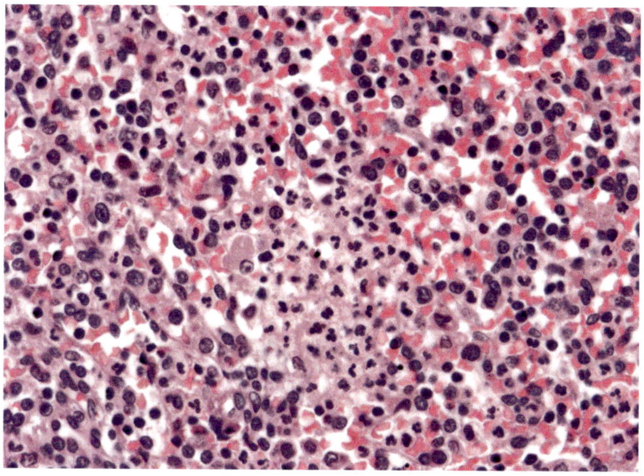

Fig. 4.21 Chronic myelogenous leukemia. In this example of CML, there are some scattered foamy macrophages. These are pseudo-Gaucher cells; macrophages that have accumulated debris from the breakdown of myeloid precursors

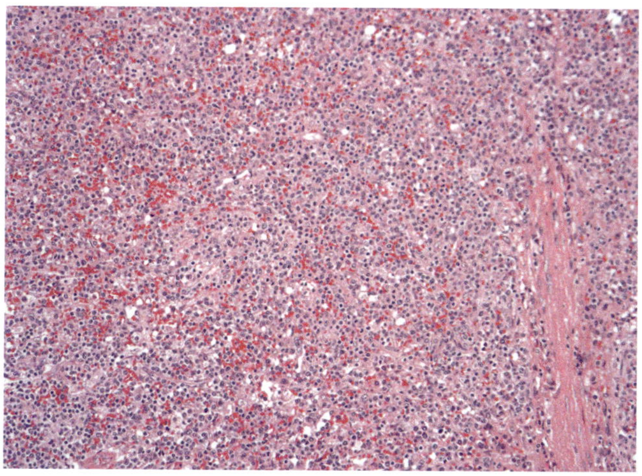

Fig. 4.22 Chronic myelogenous leukemia, blast phase. In this example, the entire splenic architecture is effaced by a proliferation of monotonous abnormal myeloblasts. No erythroid precursors or megakaryocytes are seen

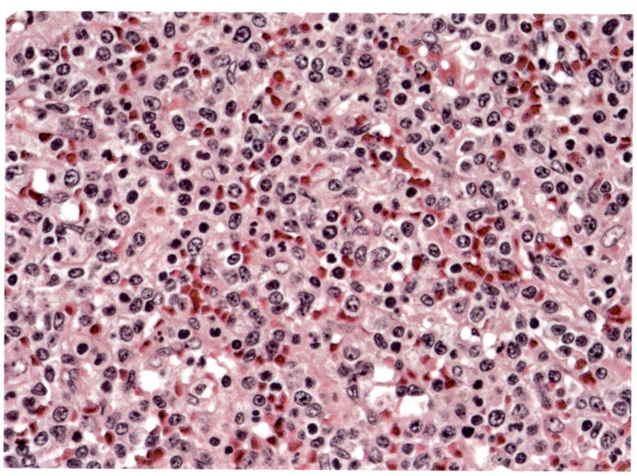

Fig. 4.23 Chronic myelogenous leukemia, blast phase. High-magnification image of blast phase of CML involving spleen. The blasts are intermediate in size with marginated, delicate chromatin and small to moderate amounts of pink cytoplasm. Rare mature granulocytes can be seen

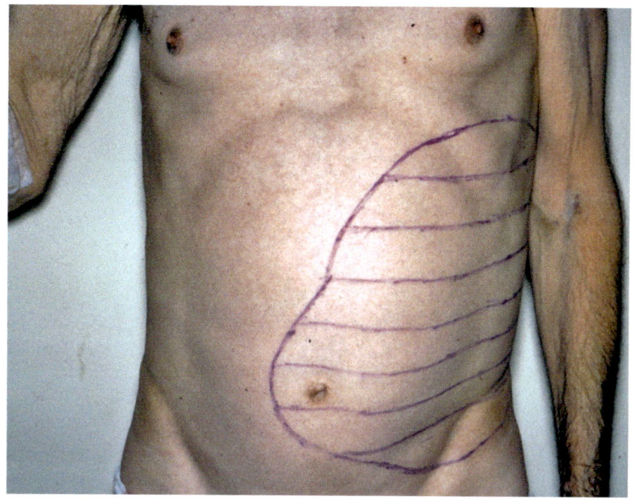

Fig. 4.24 Primary myelofibrosis. Image of a patient with primary myelofibrosis and a massively enlarged spleen (outlined in ink on the patient's abdomen). (*Image courtesy of* L. Morgenstern, Los Angeles, CA.)

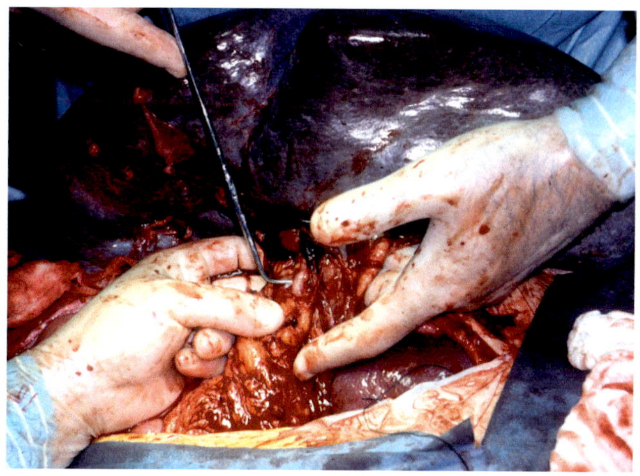

Fig. 4.25 Primary myelofibrosis. Intrasurgical image of patient in Fig. 4-24. The size of the spleen is massive. (*Image courtesy of* L. Morgenstern, Los Angeles, CA.)

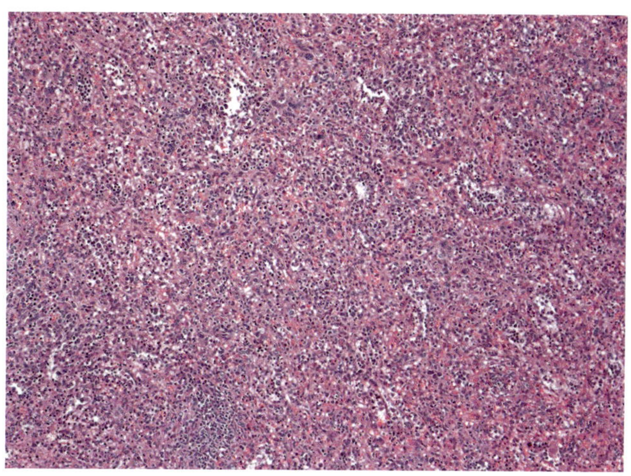

Fig. 4.26 Primary myelofibrosis. Low-magnification image of spleen with primary myelofibrosis, a myeloproliferative neoplasm. White pulp areas are inconspicuous, and there is marked expansion of red pulp areas

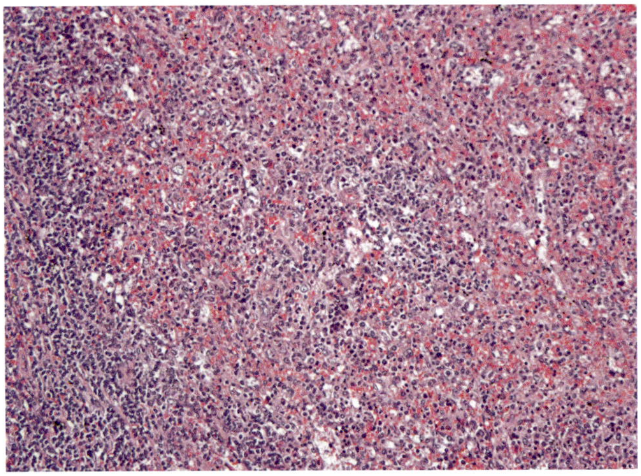

Fig. 4.27 Primary myelofibrosis. In this low-magnification image of primary myelofibrosis, small dark nuclei are seen within sinuses. As in other cases, there is massive expansion of red pulp

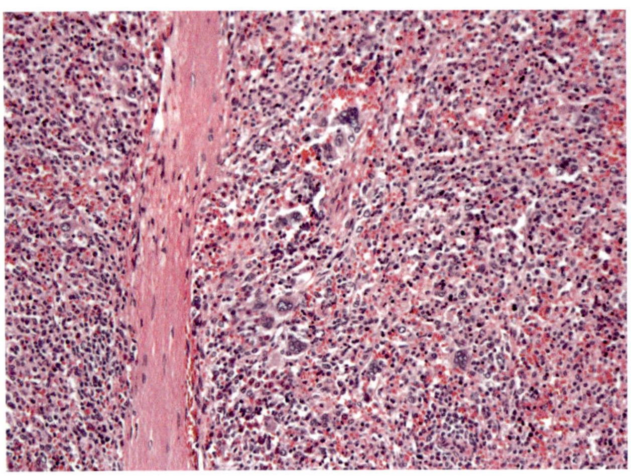

Fig. 4.28 Primary myelofibrosis. Intermediate magnification of primary myelofibrosis in spleen. A splenic fibrous trabeculae is present. Adjacent is red pulp with a marked increase in cellular elements. There are numerous megakaryocytes and the nucleated cells are a combination of red blood cell precursors and myeloid precursors

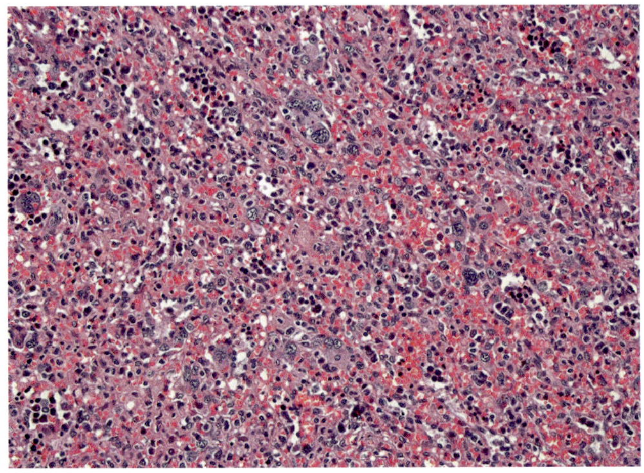

Fig. 4.29 Primary myelofibrosis. Another image of primary myelofibrosis with increased scattered megakaryocytes, including some clusters. Erythroid precursors are identified by their round nuclei, and many are quite dark blue/purple. Myeloid precursors have more open chromatin, irregular nuclei, and some pink cytoplasm

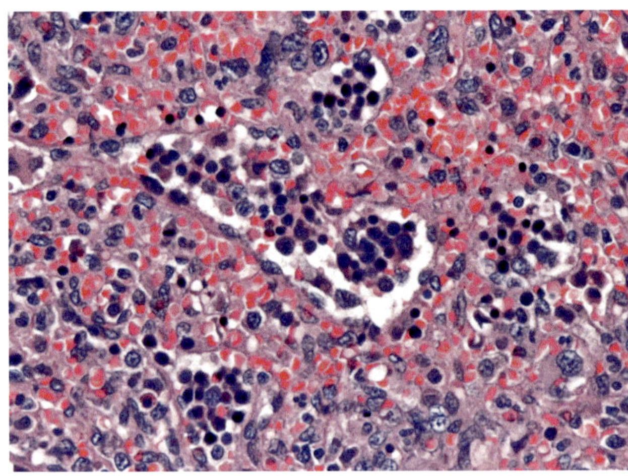

Fig. 4.30 Primary myelofibrosis. In this high-magnification image of primary myelofibrosis, large clusters of erythroid precursors are seen with splenic sinuses. There are also immature myeloid elements within splenic cords

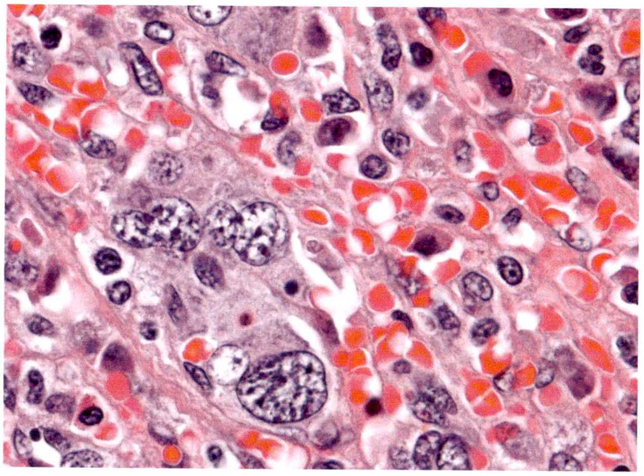

Fig. 4.31 Primary myelofibrosis. High magnification of primary myelofibrosis highlighting clusters of abnormal-appearing megakaryocytes. The neoplastic myeloid proliferation associated with primary myelofibrosis typically includes megakaryocytes, myeloid, and erythroid elements

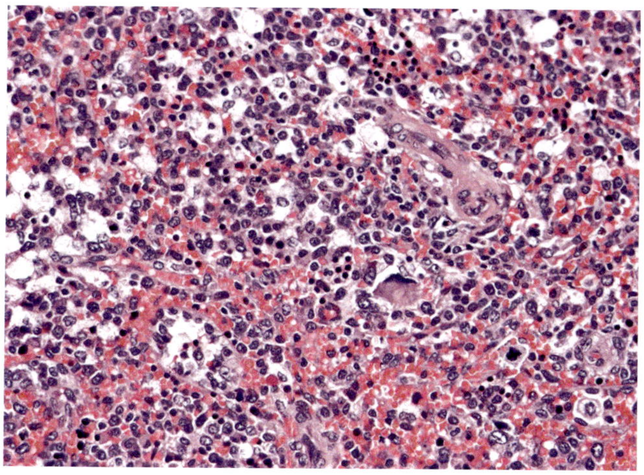

Fig. 4.32 Primary myelofibrosis. This example of primary myelofibrosis shows trilineage hematopoietic elements and numerous foamy macrophages. These are comparable to pseudo-Gaucher cells, which are filled with cellular debris from the breakdown of the abnormal myeloid elements. It is not uncommon to see this finding focally in the spleens of patients with myeloproliferative neoplasms

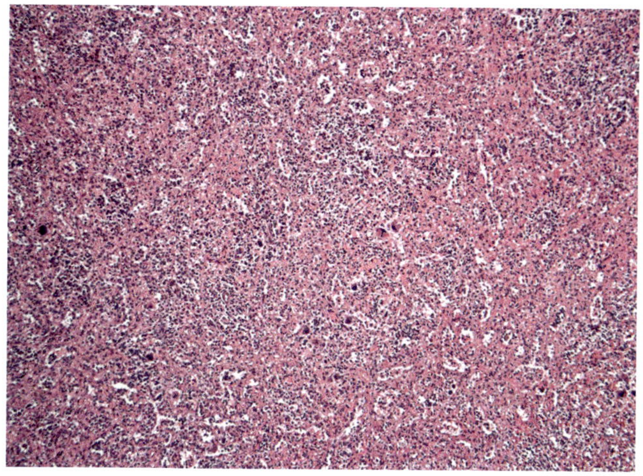

Fig. 4.33 Polycythemia vera. Low-magnification image of enlarged spleen of a patient with polycythemia vera. White pulp structures are absent and there is a marked expansion of the red pulp. Even at low magnification, the increased red pulp cells (myeloid and erythroid precursors) and megakaryocytes are apparent

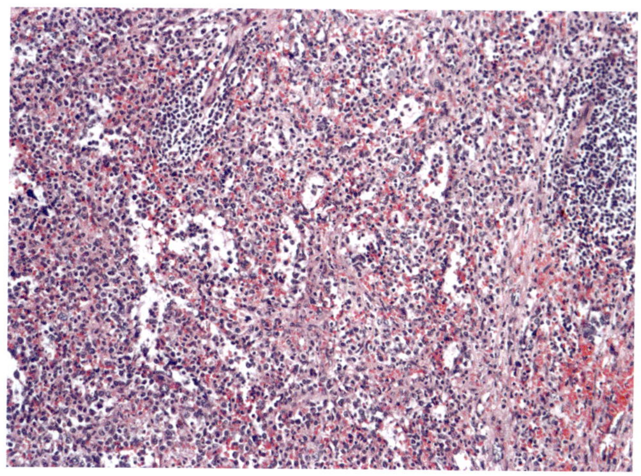

Fig. 4.34 Polycythemia vera. Intermediate magnification of polycythemia vera. Preserved white pulp structures (*right*) are seen. There are many cells within the cords and sinuses. Numerous myeloid precursors and many erythroid precursors are present

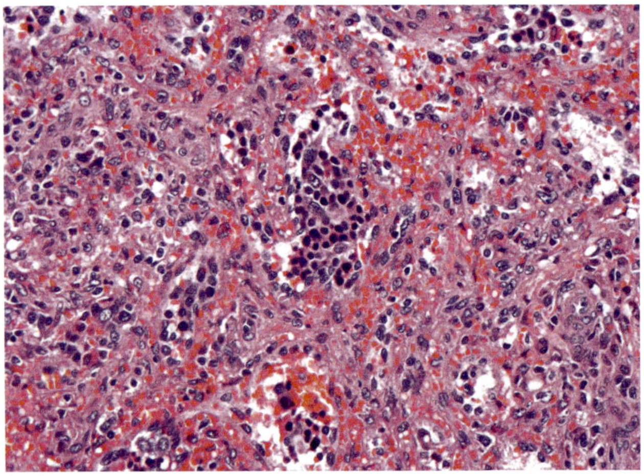

Fig. 4.35 Polycythemia vera. The neoplastic myeloid proliferations in the spleen consists of a combination of myeloid erythroid and megakaryocytic elements. In this example, prominent islands of erythroid precursors are seen within sinuses

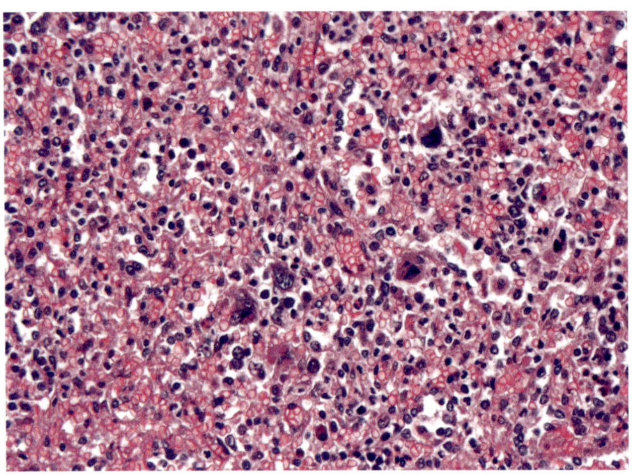

Fig. 4.36 Polycythemia vera. Megakaryocytes are seen within the red pulp in this example of polycythemia vera. They are present both within sinuses and in cords. The megakaryocyte nuclei are enlarged and hyperchromatic

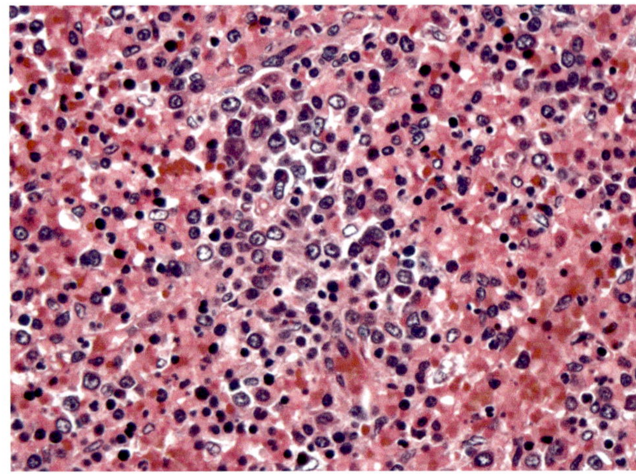

Fig. 4.37 Polycythemia vera. A cluster of immature myeloid precursors in the red pulp in polycythemia vera. Although not specific, if these cells were positive for CD34, this would be concerning for a blastic transformation

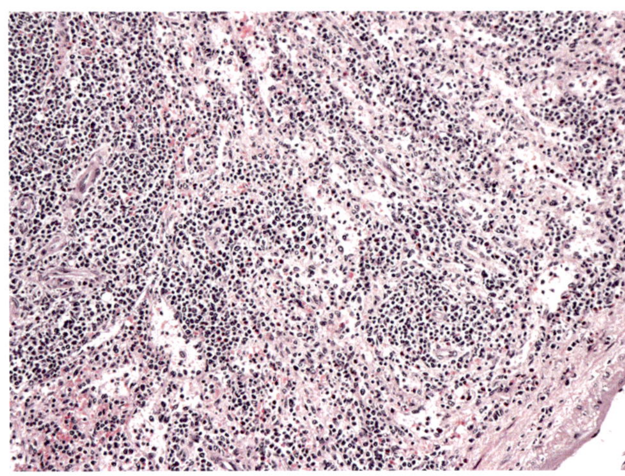

Fig. 4.38 Chronic eosinophilic leukemia (CEL). CEL involving the spleen. In this intermediate magnification image, the red pulp structures appear to have increased cellular components. Even at this magnification, the brick-red/orange color of eosinophils is appreciated

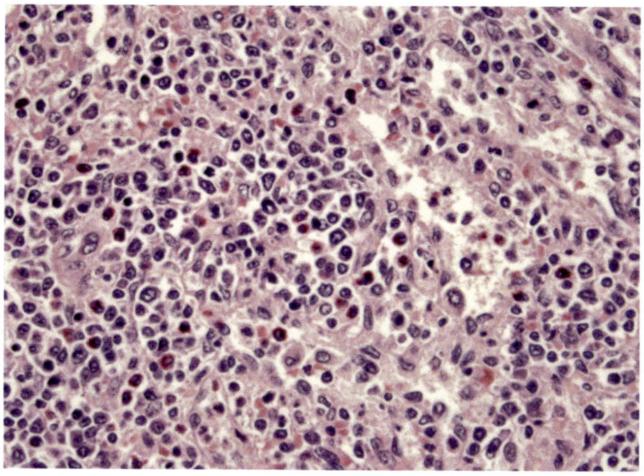

Fig. 4.39 Chronic eosinophilic leukemia. In this high-magnification image of CEL involving the spleen, there are increased eosinophils and some other immature myeloid elements. These findings are not entirely specific for CEL, as increased eosinophils could be seen in chronic myelogenous leukemia, as well as some reactive and infections conditions

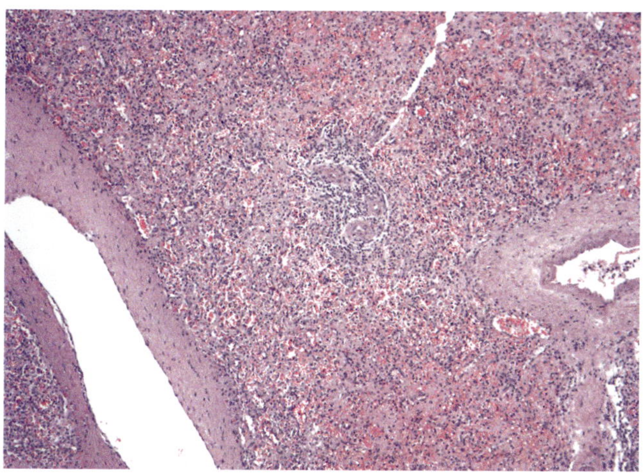

Fig. 4.40 Essential thrombocythemia. Splenomegaly is extremely rare is essential thrombocythemia. This example shows some depletion of white pulp structures and some neoplastic myeloid proliferation, including some megakaryocytes

4.4 Myelodysplastic Syndrome and Myelodysplastic/Myeloproliferative Disorders

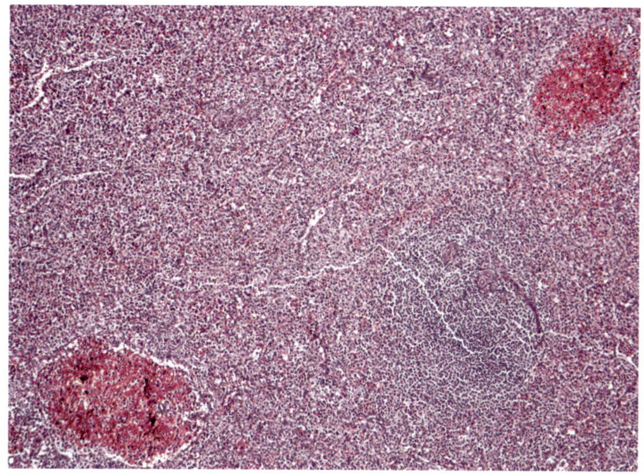

Fig. 4.41 Chronic myelomonocytic leukemia (CMML). Low-magnification image of spleen with involvement by CMML. White pulp structures are diminished with massive expansion of the red pulp. The red pulp has a significant increase in cellular components. In this case, they are neoplastic myeloid elements

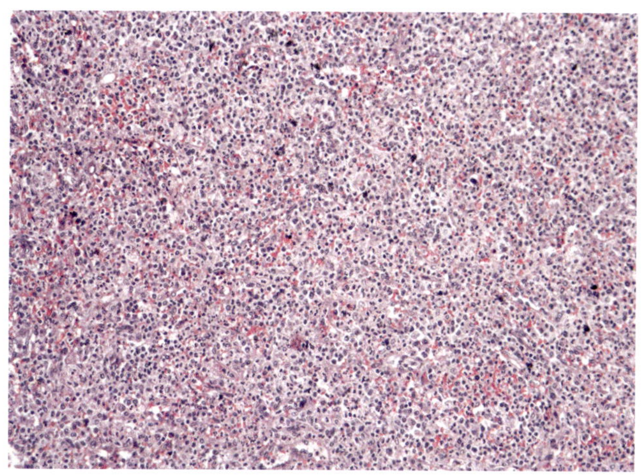

Fig. 4.42 Chronic myelomonocytic leukemia. In this example of CMML, both cords and sinuses are filled with immature myeloid elements, with some scattered erythroid elements. There is some focal pigment present, which is an artifact due to fixation

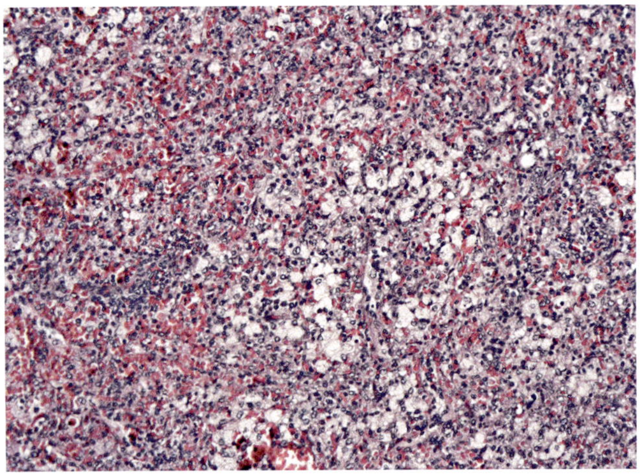

Fig. 4.43 Chronic myelomonocytic leukemia. Intermediate magnification of CMML with many foamy macrophages. As in other myeloid disorders, these macrophages have a pseudo-Gaucher appearance

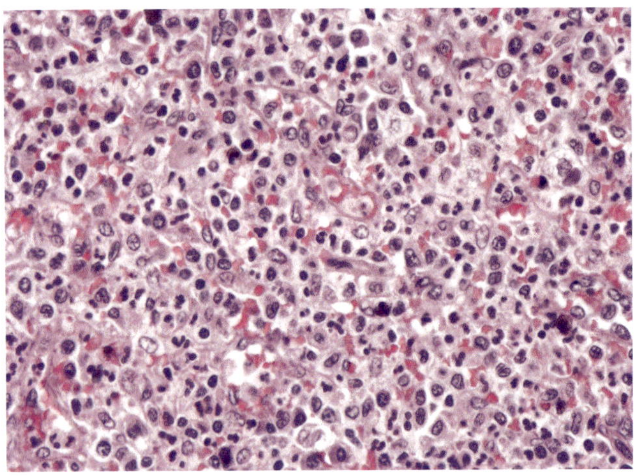

Fig. 4.44 Chronic myelomonocytic leukemia. High magnification of CMML, which shows a mixture of mature and immature myeloid elements. Monocytes are difficult to distinguish from immature granulocytic cells. Monocytic forms are present in variable numbers in CMML and may not be the predominant population seen

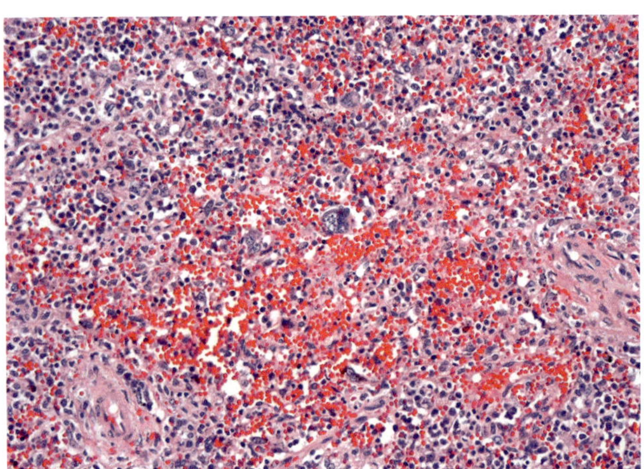

Fig. 4.45 Myelodysplastic syndrome. In typical cases of myelodysplastic syndromes, splenic involvement is rare. In this case of refractory anemia with excess blasts, there are increased neoplastic myeloid elements in the red pulp. Included are erythroid precursors, immature myeloid elements, and dysplastic-appearing megakaryocytes

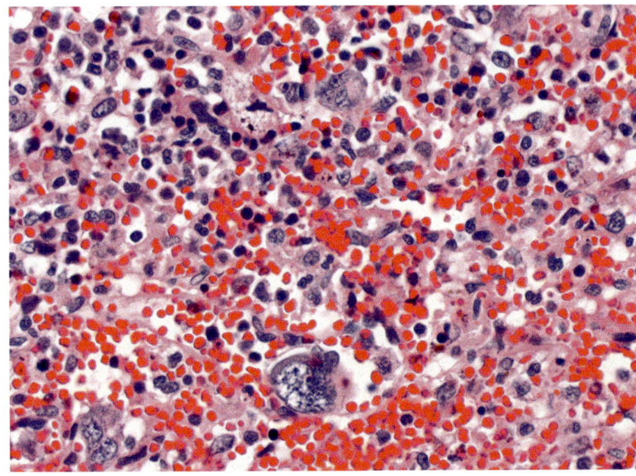

Fig. 4.46 Myelodysplastic syndrome. Higher-magnification image highlighting enlarged and dysplastic megakaryocytes in the neoplastic myeloid proliferation of a myelodysplastic syndrome (MDS) in the spleen. Specific cellular features do not distinguish MDS involvement from that of a myeloproliferative neoplasm in the spleen

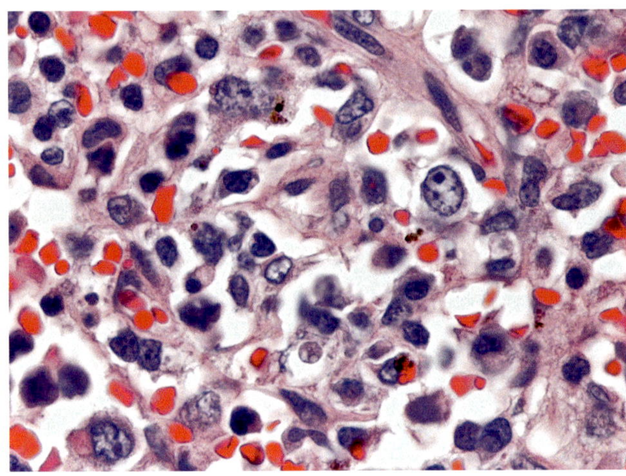

Fig. 4.47 Myelodysplastic syndrome. Clusters of immature myeloid elements within sinuses of splenic red pulp. Immature myeloid cells with slightly irregular nuclei, marginated chromatin, and small or moderate amounts of pink, sometimes granular, cytoplasm

4.5 Mastocytosis

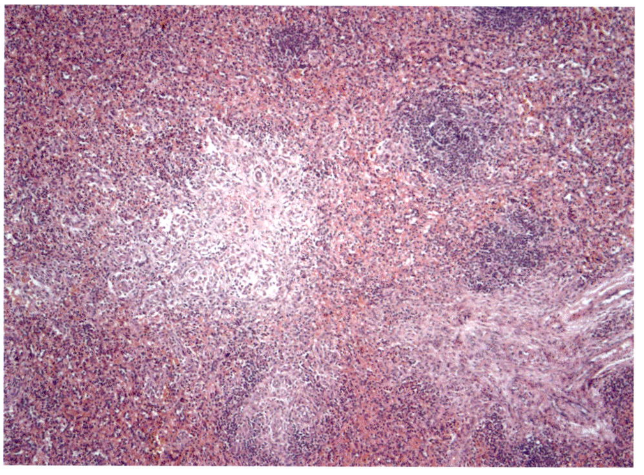

Fig. 4.48 Systemic mastocytosis. Low-magnification image of systemic mastocytosis in the spleen. The distribution of abnormal mast cells in spleen is typically perivascular. The mast cells are present in fibrotic lesions, which are often hypocellular with a mixed cellular infiltrate

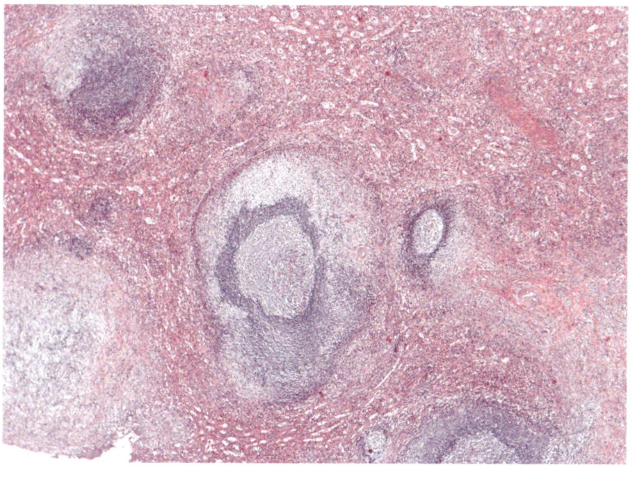

Fig. 4.49 Systemic mastocytosis. In occasional cases of systemic mastocytosis, abnormal mast cells are in a parafollicular distribution, mimicking marginal zone hyperplasia. In this case, the small follicle (*middle right*) has an abnormal proliferation of mast cells that merges into a more fibrotic lesion (*far right*)

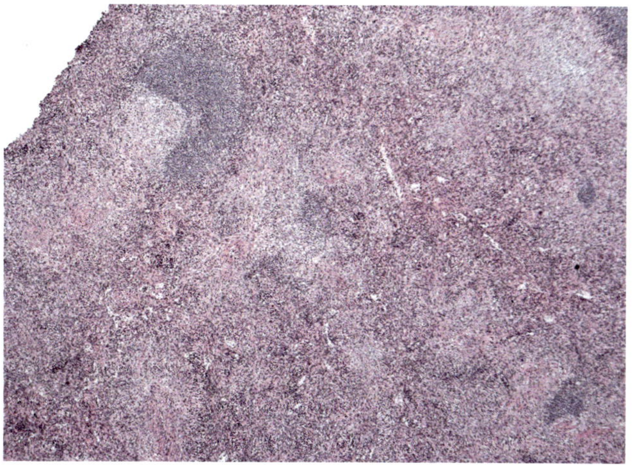

Fig. 4.50 Systemic mastocytosis. When there is more extensive involvement of systemic mastocytosis, there are larger and more confluent pale and fibrotic areas in the spleen

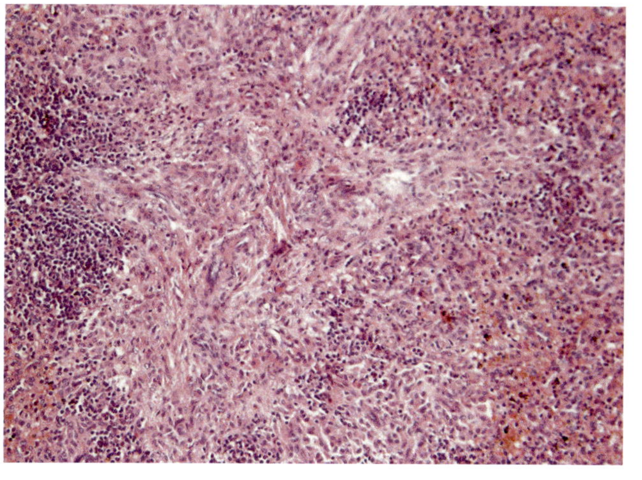

Fig. 4.51 Systemic mastocytosis. At intermediate magnification, a hematoxylin and eosin–stained section can be difficult to identify correctly. The individual mast cells are inconspicuous. Mast cells are spindle shaped, with bland nuclei and simulated fibroblasts

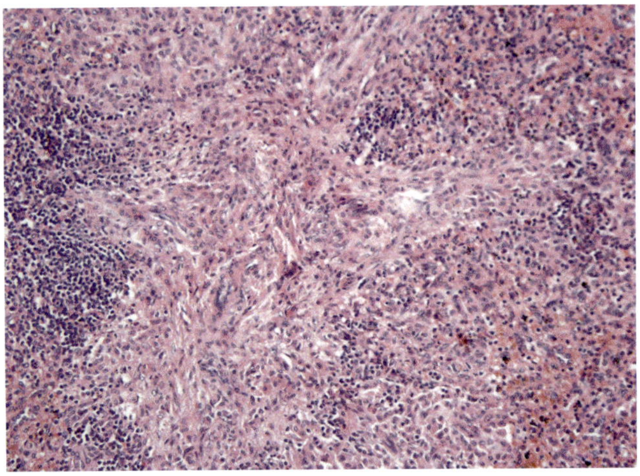

Fig. 4.52 Systemic mastocytosis. High-magnification image of a spindled region of systemic mastocytosis. Most spindle-shaped cells are mast cells. They have delicately granular cytoplasm. These granules may be visualized by a Giemsa stain, but immunohistochemical staining is more accurate, as abnormal mast cells may be hypogranular

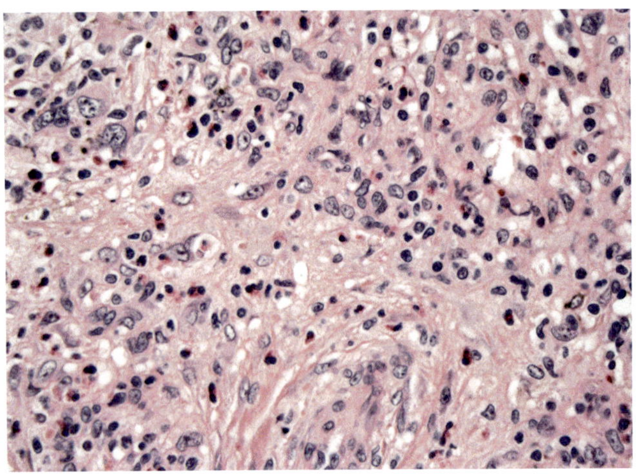

Fig. 4.53 Systemic mastocytosis. High magnification of systemic mastocytosis. In many cases, there are many accompanying eosinophils in mast cell lesions, as seen here

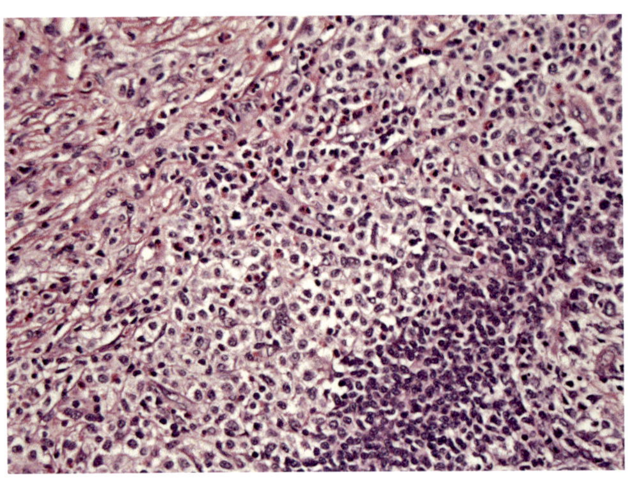

Fig. 4.54 Systemic mastocytosis. Although abnormal mast cells are often spindle shaped, in other circumstances they may be round, with increased pale cytoplasm, as seen here. This region of clustered mast cells merges with a more spindle-shaped region (*left*)

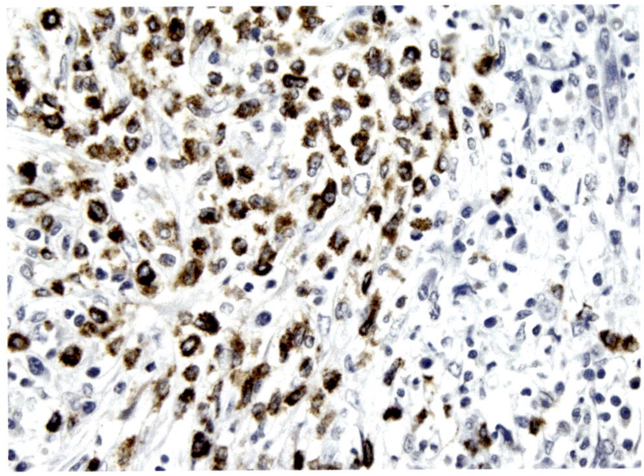

Fig. 4.55 Systemic mastocytosis, tryptase stain. Immunohistochemical staining for tryptase is the most specific single marker for identifying mast cells. However, both normal and neoplastic mast cells are positive for tryptase

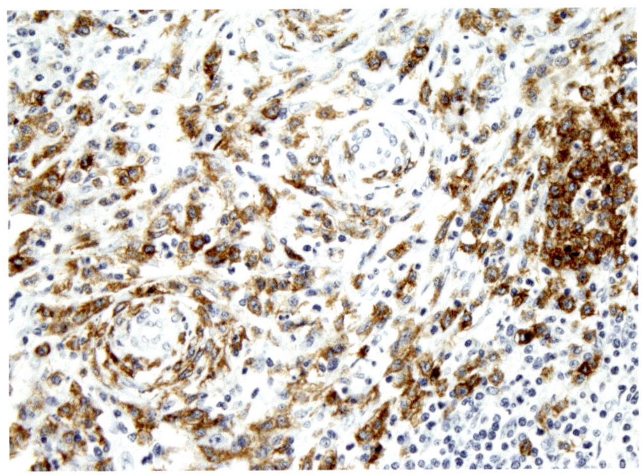

Fig. 4.56 Systemic mastocytosis, CD117 stain. Although both neoplastic and normal mast cells are strongly positive for CD117, normal mast cells are round and not present in clusters. In this case of systemic mastocytosis, the mast cells are markedly increased, spindled, and present in clusters

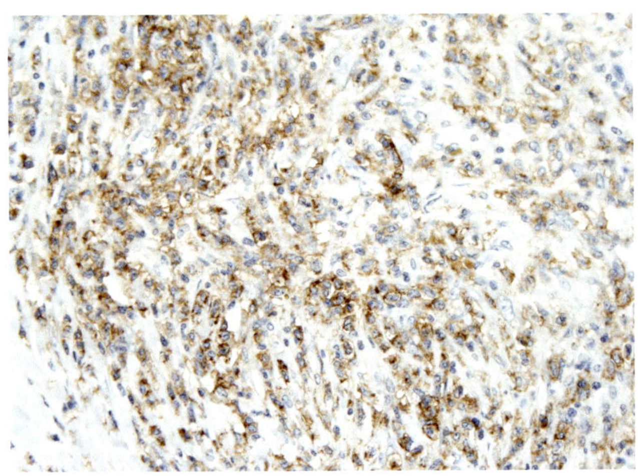

Fig. 4.57 Systemic mastocytosis, CD25 stain. In contrast to normal mast cells, neoplastic mast cells of systemic mastocytosis are often positive for CD25. This case shows strong expression of CD25 in the mast cells

4.6 Histiocytic Sarcoma

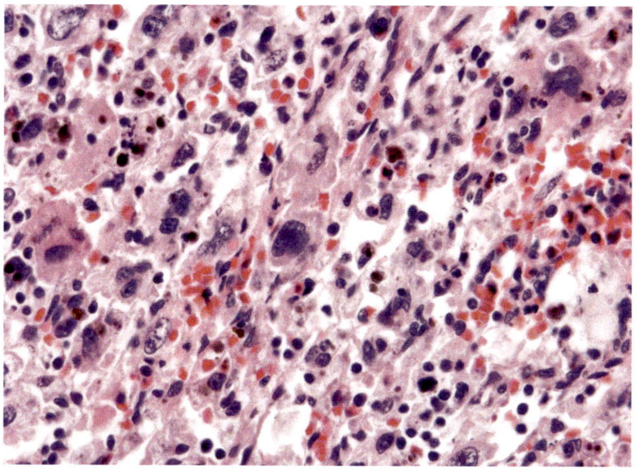

Fig. 4.58 Histiocytic sarcoma. Histiocytic sarcoma is a rare neoplasm in any site. It can be seen in the spleen, and as in this case, the degree of abnormality of the histiocytic cells may suggest other, nonhematopoietic malignancies. In this case, the abnormal histiocytes are large in size with significant pleomorphism, occasional multinucleated cells, and abnormal hyperchromatic nuclei. In this case, there is some hemosiderin deposition, which may be due to remote hemorrhage

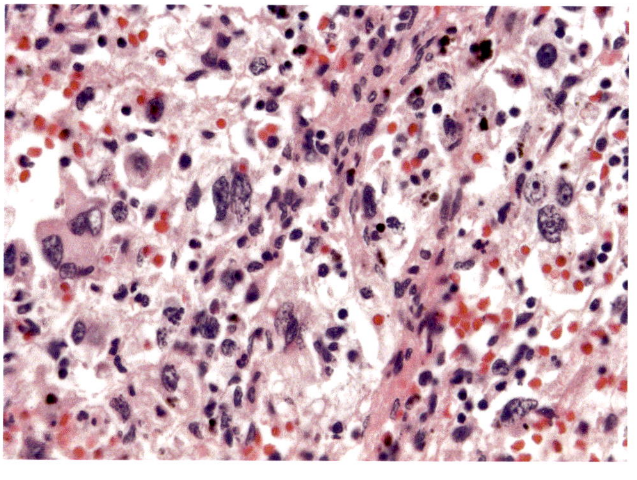

Fig. 4.59 Histiocytic sarcoma. Another example of histiocytic sarcoma in the spleen. Typically, these are masses, with no specific distribution in red or white pulp. This case shows highly pleomorphic cells, many with large amounts of pale or clear cytoplasm

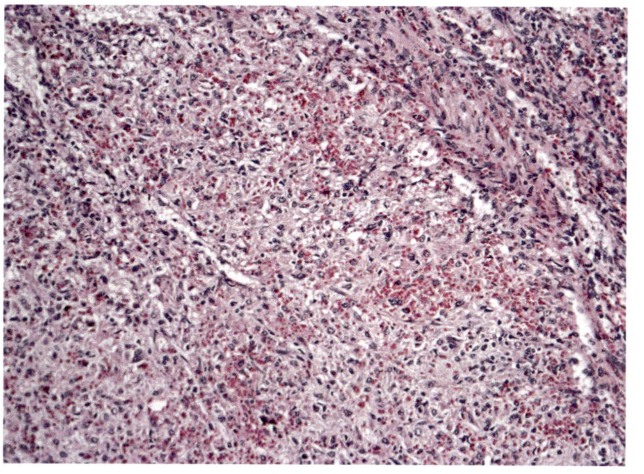

Fig. 4.60 Histiocytic sarcoma. In this example of histiocytic sarcoma, there is a more solid appearance, with mostly intermediate to large cells. In this case, exclusion of other hematolymphoid neoplasms, including lymphoma would be critical

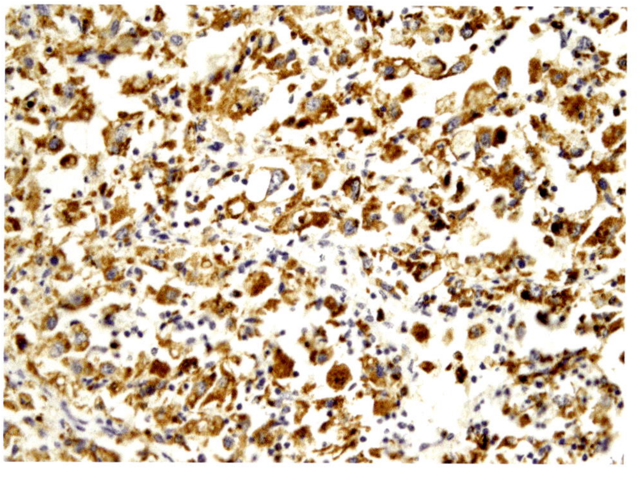

Fig. 4.61 Histiocytic sarcoma, CD68 stain. CD68 staining in a histiocytic sarcoma. Although CD68 would be expected to be positive in almost all cases of histiocytic sarcoma, it is not entirely specific for that diagnosis. Usually a combination of histiocytic stains such as CD68 and CD163, and a lack of stains for other neoplasms is required to confirm a diagnosis

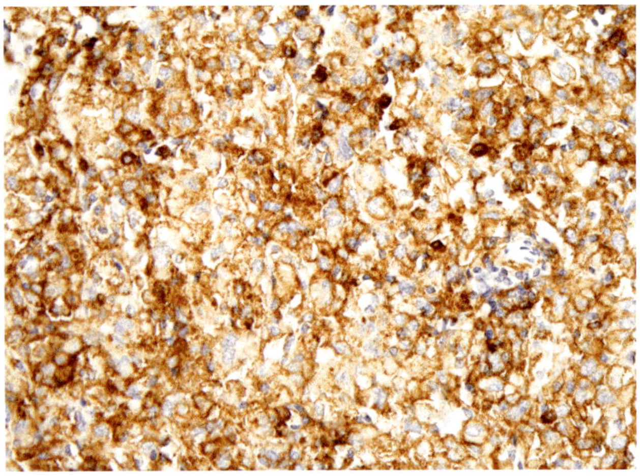

Fig. 4.62 Histiocytic sarcoma, CD33 stain. Histiocyte/macrophage lineage cells are usually derived from bone marrow myeloid progenitor cells. CD33 is positive in cells of myeloid lineage and is present in these cells of histiocytic sarcoma. Similar staining could be seen in other myeloid-derived neoplasms, such as myeloid sarcoma

5 Nonhematopoietic Lesions Including Vascular

Nonhematopoietic lesions of the spleen can be quite challenging. In this chapter, cysts of the spleen are illustrated. These include cysts of both infectious and noninfectious etiologies. Hamartoma, a benign lesion characterized by proliferation of benign splenic red pulp components, is illustrated, as well as myoid angioendothelioma, a purported variant of splenic hamartoma. Also, the difficult area of inflammatory pseudotumor (IPT) and related entities is discussed. The specific diagnosis includes "true" IPT, IPT-like follicular dendritic cell tumor, and inflammatory myofibroblastic tumor. Vascular tumors of the spleen have a broad range of histologic appearance and clinical behavior. A wide range is illustrated including SANT, littoral cell angioma, and angiosarcoma. Finally, a broad array of tumors metastatic to the spleen is illustrated.

D.P. O'Malley, *Atlas of Spleen Pathology*, Atlas of Anatomic Pathology,
DOI 10.1007/978-1-4614-4672-9_5, © Springer Science+Business Media New York 2013

5.1 Cysts

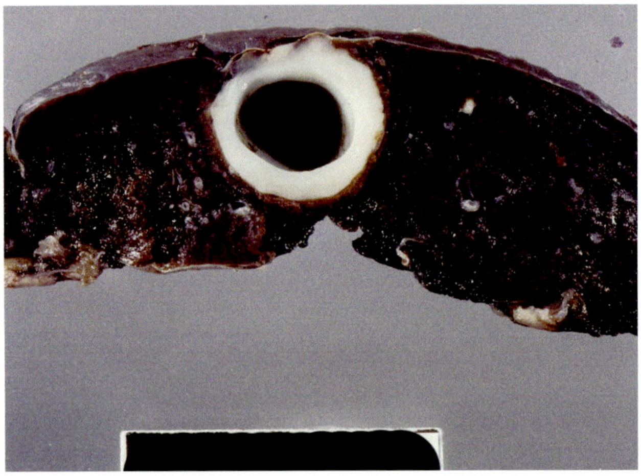

Fig. 5.1 Cyst. Gross image of a spleen section with a thick-walled cyst. Note the surrounding parenchyma is unremarkable. The wall is thickened and fibrous. The nature of the lining is not apparent in this gross examination

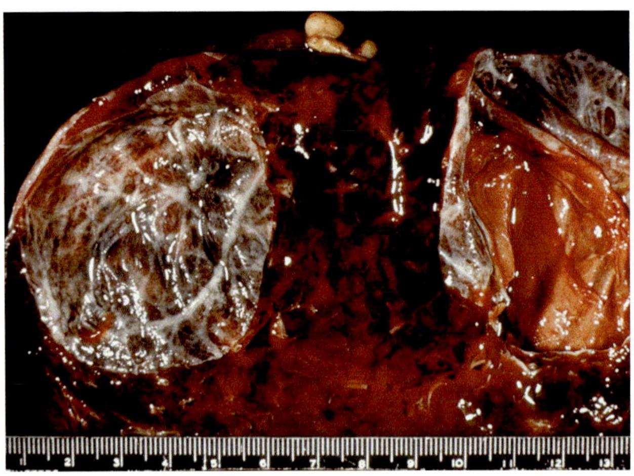

Fig. 5.2 Cyst. Gross image of a splenic cyst. The inner wall has coarse trabeculation but the wall is relatively smooth. Microscopic examination of the lining is useful to determine the type of splenic cyst. (*Image courtesy of* D. Farhi, Atlanta, GA.)

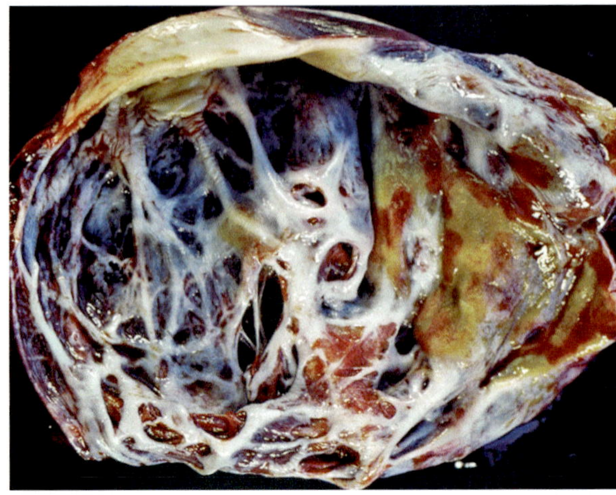

Fig. 5.3 Cyst. The wall of this splenic cyst is coarsely trabecular. By gross examination, it is unclear if there is an epithelial, mesothelial, or no lining (pseudocyst). (*Image courtesy of* R. Neiman, Harpswell, ME.)

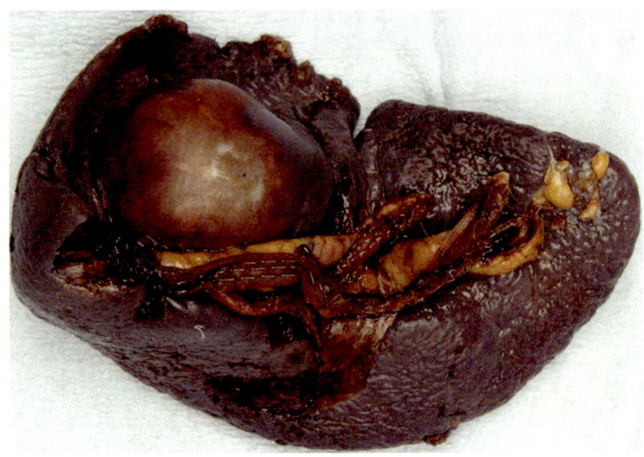

Fig. 5.4 Cyst. Gross image of a removed spleen. Note that adjacent to the hilum is the bulging surface of a cyst. (*Image courtesy of* J. Sidhu, Johnson City, NY.)

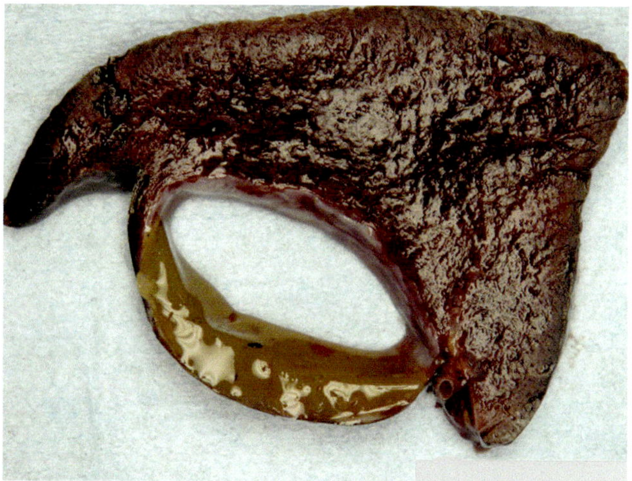

Fig. 5.5 Cyst. Cut surface of previous image showing the inner surface of a squamous lined benign cyst. The lining is shiny and smooth. The wall has a gelatinous yellow appearance, suggesting collections of macrophages. (*Image courtesy of* J. Sidhu, Johnson City, NY.)

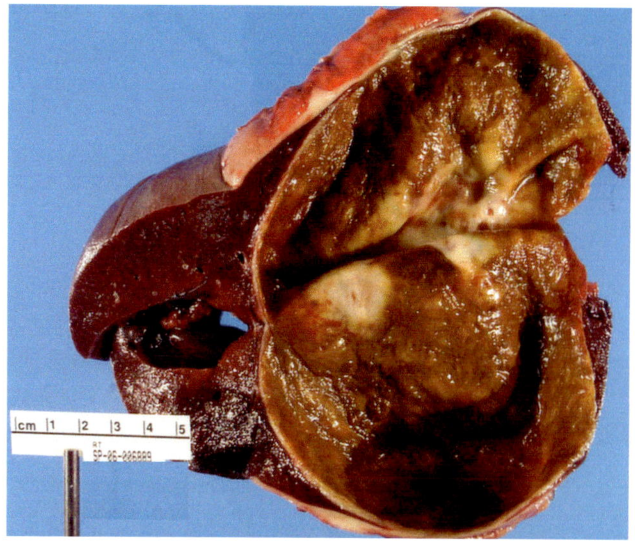

Fig. 5.7 Epithelial cyst. Gross image of an opened epithelial cyst. The wall of the cyst has a slightly rough appearance. Cysts can be generally divided in several ways: true cysts versus pseudocyst; infectious versus noninfectious; benign versus neoplastic. Each way is beneficial but in all cases the cyst wall and the contents should be evaluated. (*Image courtesy of* M. Drachenberg, Long Beach, CA.)

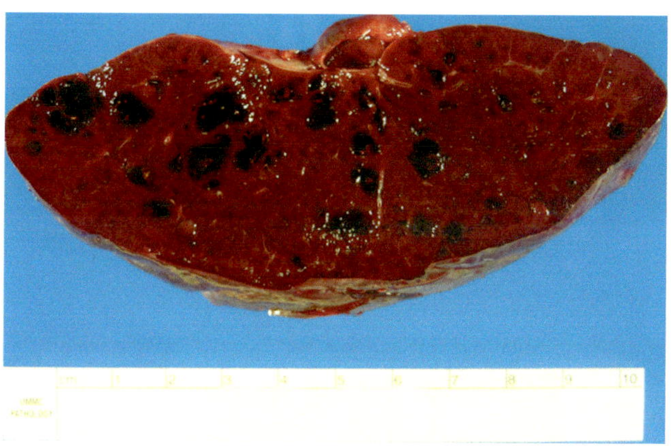

Fig. 5.6 Hemorrhagic foci. Multiple hemorrhagic foci area present in this gross sample. These cystic, blood-filled spaces could represent vascular tumors, multiple cysts, or a variety of other pathologic conditions. Microscopic examination would be required to determine the etiology. (*Image courtesy of* M. Kamionek, Boston, MA.)

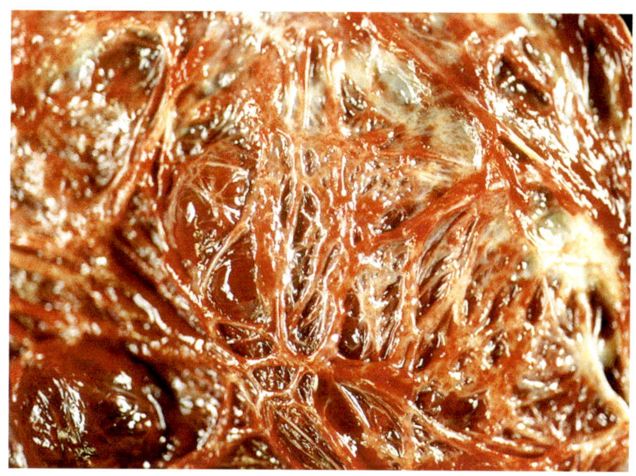

Fig. 5.8 Epithelial cyst. The wall of this epithelial cyst is coarsely trabecular. The epithelial lining is not apparent on gross examination. (*Image courtesy of* R. Neiman, Harpswell, ME.)

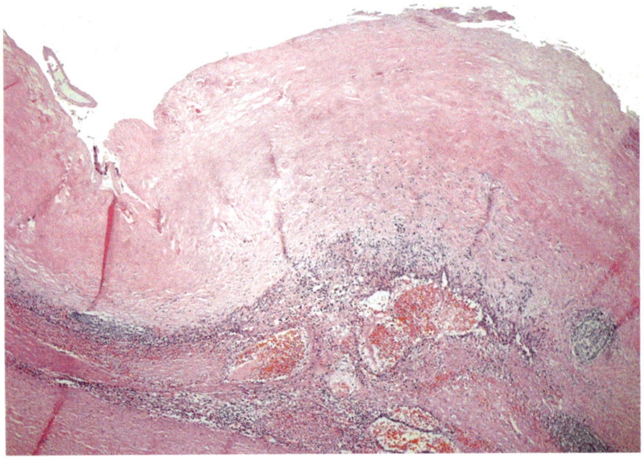

Fig. 5.9 Echinococcal cyst. The walls of infectious cysts are typically not lined with epithelium, as is the case in this echinococcal cyst. The inner wall (*upper*) is composed of paucicellular, fibrous tissue. (*Image courtesy of* T. Davis, Indianapolis, IN.)

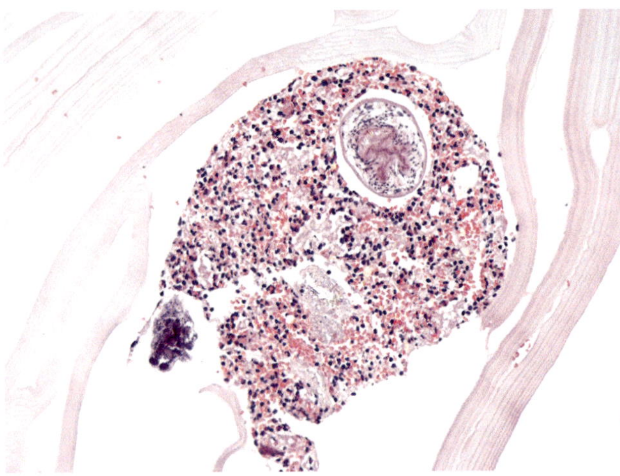

Fig. 5.10 Echinococcal cyst. Cyst contents. Many degenerated cells are present, as well as some lymphocytes and components of *Echinococcus*. (*Image courtesy of* T. Davis, Indianapolis, IN.)

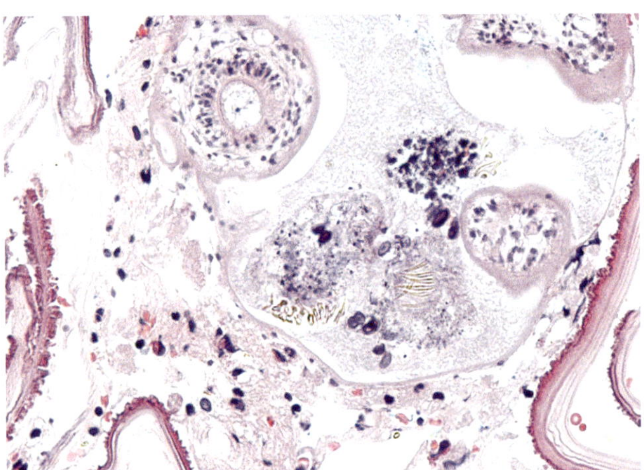

Fig. 5.11 Echinococcal cyst. High magnification of the contents of an echinococcal cyst. Note the refractile materials, which are portions of the mouthparts (*eg*, scolex) of the *Echinococcus*. (*Image courtesy of* T. Davis, Indianapolis, IN.)

5.2 Hamartoma

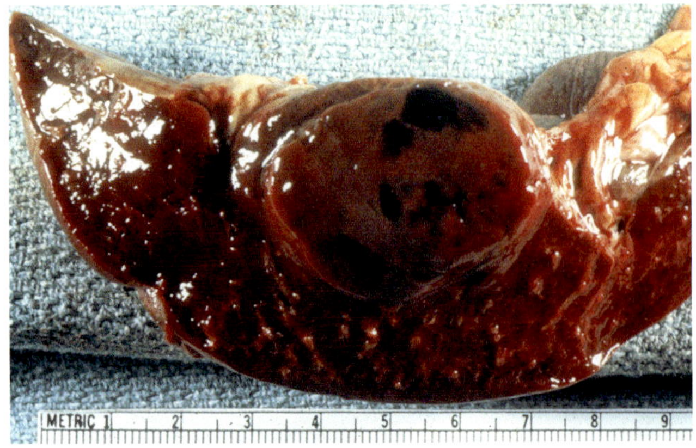

Fig. 5.12 Hamartoma. Gross image of a splenic hamartoma. The cut surface the hamartoma has a more brown-red appearance compared to the surrounding uninvolved spleen, which is more red. (*Image courtesy of* L. Morgenstern, Los Angeles, CA.)

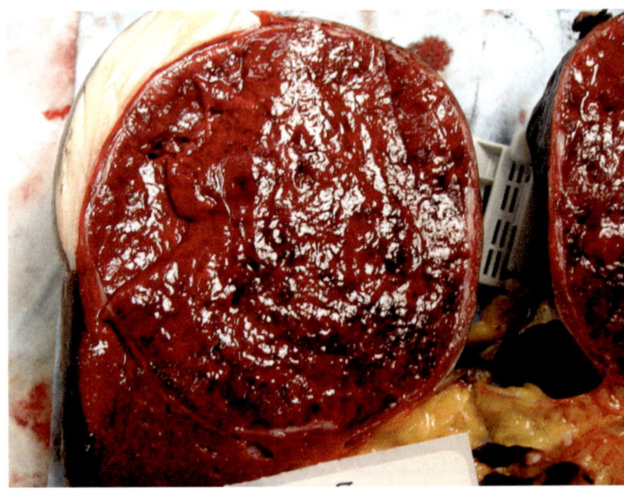

Fig. 5.13 Hamartoma with bizarre stromal cells. Gross image of splenic hamartoma with bizarre stromal cells (*see* microscopic images below). It is relatively well circumscribed and compresses the normal splenic parenchyma seen at the lower border. Often hamartomas will have a bulging cut surface. Like normal red pulp, hamartomas typically have a deep red, beefy appearance. (*Images courtesy of* M. Argyres, Denver, CO.)

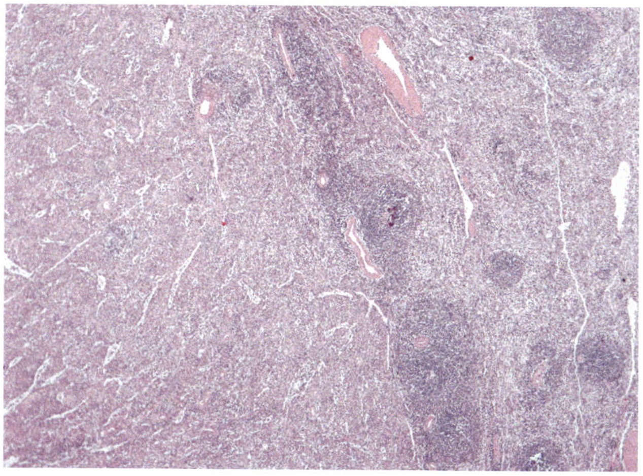

Fig. 5.14 Hamartoma. Splenic hamartoma is histologically composed of predominantly splenic red pulp structures, without normal white pulp architecture. In this case, there is a vague border between uninvolved spleen (*right*) with white pulp nodules. The splenic hamartoma (*left*) is composed of disorganized red pulp structures. Sinuses are patent

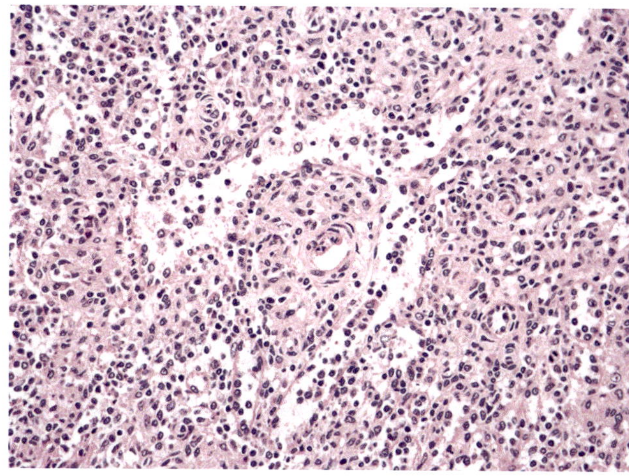

Fig. 5.16 Hamartoma. High magnification of splenic hamartoma. Sinuses are apparently normal, with histiocytes and lymphocytes within the cords. Some lymphocytes are seen within the sinuses. These are predominantly T-cells

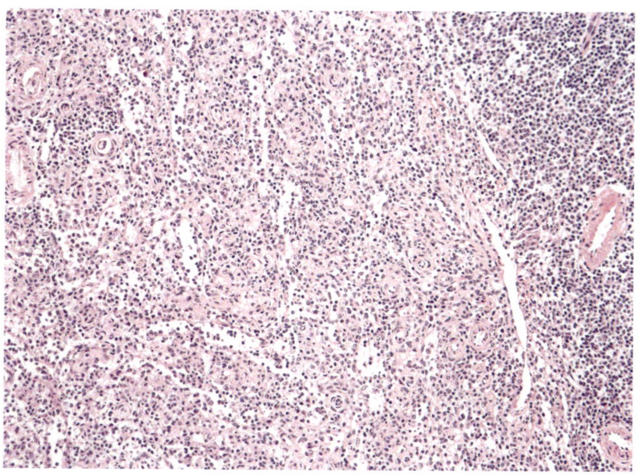

Fig. 5.15 Hamartoma. Intermediate magnification of splenic hamartoma. Without specific indications, splenic hamartoma cannot be distinguished from normal red pulp. Its presentation as a mass and a lack of *organized* white pulp structures is key to the diagnosis

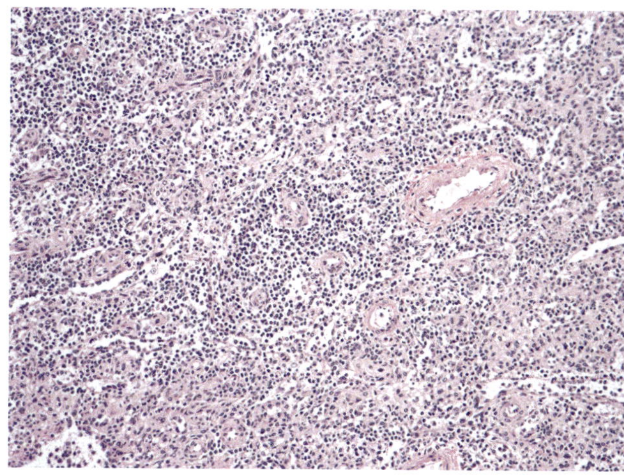

Fig. 5.17 Hamartoma. Although not present in normal amounts or with a normal distribution, lymphocytes are not completely absent from splenic hamartoma, as in this case. The lymphocytes present are typically scattered in red pulp structures or present in a periarteriolar distribution

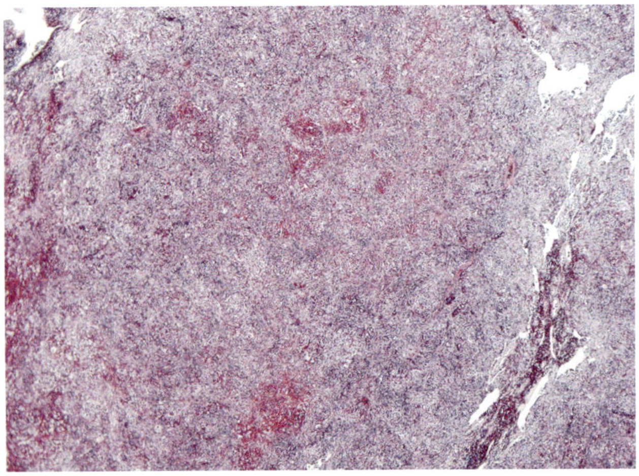

Fig. 5.18 Hamartoma. Another example of splenic hamartoma. No normal white pulp structures are seen. The differential diagnosis includes other vascular lesions of spleen including hemangioma, sclerosing angiomatoid nodular transformation (SANT), and littoral cell angioma. Inflammatory pseudotumor and related entities may also be considered in the differential diagnosis of hamartoma

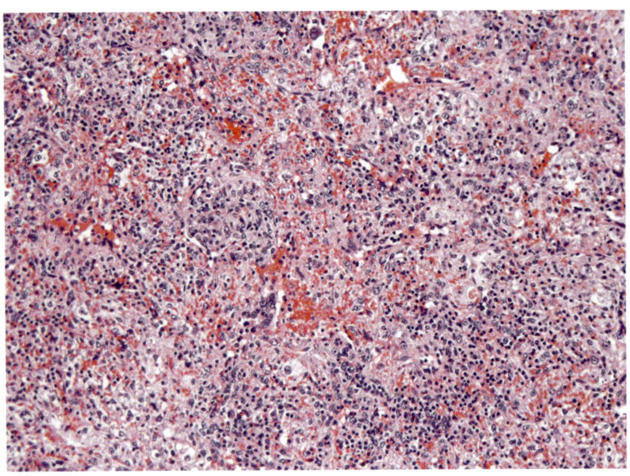

Fig. 5.19 Hamartoma. Not uncommonly, extramedullary hematopoiesis (EMH) is seen within splenic hamartoma. In this case, there are some erythroid and myeloid precursor cells present. The colonization by EMH may be a result of the "permissive" microenvironment in the hamartoma

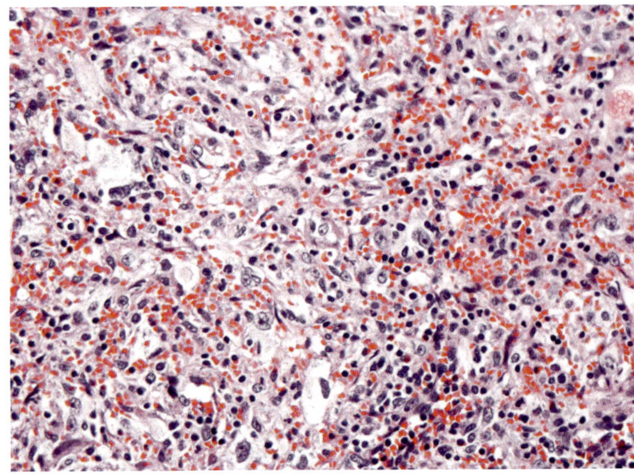

Fig. 5.20 Hamartoma. In this example, some of the stromal cells of the hamartoma are enlarged, binucleated, and have occasional prominent nucleoli. As seen below, these changes can become quite dramatic

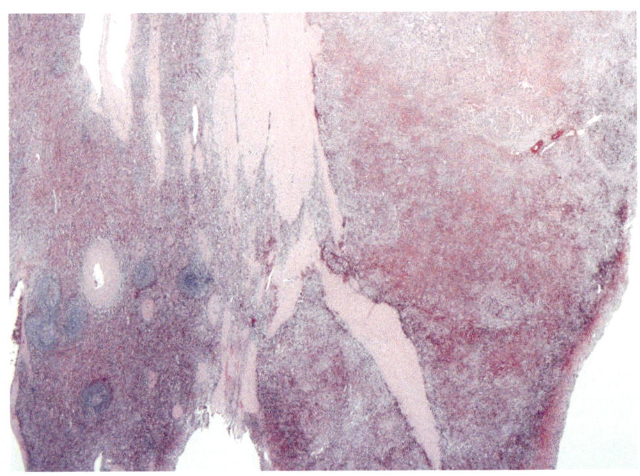

Fig. 5.21 Hamartoma with bizarre stromal cells. An unusual variant, hamartoma with bizarre stromal cells has a similar gross appearance of splenic hamartoma, and has the typical, ill-defined border with normal splenic parenchyma. In this case, there is a fibrous pseudocapsule present. Normal tissue with white pulp is present at left, with the hamartomatous tissue present at right. (*Image courtesy of* M. Argyres, Denver, CO.)

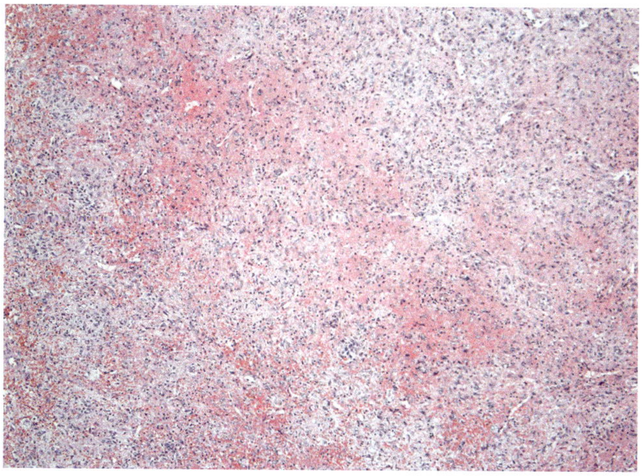

Fig. 5.22 Hamartoma with bizarre stromal cells. An intermediate magnification of splenic hamartoma with bizarre stromal cells that is paucicellular, with some ill-defined vascular spaces. There are increased stromal elements (cordal component). (*Image courtesy of* M. Argyres, Denver, CO.)

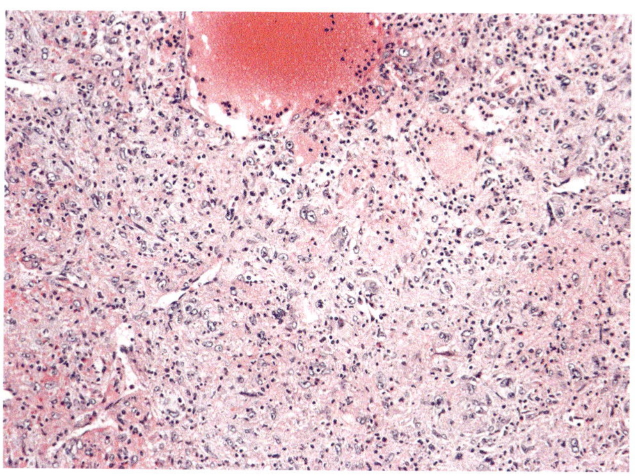

Fig. 5.23 Hamartoma with bizarre stromal cells. At intermediate-high magnification, the atypia of the stromal cells can be seen in splenic hamartoma with bizarre stromal cells. The large cells are multinucleated, with increased amounts of cytoplasm. They have open, marginated chromatin. (*Image courtesy of* M. Argyres, Denver, CO.)

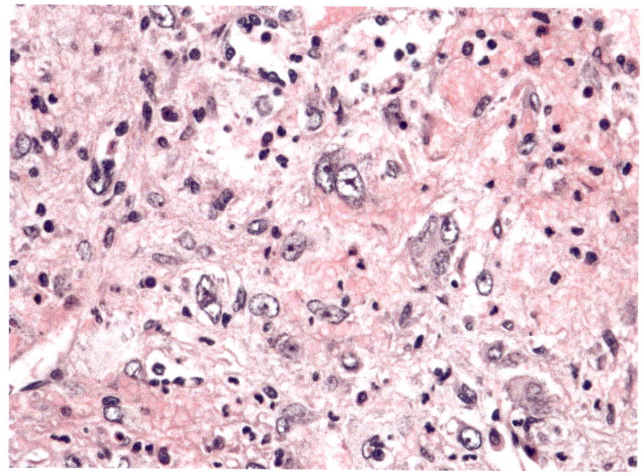

Fig. 5.24 Hamartoma with bizarre stromal cells. High magnification of the abnormal stromal cells in splenic hamartoma with bizarre stromal cells. The changes in these cells is thought to be comparable with "ancient change" in schwannomas. The large cells, although ominous in appearance, are benign. However, careful evaluation is necessary to exclude the possibility of a neoplasm. The cells lack mitotic figures. (*Image courtesy of* M. Argyres, Denver, CO.)

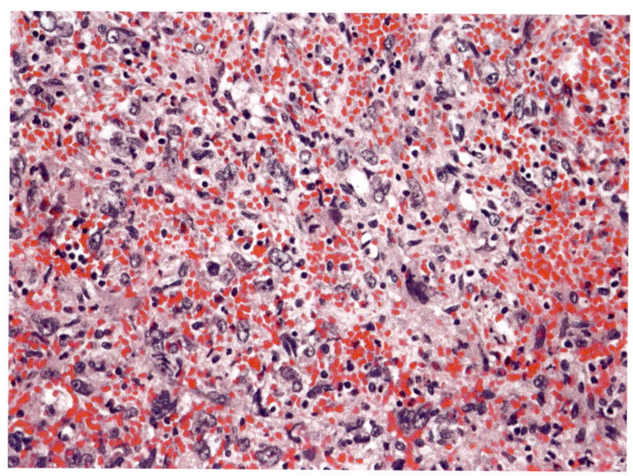

Fig. 5.25 Hamartoma with bizarre stromal cells. In this area of splenic hamartoma with bizarre stromal cells, the combination of atypical cells and vascular spaces raises the possibility of a vascular neoplasm such as angiosarcoma. Splenic hamartoma with bizarre stromal cells can be a diagnostic pitfall. (*Image courtesy of* M. Argyres, Denver, CO.)

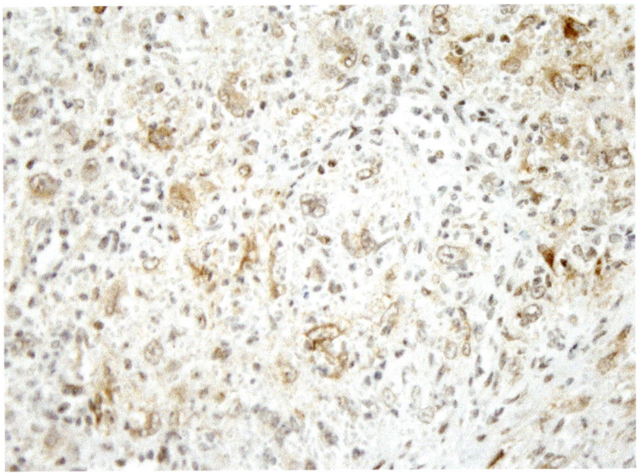

Fig. 5.26 Hamartoma with bizarre stromal cells, glypican stain. Extensive immunohistochemical evaluation of splenic hamartoma with bizarre stromal cells will often reveal that the large cells are mostly unreactive with a large panel of markers (author's experience). Although by no means specific, staining for glypican was identified in this case. Stains for keratin, S-100, histiocytic markers (CD68, CD163), and muscle markers are negative. (*Images courtesy of* M. Argyres, Denver, CO.)

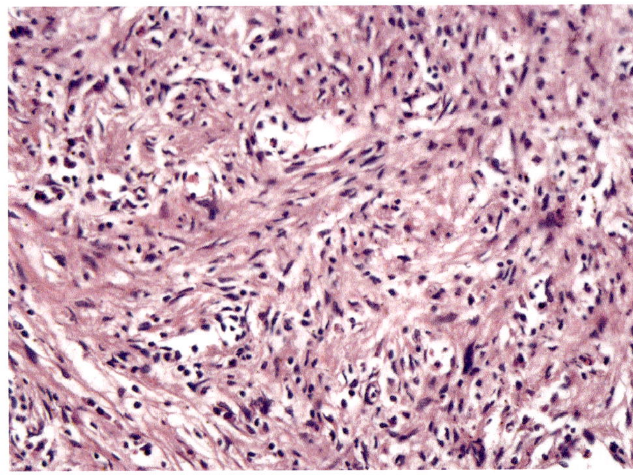

Fig. 5.28 Myoid angioendothelioma. High magnification of myoid angioendothelioma. The majority of cells are spindle shaped, with some interspersed vascular spaces. These cells are myofibroblasts with a bland cytologic appearance

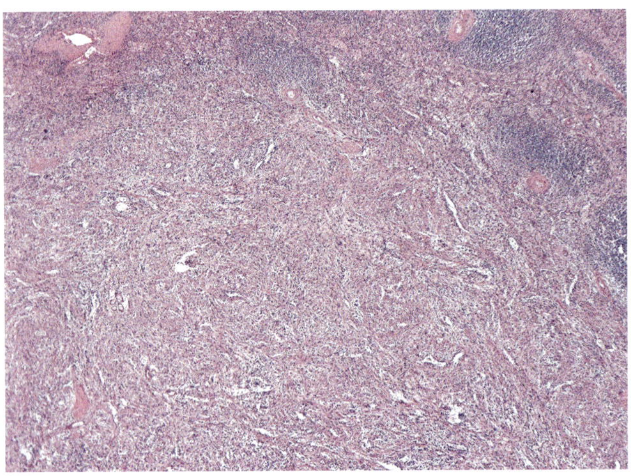

Fig. 5.27 Myoid angioendothelioma. Myoid angioendothelioma is thought to represent a variant of splenic hamartoma. Normal spleen is present (*upper right*) and, as with other hamartomas, well-formed white pulp structures are absent. However, in contrast to classic hamartoma, myoid angioendothelioma is composed of predominantly spindle cells

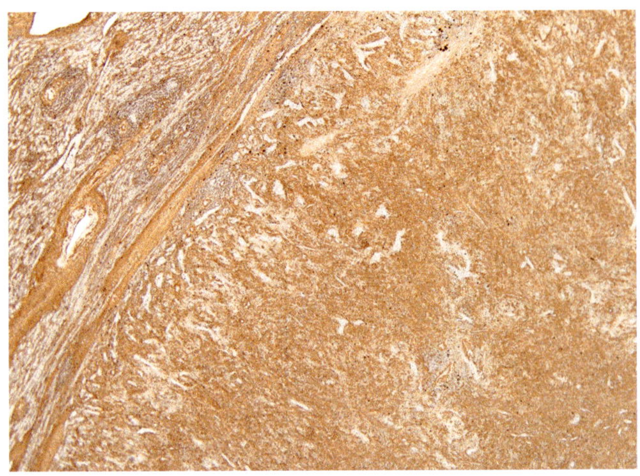

Fig. 5.29 Myoid angioendothelioma, smooth muscle actin stain. Myoid angioendothelioma myofibroblasts stain strongly for smooth muscle actin

5.3 Inflammatory Pseudotumor

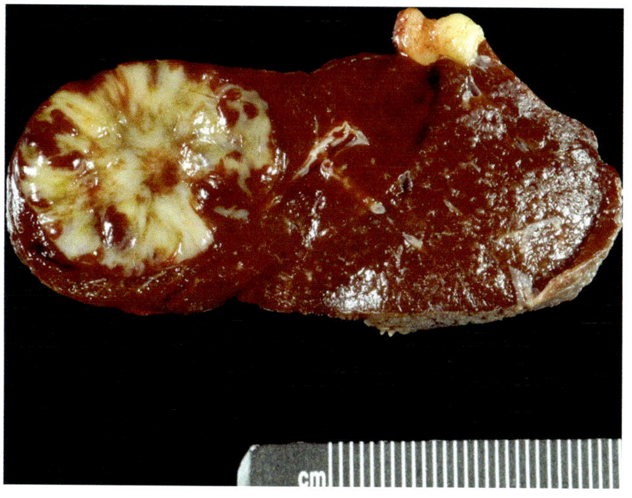

Fig. 5.30 Inflammatory pseudotumor. Gross image of an inflammatory pseudotumor of spleen. This example is small, with yellow-tan fibrotic bands appearing to emanate from a central point. This mass partly compresses the surrounding splenic parenchyma. Note that this gross appearance is indistinguishable from other histologic mimics of IPT. (*Image courtesy of* C. Bueso-Ramos, Houston, TX.)

Fig. 5.31 Inflammatory pseudotumor. Gross image of a fixed section spleen with inflammatory pseudotumor (IPT). The IPT (*upper right*) has a relatively well-circumscribed border grossly but this may be less well-defined by histology. The gross appearance is somewhat variegated in color. (*Image courtesy of* D. Farhi, Atlanta, GA.)

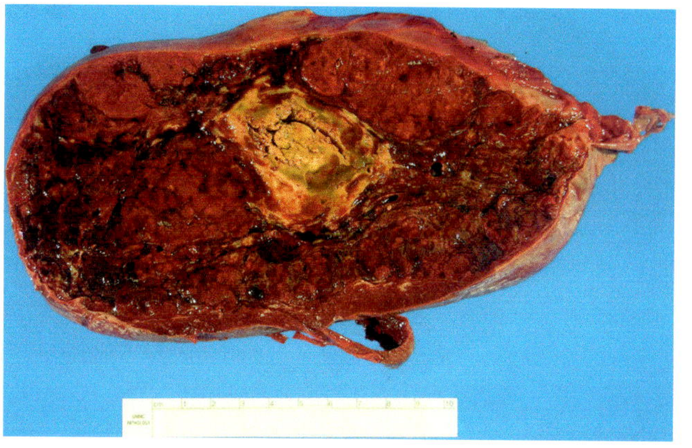

Fig. 5.32 Inflammatory pseudotumor. In this example of IPT, there is evidence of central necrosis, with subsequent splenic rupture. Necrosis is not typical of IPT, but can be seen when there is vascular compromise. (*Image courtesy of* M. Kamionek, Boston, MA.)

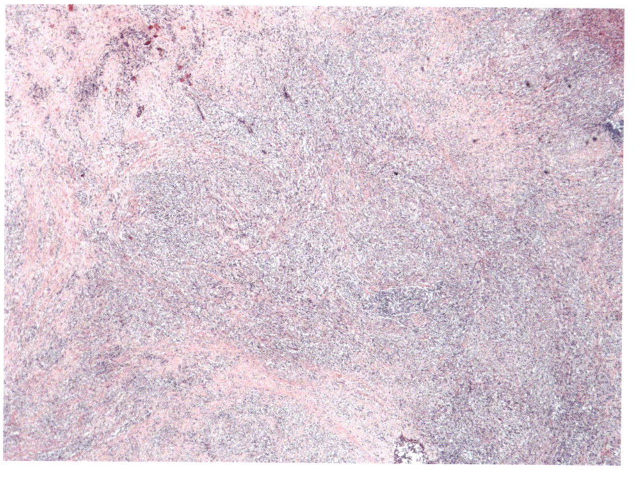

Fig. 5.33 Inflammatory pseudotumor. Low-magnification image of IPT. Note the overall hypocellularity and sclerosis. There is no capsule but the lesion appears well-circumscribed

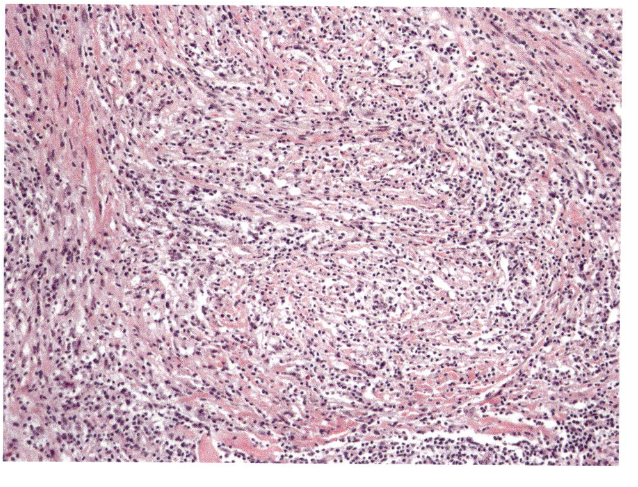

Fig. 5.34 Inflammatory pseudotumor. High magnification of IPT with sclerosis, small lymphocytes, plasma cells, and histiocytes. Spindle cells are a combination of fibroblasts and other stromal elements

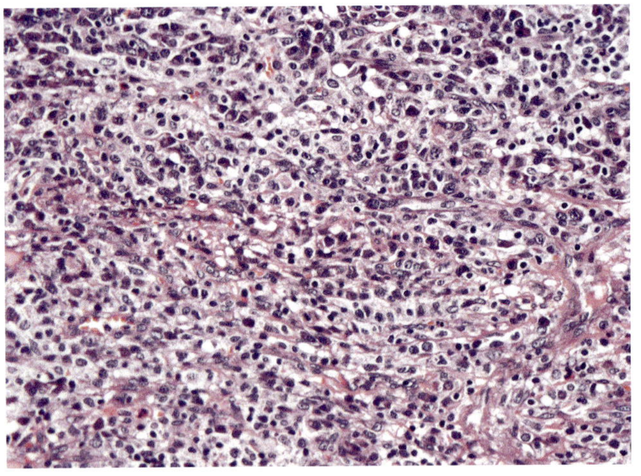

Fig. 5.35 Inflammatory pseudotumor. Inflammatory pseudotumor (IPT) is composed of spindle cells with an admixture of mixed inflammatory elements. IPT will lack features specific for other similar entities, including inflammatory myofibroblastic tumor, follicular dendritic cell tumor, and other spindle cell neoplasms

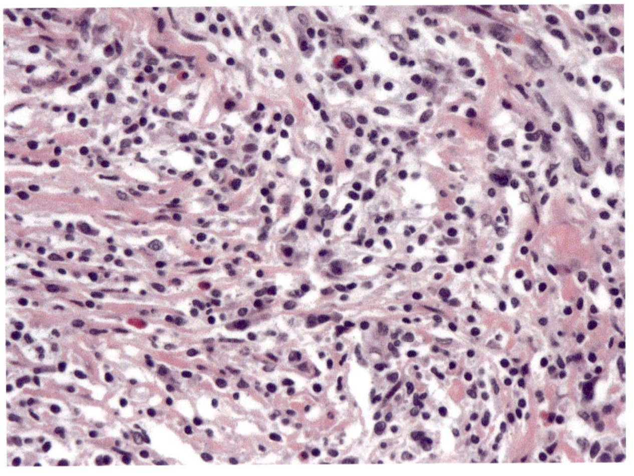

Fig. 5.36 Inflammatory pseudotumor. This example of IPT shows many plasma cells. Lymphoid and plasma cell neoplasms should be excluded. IPT may be associated with infections or of unknown etiology

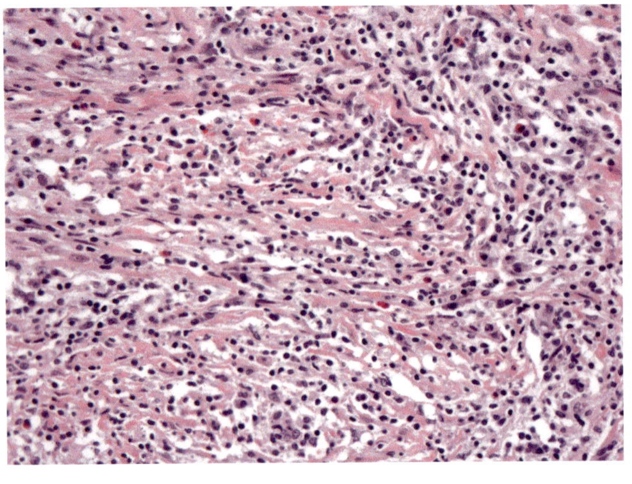

Fig. 5.37 Inflammatory pseudotumor. Another high-magnification image of IPT. Bland spindle cells, plasma cells, small lymphocytes, and sclerosis are typical findings

5.4 Follicular Dendritic Cell Tumor

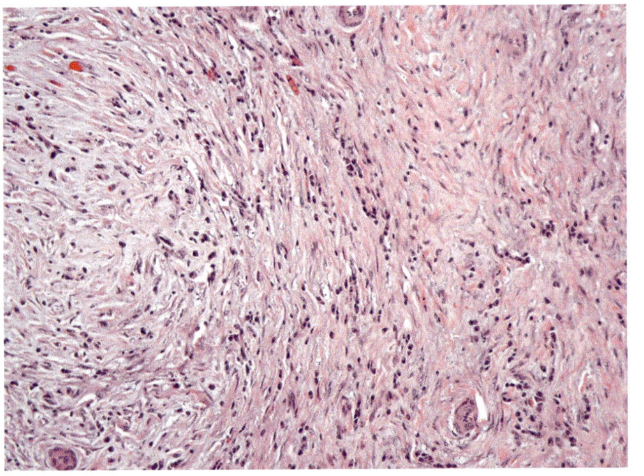

Fig. 5.38 Inflammatory pseudotumor-like follicular dendritic cell tumor. IPT-like follicular dendritic cell tumor has similar low-magnification features to IPT. There are bland spindle cells, a nonencapsulated mass with mixed inflammatory elements, and sclerosis

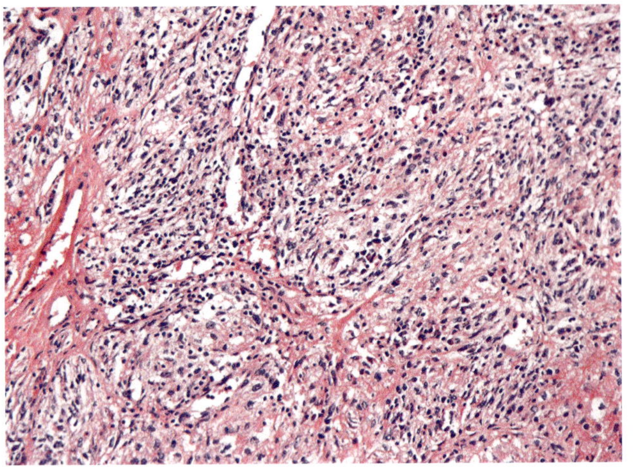

Fig. 5.39 Inflammatory pseudotumor-like follicular dendritic cell tumor. Some cases of IPT-like follicular dendritic cell tumor have fewer spindle cells and more inflammatory elements. In contrast to IPT, the spindled elements are follicular dendritic cells, which must be identified by immunohistochemistry

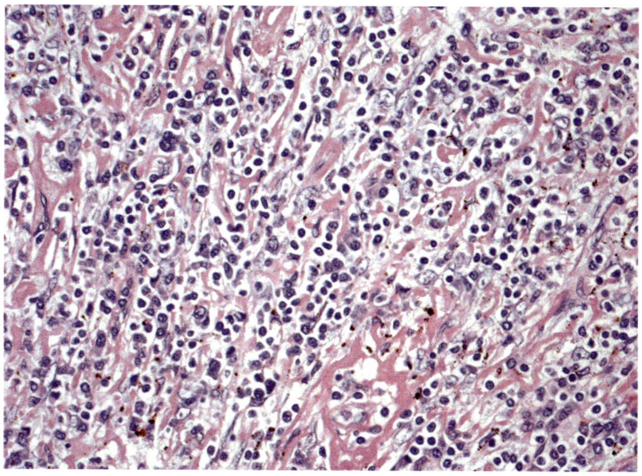

Fig. 5.40 Inflammatory pseudotumor-like follicular dendritic cell tumor. High magnification of IPT-like follicular dendritic cell tumor shows the polymorphous nature of the inflammatory infiltrate. The follicular dendritic cells are inconspicuous spindled cells

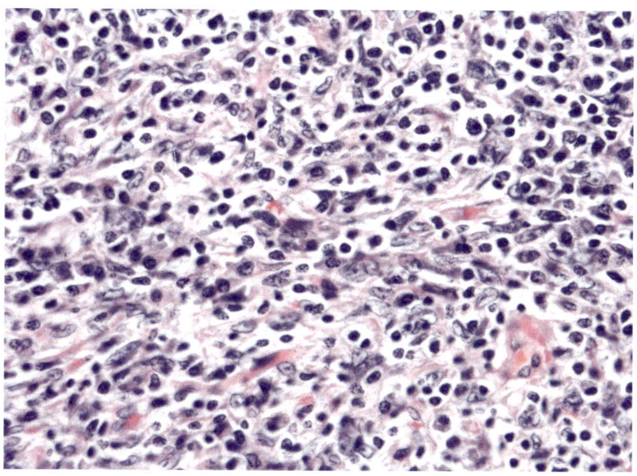

Fig. 5.41 Inflammatory pseudotumor-like follicular dendritic cell tumor. In this high magnification of IPT-like follicular dendritic cell tumor, some follicular dendritic cells are enlarged with atypical-appearing nuclei

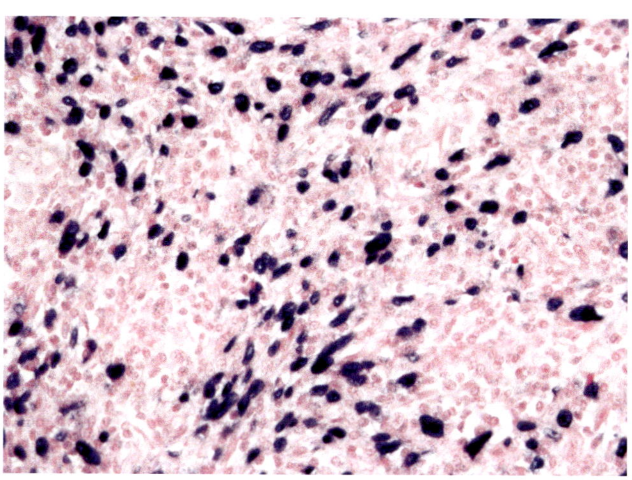

Fig. 5.42 Inflammatory pseudotumor-like follicular dendritic cell tumor, Epstein-Barr virus (EBV) in situ stain. In situ staining for EBV in IPT-like follicular dendritic cell tumor showing many positive spindle cells. In nearly all cases, the follicular dendritic cells are positive for EBV by in situ stains. Also, follicular dendritic cell markers, such as CD21, CD35, D2-40, and others would be positive

5.5 Other Stromal Tumors

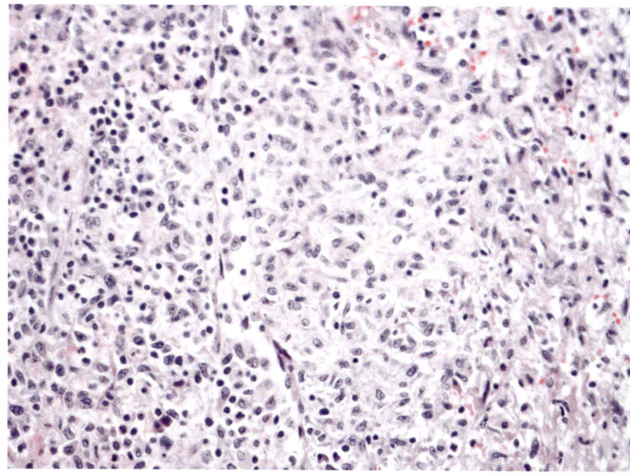

Fig. 5.43 Indeterminate tumor. A rare accessory/dendritic cell tumor of the spleen, indeterminate cells are alleged precursors of Langerhans cells. (Image courtesy of R. Agrawal, Aliso Viejo, CA.)

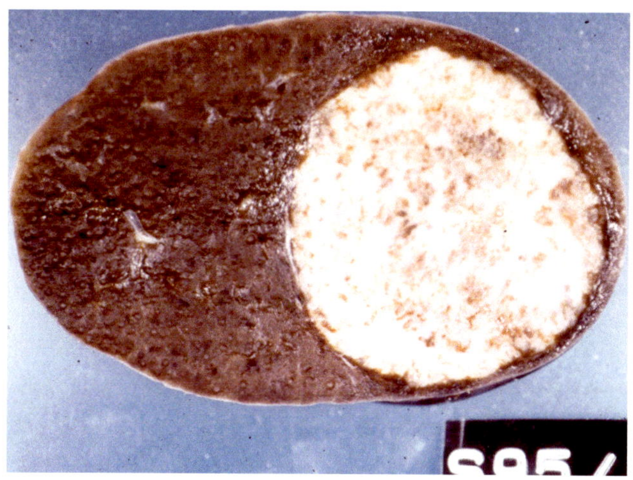

Fig. 5.44 Inflammatory myofibroblastic tumor. Gross image of an inflammatory myofibroblastic tumor of spleen. Compared to the adjacent uninvolved spleen, the tumor appears to be firm white-tan tissue with some slight variation in color (possible areas of blood). Although appearing relatively well-circumscribed, no true capsule would be seen by histology. (*Image courtesy of* D. Farhi, Atlanta, GA.)

5.6 Splenic Artery Aneurysm

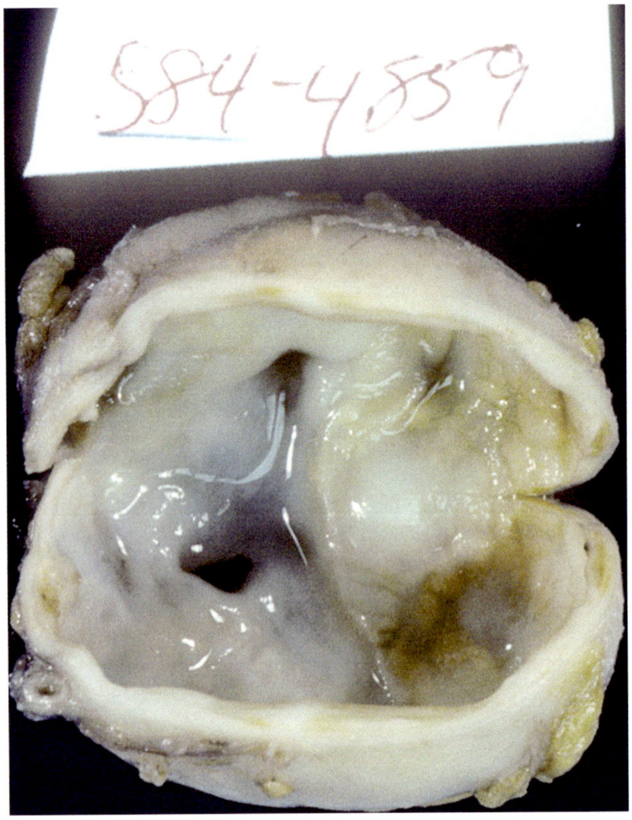

Fig. 5.45 Splenic artery aneurysm. Gross image of a ruptured splenic artery aneurysm. In this case, the aneurysmal cavity has been sectioned to reveal the interior. Splenic artery aneurysms are quite rare but are more common in women. (*Image courtesy of* D. Farhi, Atlanta, GA.)

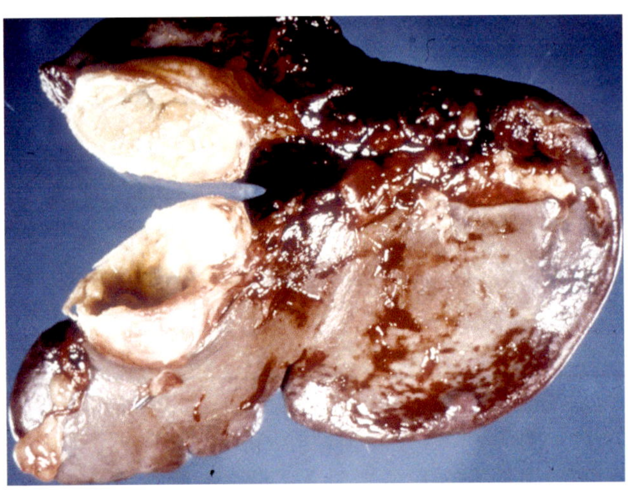

Fig. 5.46 Splenic artery aneurysm. Gross image of dissected splenic artery aneurysm (SAA). SAA are more common in women and more frequently rupture during pregnancy, possibly due to weakening of the arterial media secondary to hormonal changes of pregnancy. (*Image courtesy of* D. Farhi, Atlanta, GA.)

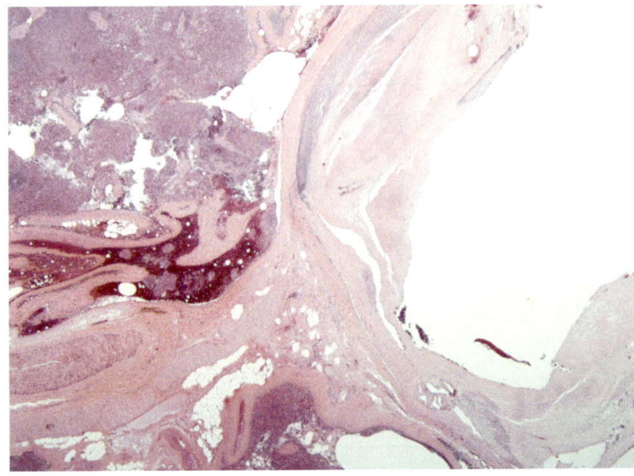

Fig. 5.47 Splenic artery aneurysm. Microscopic image of a splenic artery aneurysm. Normal splenic parenchyma is present at upper left. Note the ballooned and thinned wall of the artery. Elastin staining would show thinning and disruption of the regular elastic fibers of the artery

5.7 Benign Vascular Lesions

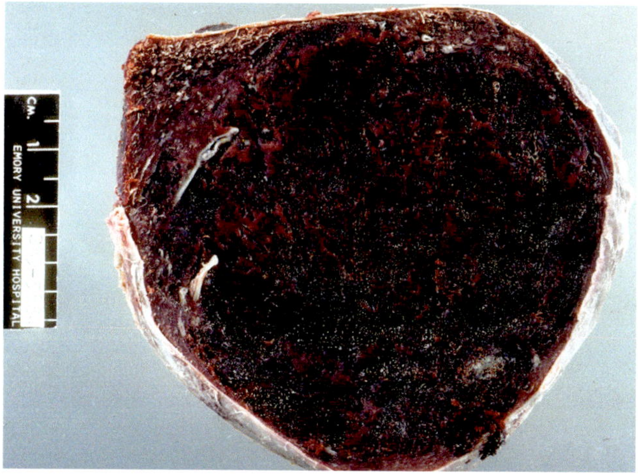

Fig. 5.48 Peliosis. Gross image of splenic peliosis. Although some normal splenic parenchyma is present at the periphery, the majority of tissue present has a dark, purple, congested appearance. The spongy appearance of the red pulp is accentuated by numerous blood-filled spaces. Peliosis occurs most frequently in the liver and less commonly in both liver and spleen, but cases of isolated splenic peliosis have been reported. The exact etiology is unknown but the pathogenesis of peliosis is speculated to be underlying vascular malformations, made apparent by local alterations of vascular pressure. (*Image courtesy of* D. Farhi, Atlanta, GA.)

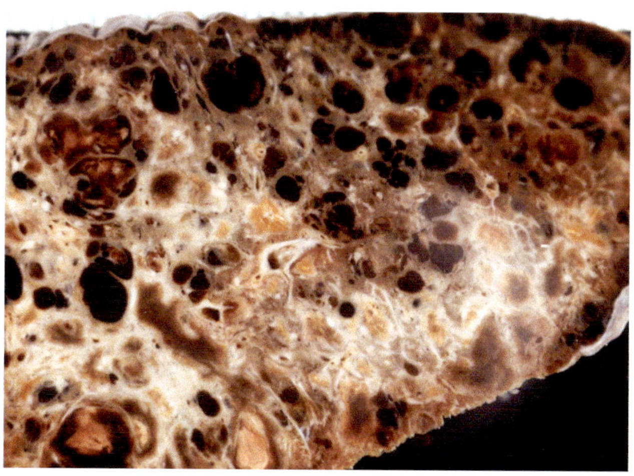

Fig. 5.49 Peliosis. Another gross image of peliosis. The splenic parenchyma is replaced by a confluence of (empty) cystic spaces, imparting a Swiss cheese–like appearance. (*From* O'Malley et al. [1]; with permission.)

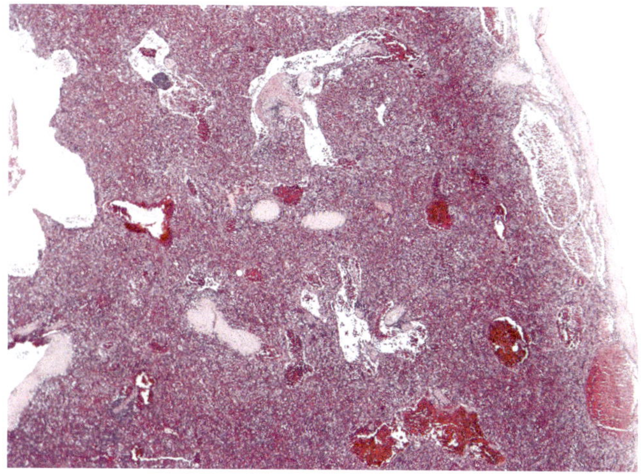

Fig. 5.50 Peliosis. Low-magnification histologic image of splenic peliosis. Numerous dilated cystic, blood-filled spaces are seen. (*Image courtesy of* L. Weiss, Aliso Viejo, CA)

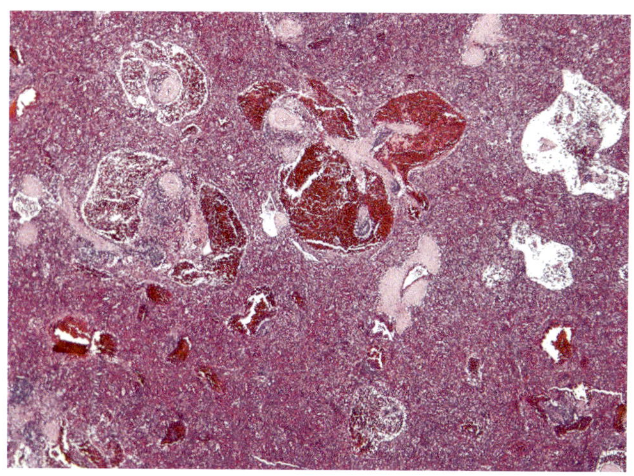

Fig. 5.51 Peliosis. Higher magnification. The lining of the blood-filled spaces is bland. It likely represents expansion of normal splenic sinuses. (*Image courtesy of* L. Weiss, Aliso Viejo, CA.)

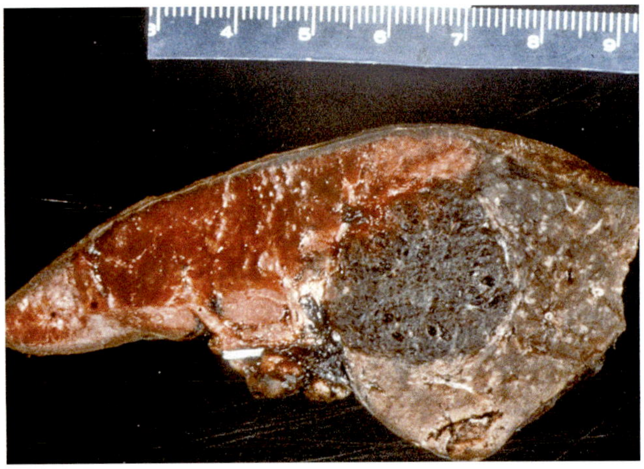

Fig. 5.52 Hemangioma. Gross image of a splenic hemangioma. In this case, the hemangioma is a relatively well-circumscribed nodular lesion in the central portion of the spleen, the left side of the sample is unfixed, whereas the right side is partly fixed. The hemangioma has a spongy, red-brown appearance. Some blood-filled, cystic spaces are seen. (*Image courtesy of* D. Farhi, Atlanta, GA.)

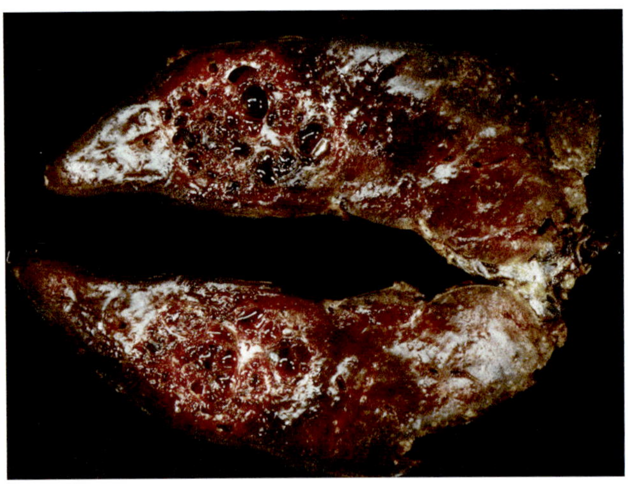

Fig. 5.53 Hemangioma. Another gross image of splenic hemangioma. Based on the size of some spaces, this is at least partly a cavernous hemangioma. Hemangiomas may be single or multiple masses and may have irregular borders. They are not typically encapsulated and merge imperceptibly into adjacent normal spleen. (*Image courtesy of* D. Farhi, Atlanta, GA.)

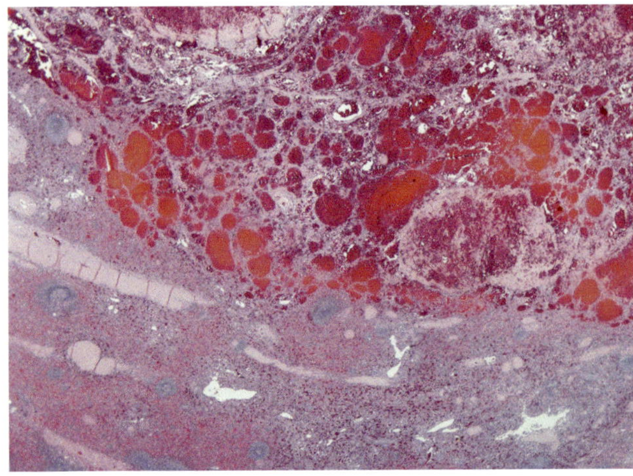

Fig. 5.54 Hemangioma. Low-magnification image of splenic hemangioma. These vascular lesions are often incidental findings on radiologic examination

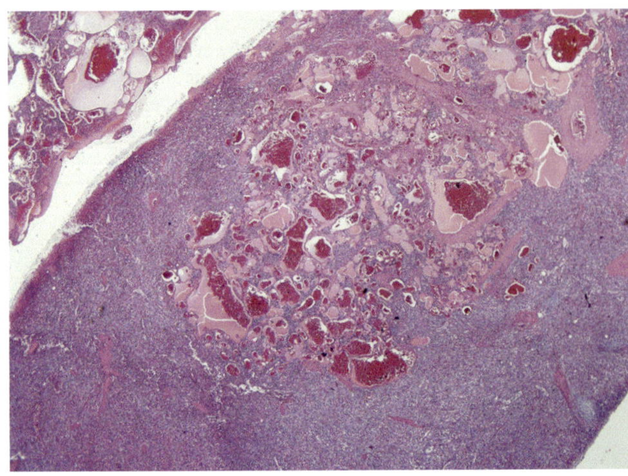

Fig. 5.55 Hemangioma. Splenic hemangiomas often have a random distribution in the spleen. Although they are most often in the red pulp, they may surround or extend into white pulp areas

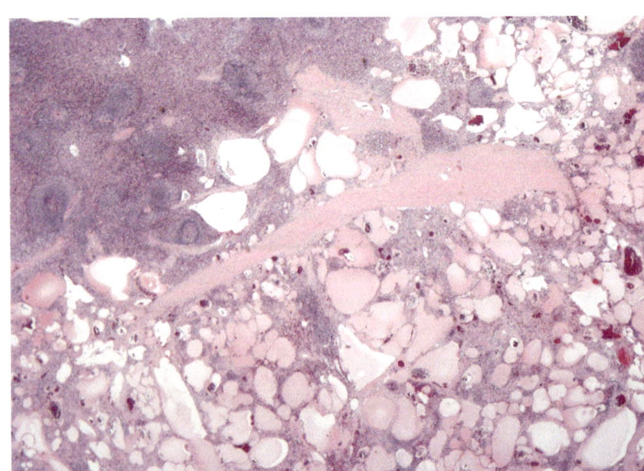

Fig. 5.56 Hemangioma, cavernous. Although the distinction can be arbitrary, a cavernous hemangioma typically has larger caliber vascular spaces compared to the smaller vascular spaces of capillary hemangioma

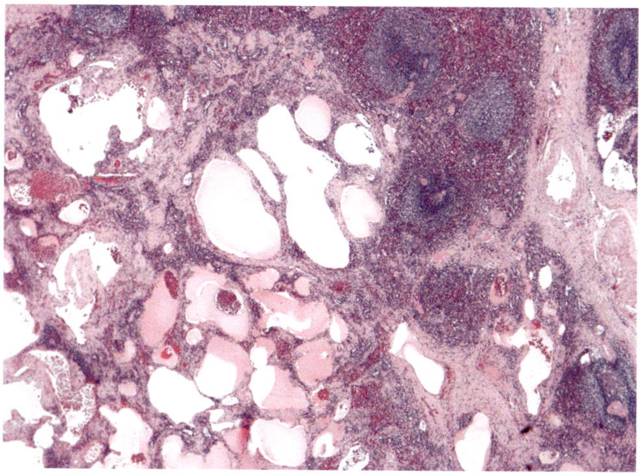

Fig. 5.57 Hemangioma, cavernous. Another low-magnification image of a cavernous hemangioma of spleen. Hemangiomas are unencapsulated and merge into adjacent normal splenic parenchyma (*upper*)

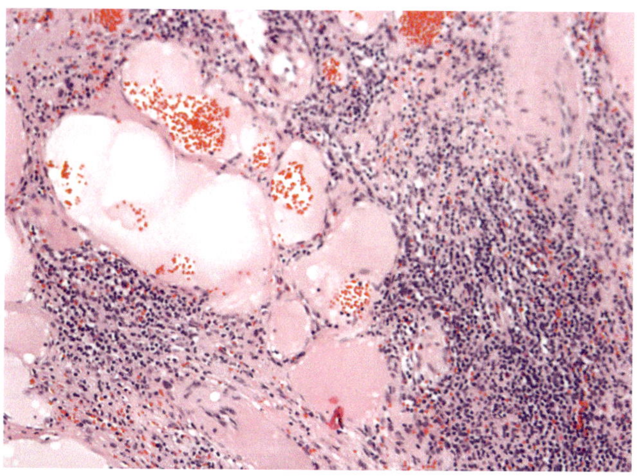

Fig. 5.58 Hemangioma, cavernous. High-magnification image of the vascular spaces of splenic hemangioma. The endothelial cells are flattened, with bland nuclei

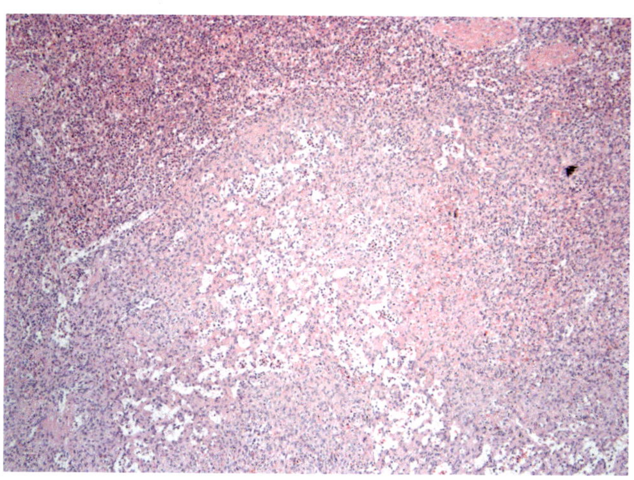

Fig. 5.59 Hemangioma, capillary. Low-magnification image of capillary hemangioma. The vascular spaces are very small in caliber and not dilated

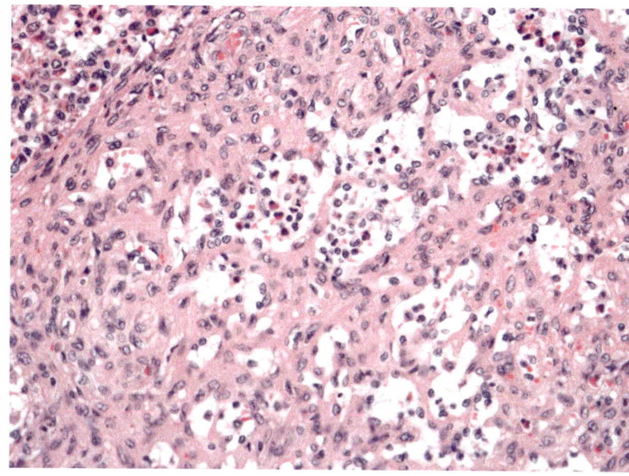

Fig. 5.60 Hemangioma, capillary. High-magnification image of capillary hemangioma. The endothelial cells are bland but the interconnected vascular spaces are small in size

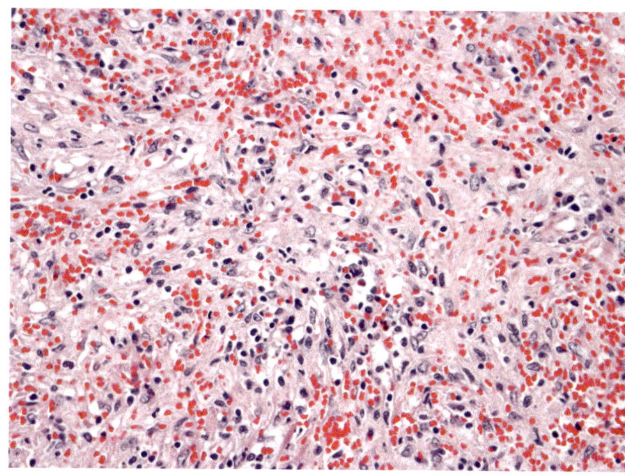

Fig. 5.61 Hemangioma. Another example of a capillary hemangioma. In this case, the vascular spaces are slightly spindle-shaped, which could cause consideration of Kaposi sarcoma in the differential diagnosis

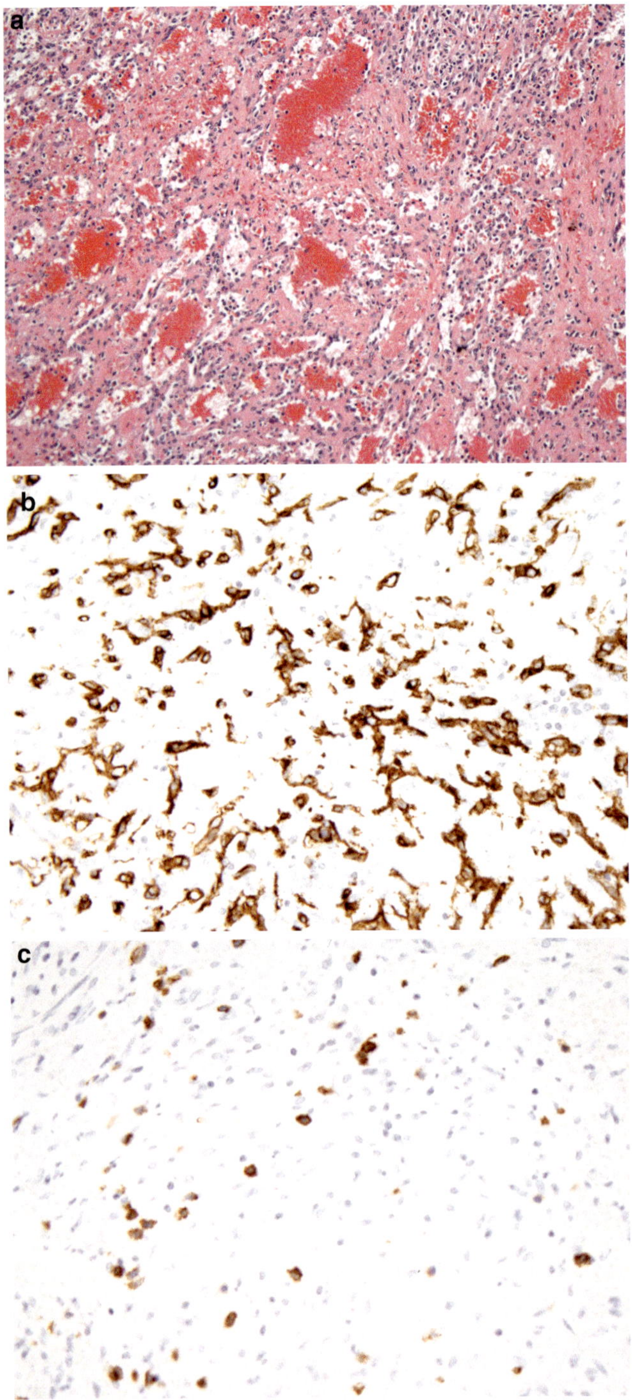

Fig. 5.62 Hemangioma, hematoxylin-and-eosin stain (H&E), CD34 stain, CD8 stain. In this hemangioma (**a**), the vascular spaces are of moderate size. CD34 staining (**b**) is positive, confirming the vascular nature of the cells. In contrast to red-pulp sinus-lining cells, hemangioma vascular elements are CD8 negative (**c**). Only cytotoxic T-cells are positive for CD8 in this area

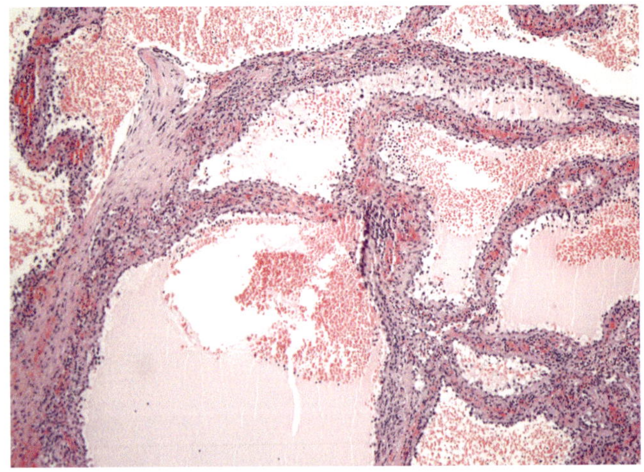

Fig. 5.63 Arteriovenous malformation/hemangioma. In this example, there are very large caliber abnormal vascular spaces present. The distinction of hemangioma of cavernous-type from an arteriovenous malformation can be difficult

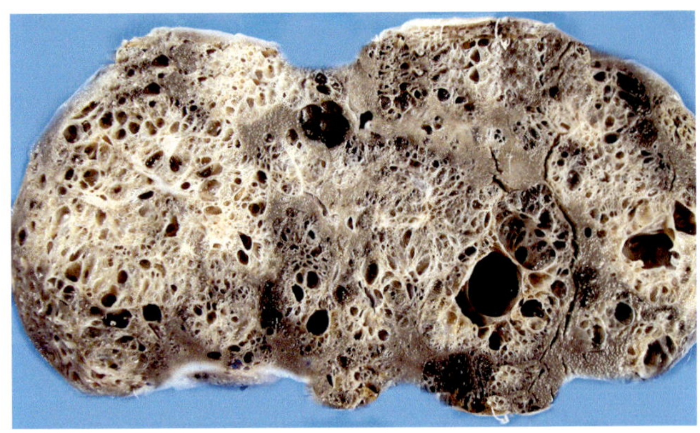

Fig. 5.64 Lymphangiomatosis. Sectioned spleen with extensive involvement by lymphangiomatosis. Numerous cystic spaces, with smooth walls are seen. The fluid filling these cysts is typically clear fluid, although brown or frankly bloody fluid could also be seen. (*From* O'Malley et al. [1]; with permission)

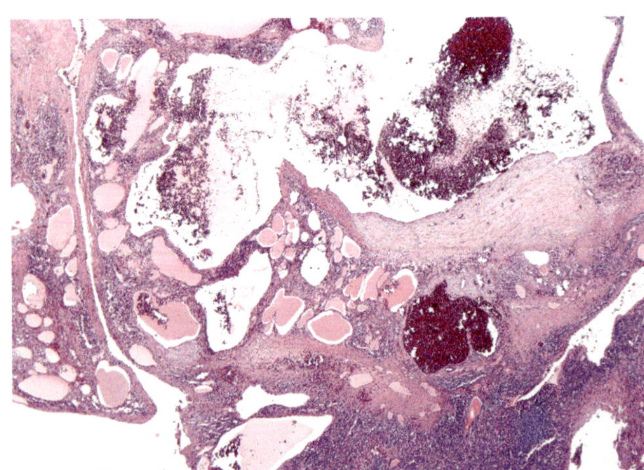

Fig. 5.65 Lymphangioma. Low-magnification image of lymphangioma of the spleen. Note the large open spaces filled with proteinaceous fluid

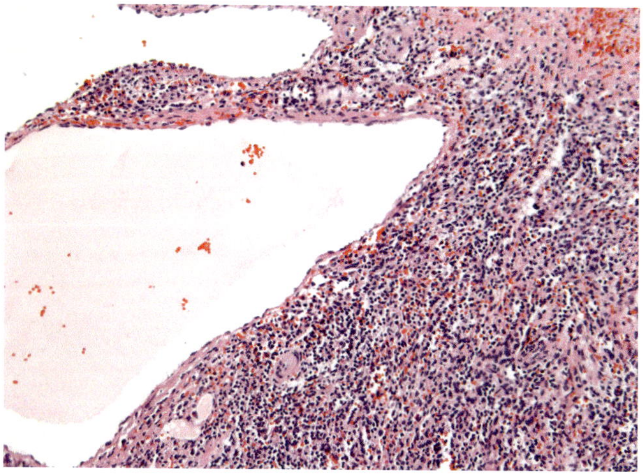

Fig. 5.66 Lymphangioma. Higher-magnification image of lymphangioma. This highlights the bland endothelial/lymphatic lining of the cystic spaces. The lesion is intimately admixed with the normal splenic parenchyma without a well-defined border or capsule

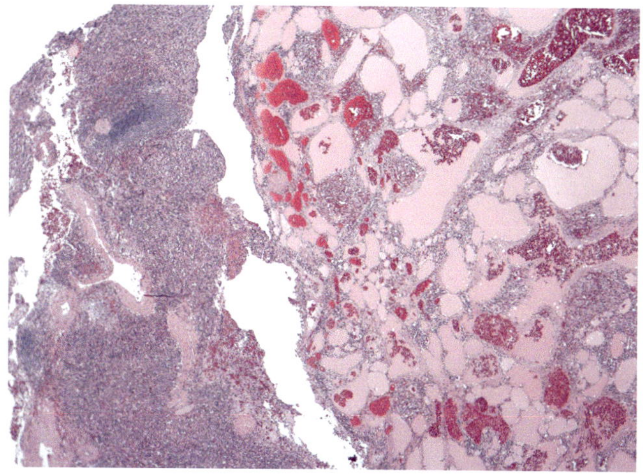

Fig. 5.68 Lymphangioma. Low-magnification image of a lymphangioma. Distinguishing lymphangioma from hemangioma can be difficult. In most cases, it is not clinically relevant. However, if necessary, staining with markers of lymphatic-associated antigens, such as D2-40, can help in the distinction

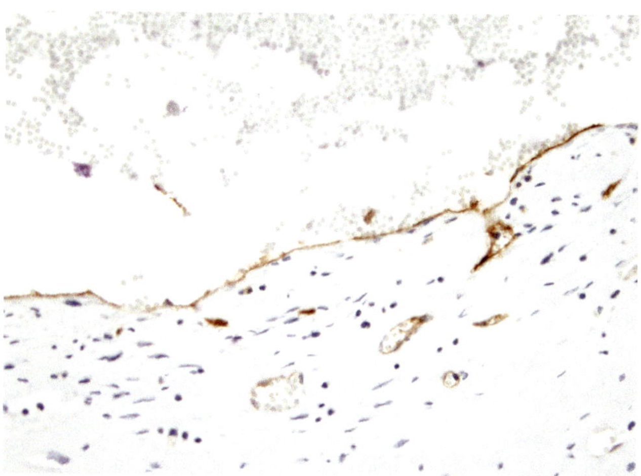

Fig. 5.67 Lymphangioma, D2-40 stain. Immunohistochemical staining for D2-40, a lymphatic endothelial marker, highlighting the lining of a lymphangioma of the spleen

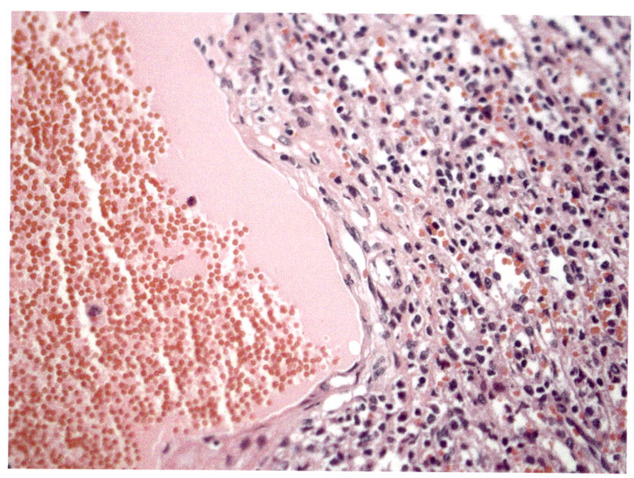

Fig. 5.69 Lymphangiomatosis. A high-magnification image of the lining of a cystic space in lymphangiomatosis. The endothelial cells are flat and attenuated, with bland nuclei. Although the fluid can be blood, there is often acellular clear fluid present (as seen at the edge of cystic space)

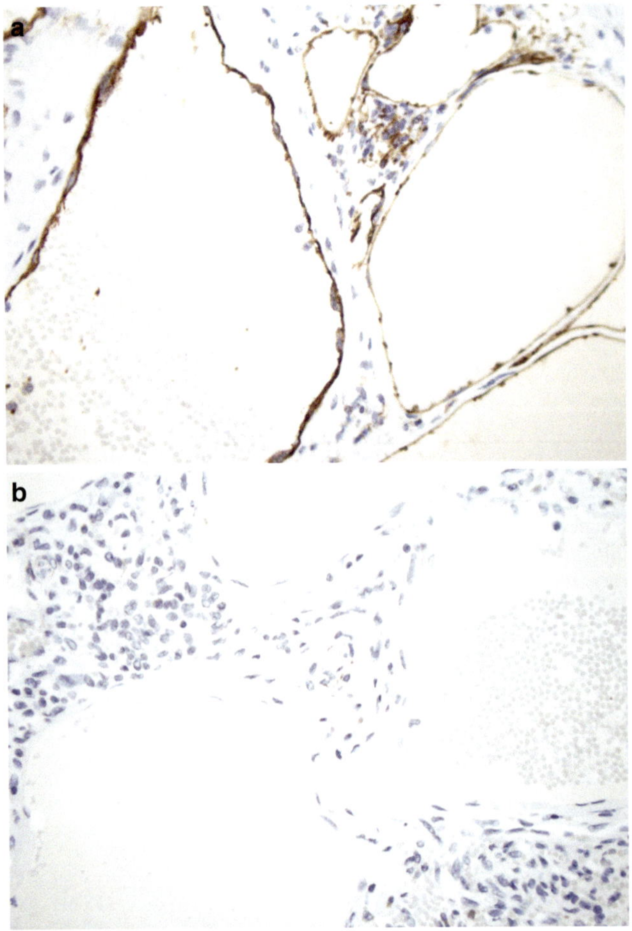

Fig. 5.70 Lymphangiomatosis, CD31 stain, pankeratin stain. An immunohistochemical stain for CD31 (**a**) confirms the endothelial nature of the lining in this example of a lymphangitic cyst. The lack of staining for pankeratin (**b**) excludes a mesothelial cyst, which could have a similar histologic appearance

5.8 Littoral Cell Angioma

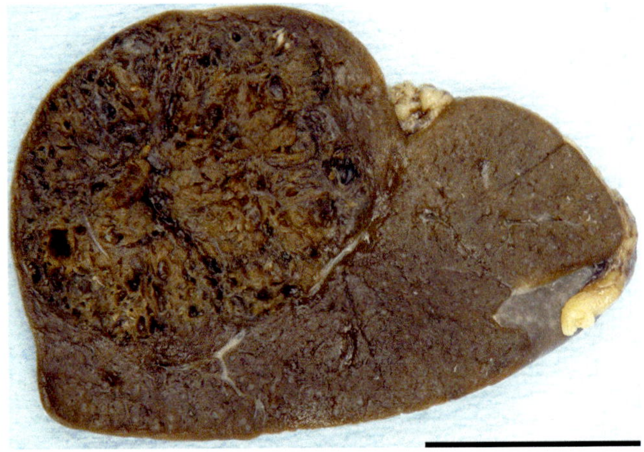

Fig. 5.71 Littoral cell angioma (LCA). Gross image of LCA (formalin fixed). Note that the tumor is relatively well-circumscribed but a true capsule is not apparent. The tumor has a trabecular appearance with empty spaces that would be filled with blood. (*Image courtesy of* J. Sidhu, Johnson City, NY.)

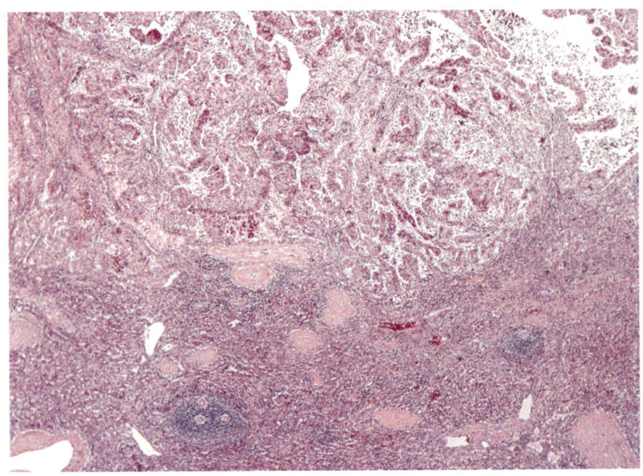

Fig. 5.72 Littoral cell angioma. LCA is a benign vascular neoplasm that occurs exclusively in splenic tissue. It is thought to be derived from neoplastic transformation of the splenic sinus lining cells (*eg*, littoral cells). It is unencapsulated and has a pushing border with uninvolved splenic parenchyma (*lower*). (*Image courtesy of* C. Fisher, London, UK.)

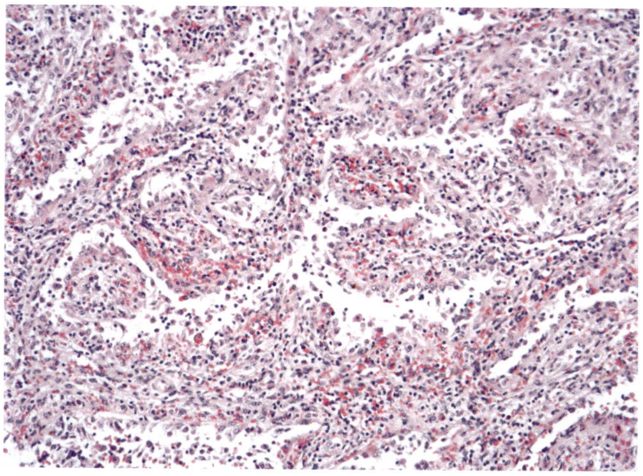

Fig. 5.73 Littoral cell angioma. Intermediate magnification of LCA. There are vascular species (mostly devoid of red blood cells). The lining cells are often plump with a "tombstone" appearance. Between the lining spaces are macrophages and stromal elements

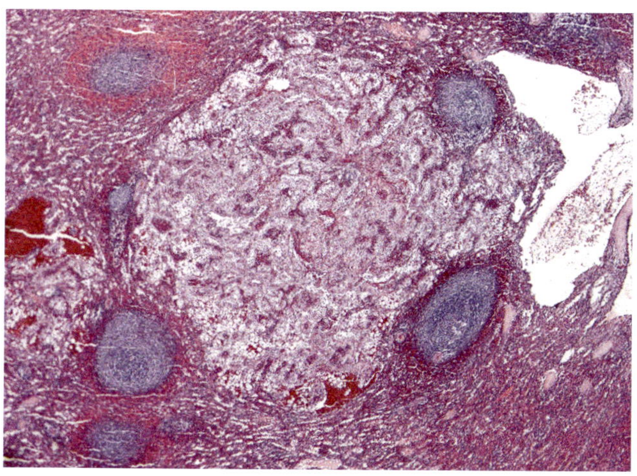

Fig. 5.74 Littoral cell angioma. A small focus of LCA surrounded by otherwise normal spleen. LCA is often discovered incidentally on radiologic examination. (*Image courtesy of* N. Crowson, Tulsa, OK.)

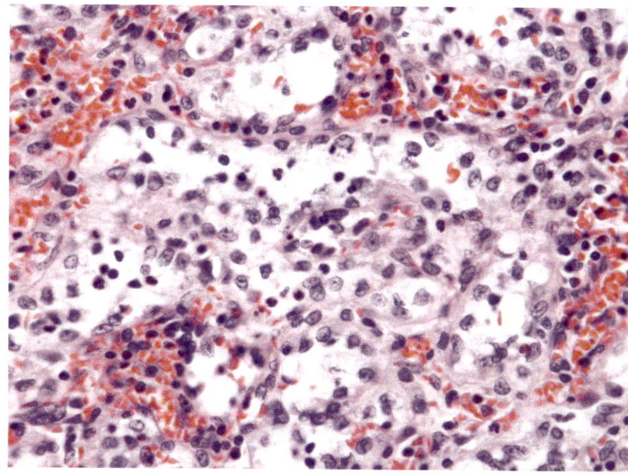

Fig. 5.75 Littoral cell angioma. High magnification of LCA. In this example, the lining cells appear plump with bland nuclear features. There are some cells in the vascular spaces and there are often macrophages, which may be filled with debris. Extramedullary hematopoiesis may be seen in LCA. (*Image courtesy of* N. Crowson, Tulsa, OK.)

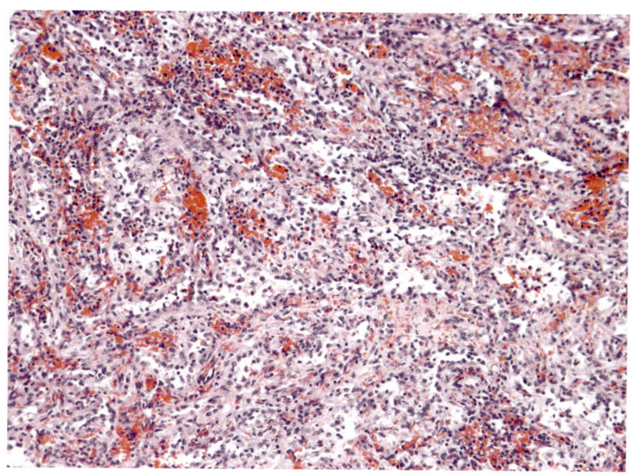

Fig. 5.76 Littoral cell angioma. Intermediate magnification of LCA. The similarity of the splenic sinuses can be appreciated but there is also more supporting stroma in the intervening spaces between vascular spaces. (*Image courtesy of* N. Crowson, Tulsa, OK)

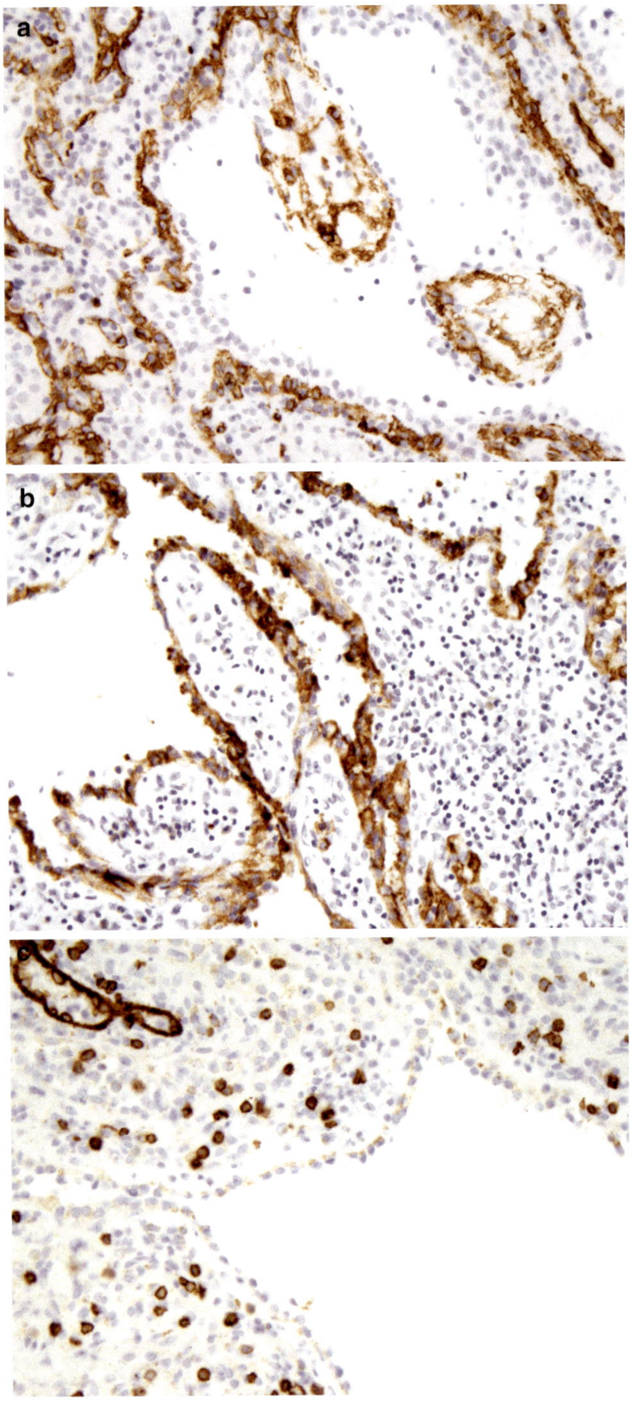

Fig. 5.77 Littoral cell angioma, CD34 stain, CD21 stain, CD8 stain. The endothelial lining cells of LCA are negative for CD34, similar to normal littoral cells (staining of capillaries beneath the endothelial lining is positive) (**a**) and are frequently positive for CD21 (**b**). In contrast to normal littoral cells, LCA do not express CD8 (**c**)

5.9 Sclerosing Angiomatoid Nodular Transformation (SANT)

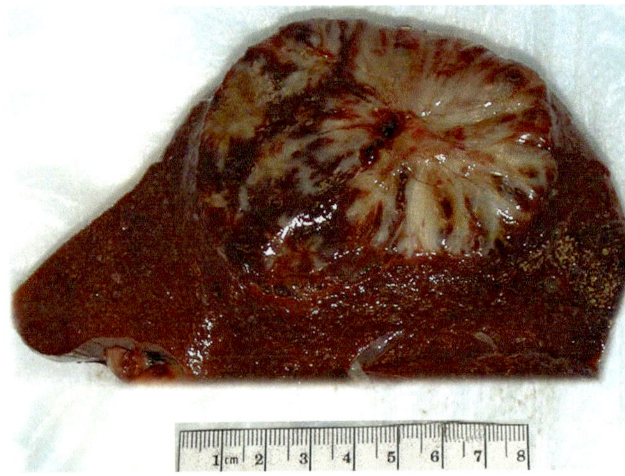

Fig. 5.78 Sclerosing angiomatoid nodular transformation (SANT). Gross image of SANT. Note the bulging cut surface, bands of tan to yellow fibrous material with foci of deep red hemorrhage. This gross appearance is similar to that seen in inflammatory pseudotumor (*see* Figs. 5-30 to 5-32). (*Image courtesy of* R. Dewar, Boston, MA.)

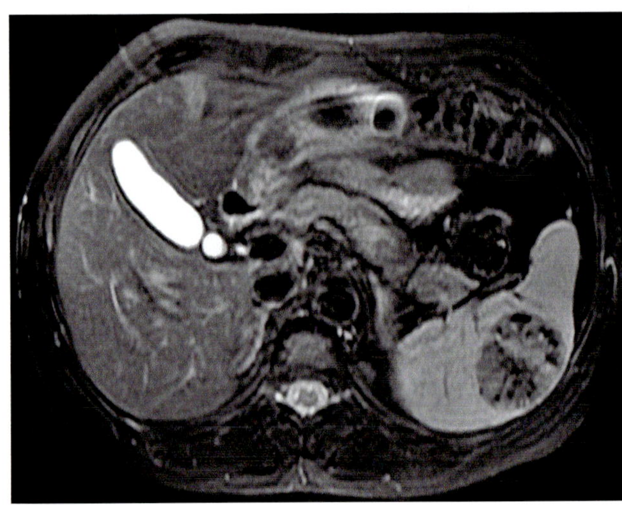

Fig. 5.79 Sclerosing angiomatoid nodular transformation. CT image of SANT from above. The spleen (*right*) shows well-circumscribed nodular masses with irregular bands in the central portion of intermediate density

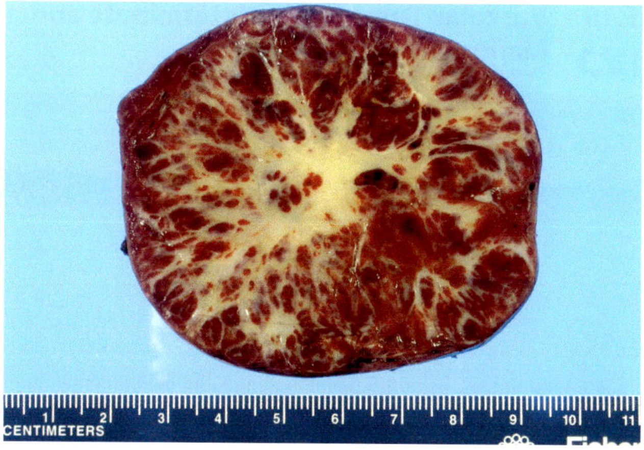

Fig. 5.80 Sclerosing angiomatoid nodular transformation. Gross image of a relatively large area of SANT. Characterized by vascular elements with stromal proliferations, the exact nature of this lesion is controversial. (*Image courtesy of* A. Kalof, Burlington, VT.)

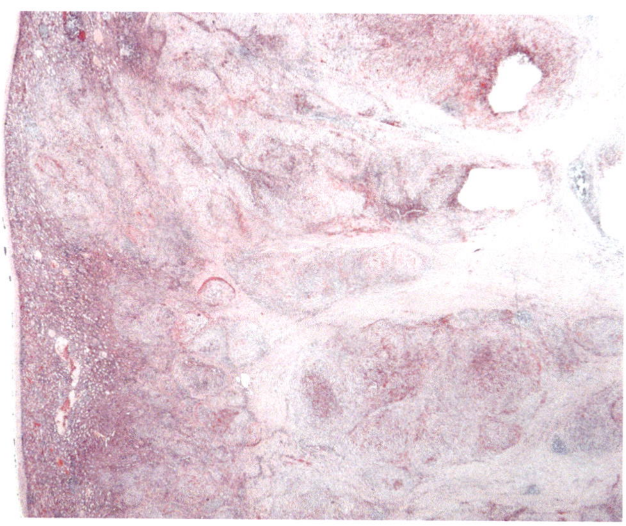

Fig. 5.81 Sclerosing angiomatoid nodular transformation. Low-magnification image of SANT. Normal splenic parenchyma (*left*) merges with the abnormal fibrovascular lesion (*right*). SANT is unencapsulated. (*Image courtesy of* N. Crowson, Tulsa, OK.)

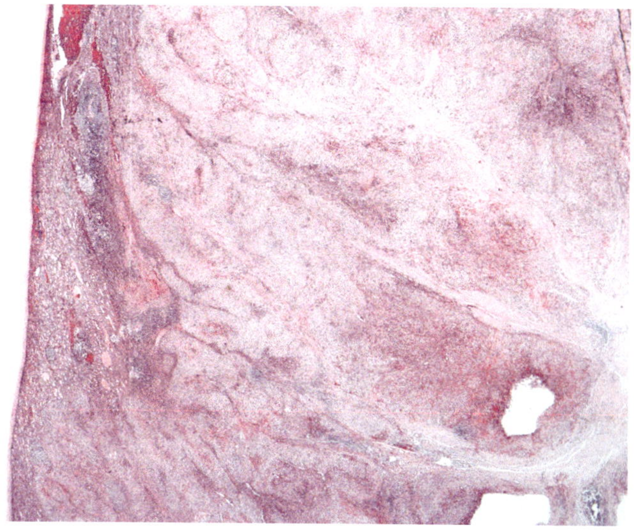

Fig. 5.82 Sclerosing angiomatoid nodular transformation. Low magnification of SANT. There is a distinct nodular pattern present. This is composed of bands of fibrous or stromal tissue with interspersed slightly more cellular nodules. (*Image courtesy of* N. Crowson, Tulsa, OK.)

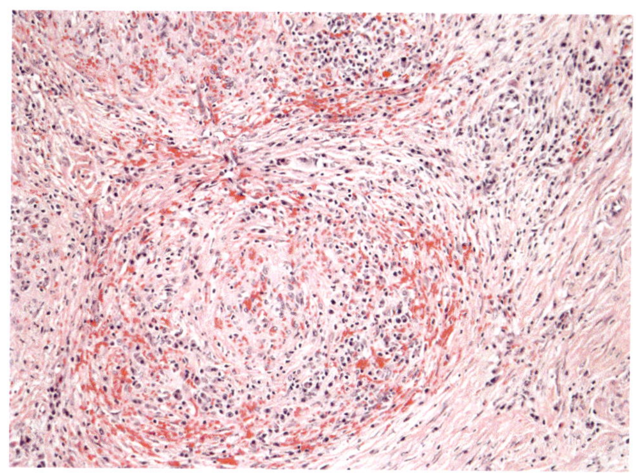

Fig. 5.83 Sclerosing angiomatoid nodular transformation. High magnification of details of SANT. The intervening spindle cells are bland. The nodules are composed of small caliber vascular elements with scattered inflammatory elements and macrophages and stromal cells. (*Image courtesy of* N. Crowson, Tulsa, OK.)

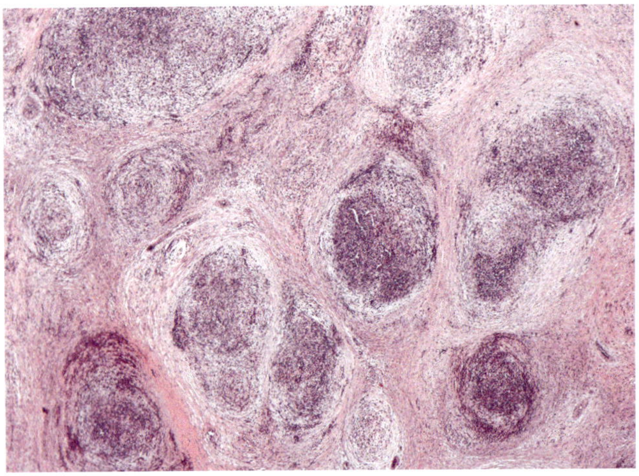

Fig. 5.84 Sclerosing angiomatoid nodular transformation. Intermediate magnification of SANT. In this case, the nodules appear more cellular with congestion and increased red blood cells. (*Image courtesy of* J. Goldblum, Cleveland, OH.)

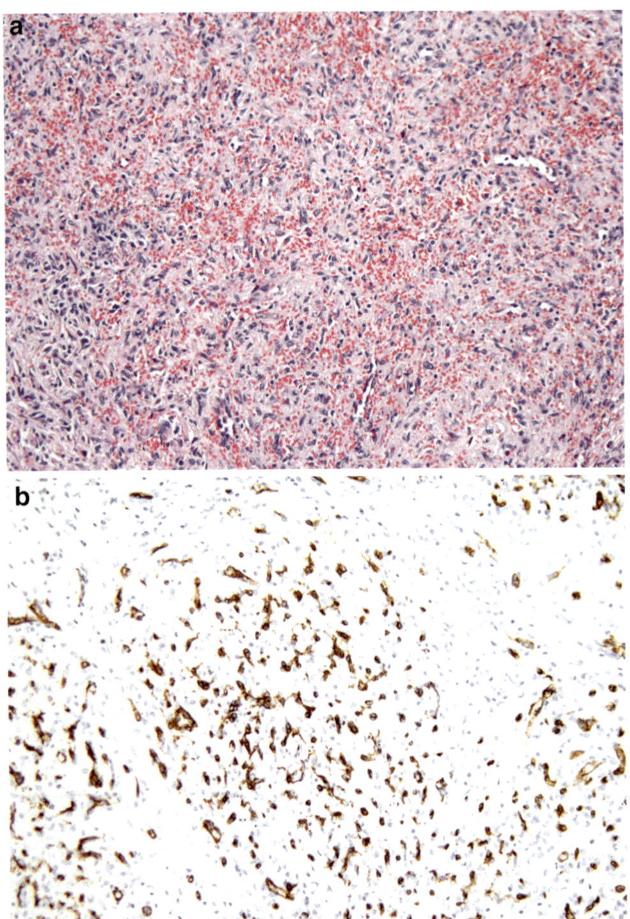

Fig. 5.85 Sclerosing angiomatoid nodular transformation, H&E (**a**), CD34 stain (**b**). High-magnification image of nodule of SANT (**a**), with CD34 staining in vascular elements (**b**)

5.10 Vascular Tumors of Indeterminate and Malignant Behavior

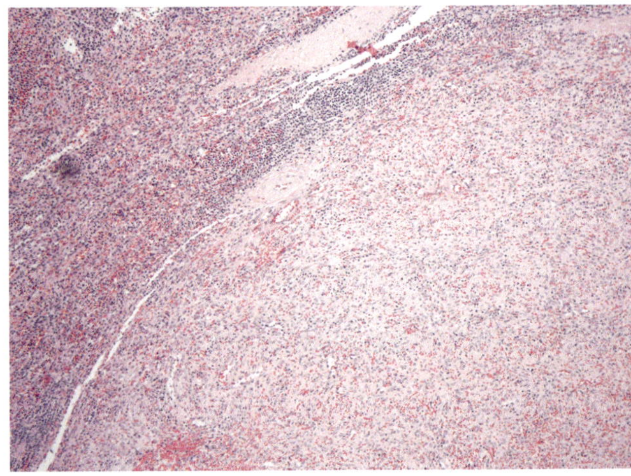

Fig. 5.86 Hemangioendothelioma. Hemangioendothelioma is thought to represent a vascular neoplasm of intermediate or uncertain biologic potential. It is more cellular with a greater degree of pleomorphism than hemangiomas. On occasion it may have epithelioid features. (*Image courtesy of* Lawrence Weiss, Aliso Viejo, CA.)

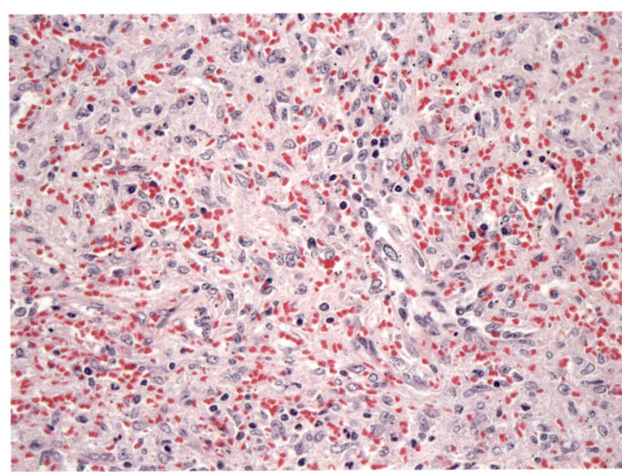

Fig. 5.87 Hemangioendothelioma. High-magnification image of hemangioendothelioma of spleen. In this case, there is clear formation of vascular spaces. The cells are plumper with more pink cytoplasm than is typical for hemangioma. However, it lacks the degree of atypia seen in angiosarcoma. (*Image courtesy of* Lawrence Weiss, Aliso Viejo, CA.)

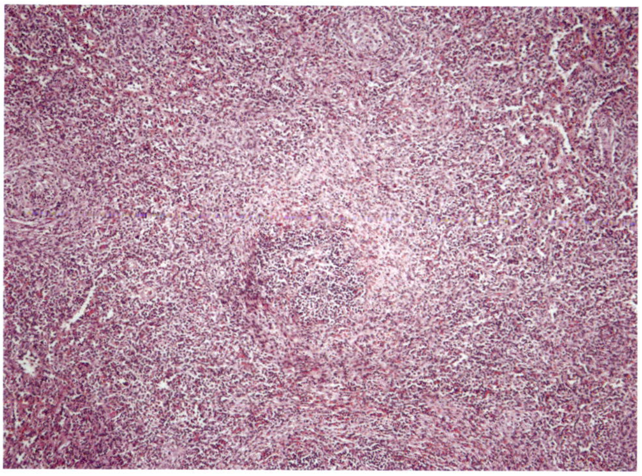

Fig. 5.88 Kaposi sarcoma (KS). Microscopic image of KS involving spleen. Note that involvement is not as prominent in some other sites, as the vascular nature of the lesion may be hidden within the background vasculature. Despite this, there is proliferation of abnormal, predominantly spindle-shaped cells, frequently in perifollicular areas. These spindle-shaped cells form irregular vascular spaces. There may be foci of intravascular red blood cells, fragmented red blood cells, pink hyaline bodies within cells, and hemosiderin deposition. (*Image courtesy of* Lawrence Weiss, Aliso Viejo, CA.)

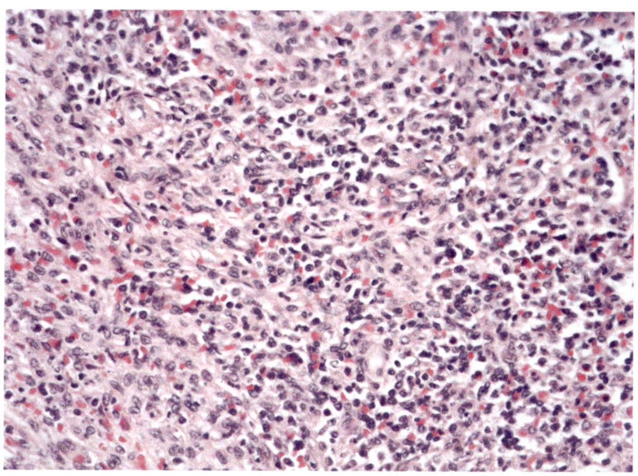

Fig. 5.89 Kaposi sarcoma. High-magnification image of KS of the spleen. Note the irregularity of vascular spaces. The endothelial cells are slightly atypical, but lack the extreme atypia of angiosarcoma

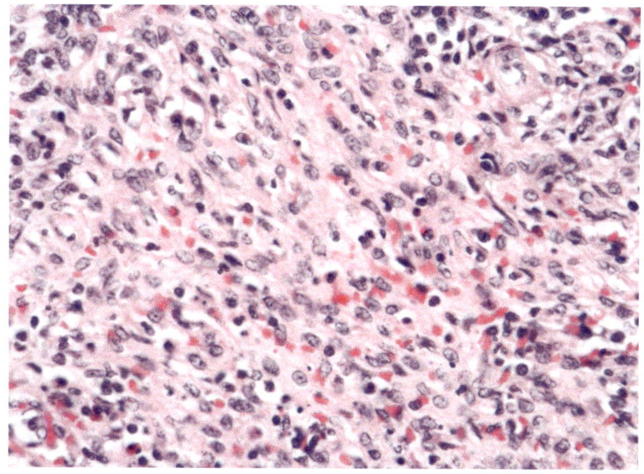

Fig. 5.90 Kaposi sarcoma. High-magnification image of KS in spleen. The irregular slit-like vascular spaces are apparent. Most of the endothelial cells are spindle shaped. Some have more round nuclei. In almost all cases, human herpesvirus 8 (HHV-8) can be identified in KS by immunohistochemical staining

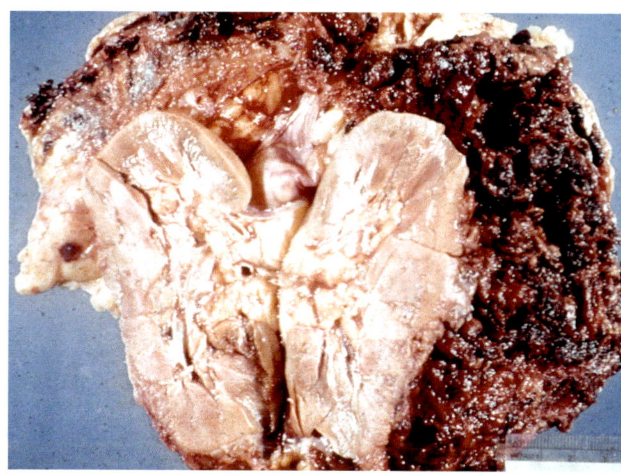

Fig. 5.91 Angiosarcoma. Gross image of angiosarcoma of spleen. Normal splenic parenchyma is not appreciable as the tumor was removed en bloc from the left upper quadrant. The tumor enveloped the kidney, which was removed along with the angiosarcoma. Note the hemorrhagic appearance of the tumor. (*Image courtesy of* D. Farhi, Atlanta, GA.)

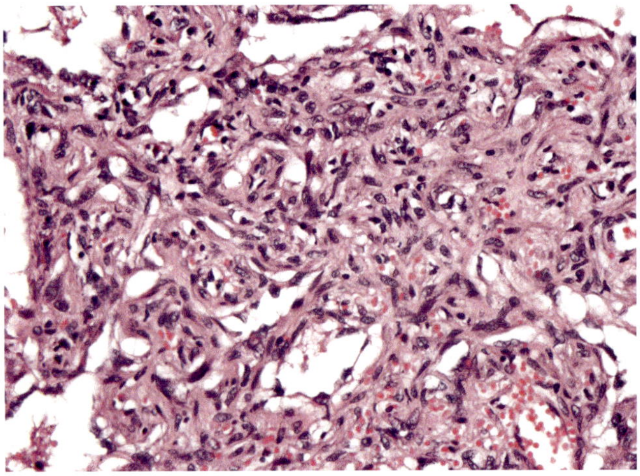

Fig. 5.92 Angiosarcoma. High-magnification image of splenic angiosarcoma. Note the spindle-shaped tumor cells and the ill-formed, anastomosing vascular spaces

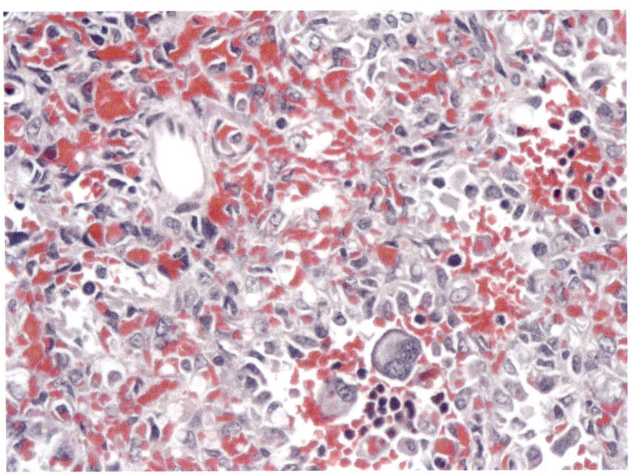

Fig. 5.93 Angiosarcoma. Another example of splenic angiosarcoma. As above, there are numerous bizarre vascular spaces. Large pleomorphic tumor cells are seen

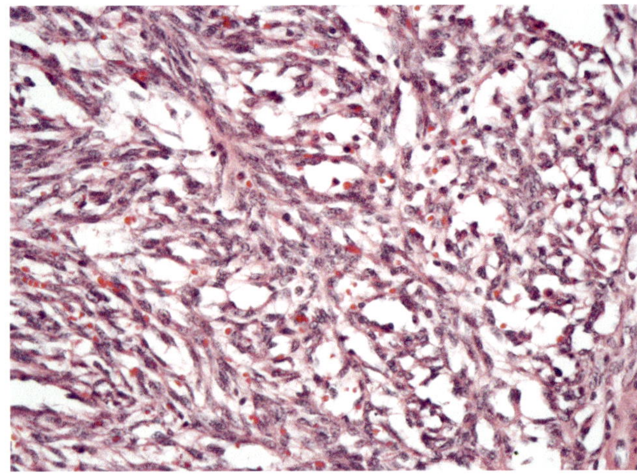

Fig. 5.94 Angiosarcoma. Another splenic angiosarcoma. In this case, the spindle cell morphology is quite pronounced. As in other cases, abnormal vascular spaces are present

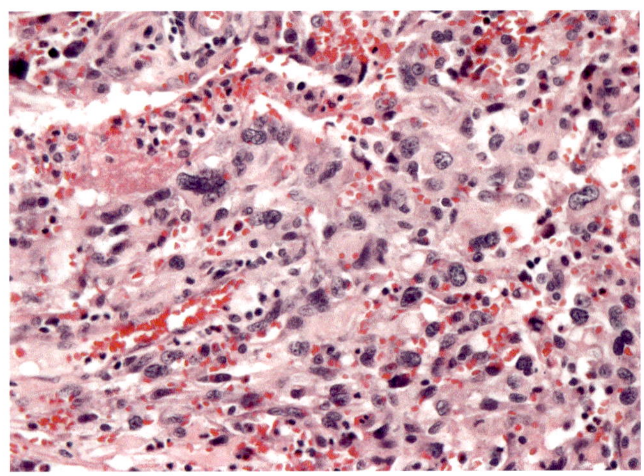

Fig. 5.95 Angiosarcoma. In this example of splenic angiosarcoma, there is significant cellular pleomorphism, but there is also clear evidence of vascular space formation. (*Image courtesy of* L. Weiss, Aliso Viejo, CA.)

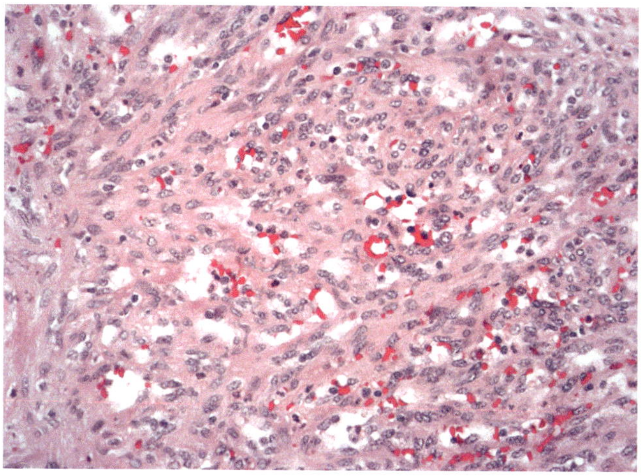

Fig. 5.96 Angiosarcoma. Histologic findings in a low-grade angiosarcoma of the spleen. There is clear evidence of vascular space formation but this case lacks the overt cytologic atypia of other cases. There is also minimal cellular pleomorphism. Distinguishing low-grade angiosarcoma from hemangioma or hemangioendothelioma can be difficult. (*Image courtesy of* L. Weiss, Aliso Viejo, CA.)

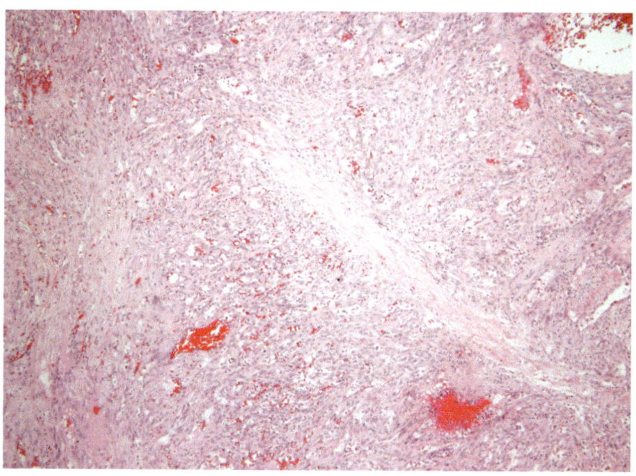

Fig. 5.97 Angiosarcoma. Another example of a low-grade angiosarcoma of spleen. In this case, the cellularity is considerably higher than expected in hemangioma. However, there is minimal cytologic atypia seen. Although the vascular spaces are clearly abnormal, they are relatively homogeneous in appearance. (*Image courtesy of* L. Weiss, Aliso Viejo, CA.)

5.11 Metastatic Tumors

Fig. 5.98 Metastatic carcinoma. Gross image of splenic metastases of carcinoma. In this image, multiple nodules of tan carcinoma are present, distributed mostly in perivascular locations. They have a solid tan appearance. In this case, this is a relatively small volume of metastatic disease, although much larger examples of metastases can be seen. (*Image courtesy of* D. Hawley, Indianapolis, IN.)

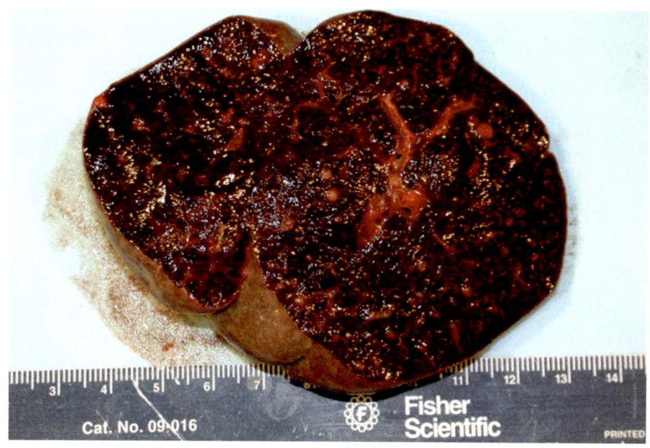

Fig. 5.99 Metastatic carcinoma. Gross image of spleen with metastatic carcinoma. The normal splenic architecture, including the small white nodules representing white pulp, are in contrast to the yellow-tan appearance of the metastatic carcinoma. (*Image courtesy of* D. Hawley, Indianapolis, IN.)

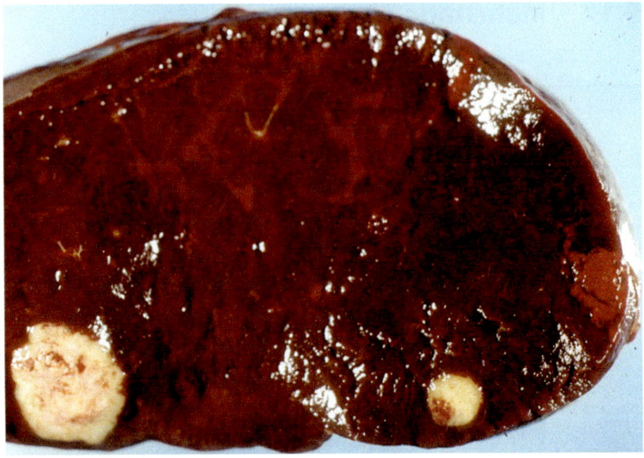

Fig. 5.100 Metastatic adenosquamous carcinoma, infarct. Gross image of spleen with two nodules of metastatic adenosquamous carcinoma (yellow-tan nodules). Contrast these to the area of infarction (*middle right*) with a pale orange-red appearance. The infarct is wedge-shaped

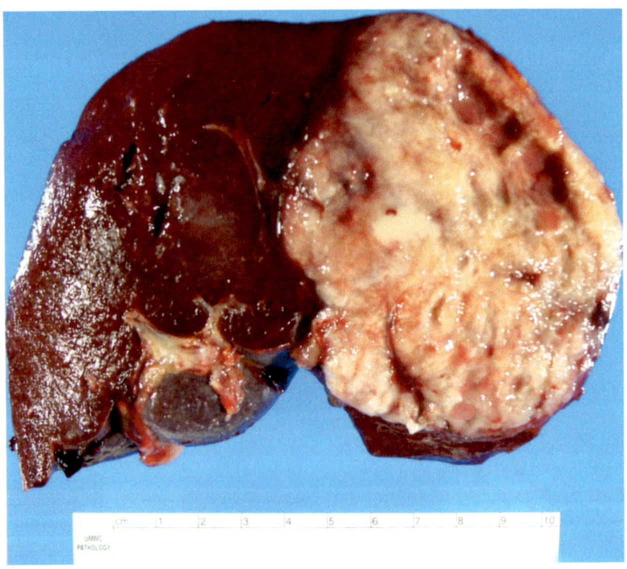

Fig. 5.101 Metastatic adenocarcinoma. Gross image of a metastasis of pulmonary adenocarcinoma to spleen. There is a single mass present with a soft, fleshy mass with a somewhat variegated appearance. There are likely some areas of necrosis and some focal hemorrhage present. (*Image courtesy of* M. Kamionek, Boston, MA.)

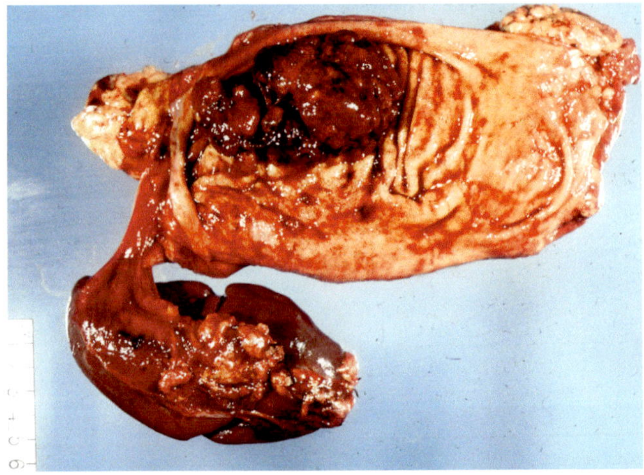

Fig. 5.102 Metastatic adenocarcinoma. Gross image show a composite resection of stomach and spleen. The opened stomach shows a large fungating adenocarcinoma. The adjacent spleen shows large aggregates of metastatic carcinoma. At the upper pole of the spleen (sutures) there are nodules of tumor. In this case, given the proximity of the two organs, it is likely that the tumor involved the spleen through direct extension rather than true metastatic disease. (*Image courtesy of* D. Farhi, Atlanta, GA.)

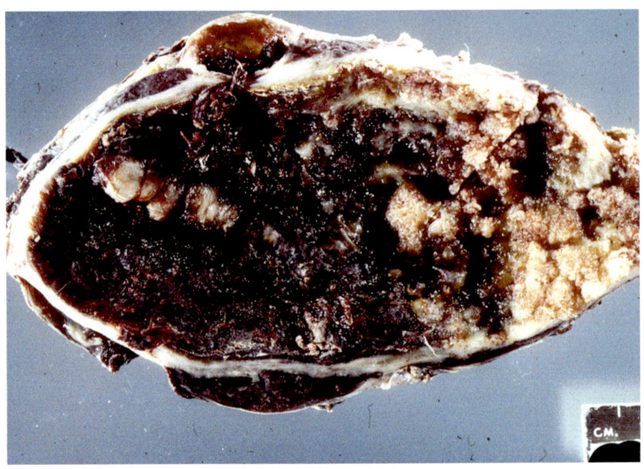

Fig. 5.103 Metastatic adenocarcinoma. This gross image shows a spleen with several nodules of metastatic tumor and diffuse involvement of one pole (*right*). There is also perisplenitis (thickening of the capsule)

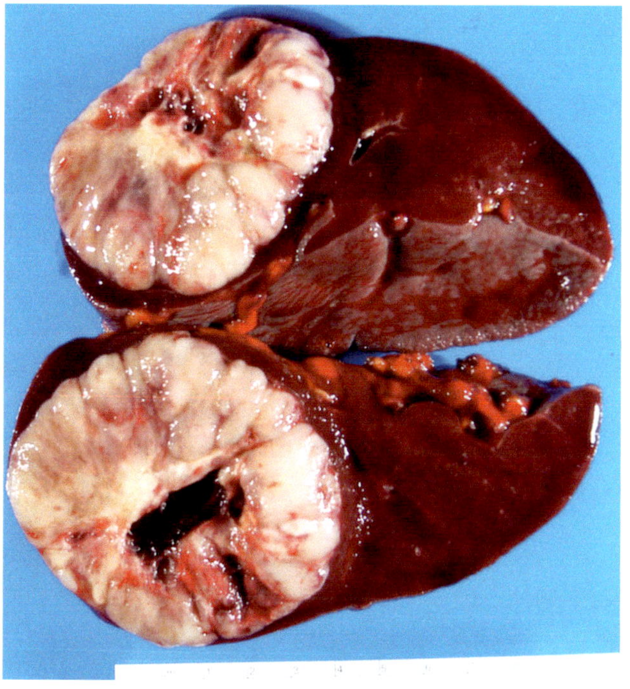

Fig. 5.104 Metastatic melanoma. Gross image of metastatic melanoma. A large fleshy mass with central necrosis. No obvious pigmentation is present to suggest that this is melanoma. (*Image courtesy of* M. Kamionek, Boston, MA.)

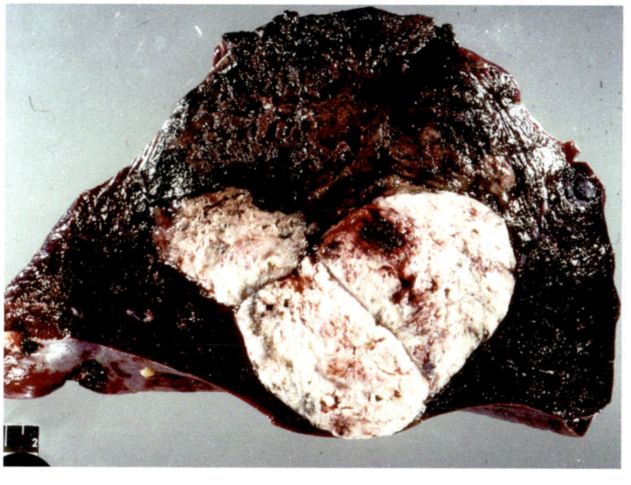

Fig. 5.105 Metastatic melanoma. Another gross image of metastatic melanoma. Three massive, pale, fleshy nodules displace normal splenic parenchyma. (*Image courtesy of* D. Farhi, Atlanta, GA.)

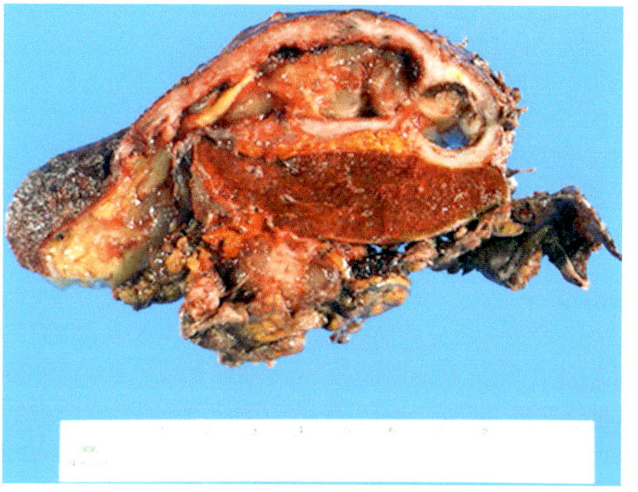

Fig. 5.106 Pseudomyxoma peritonei. An en bloc resection of abdominal contents with spleen at left. A disseminated mucin-producing adenocarcinoma of the abdominal cavity was present, entrapping many of the abdominal organs. (*Image courtesy of* M. Kamionek, Boston, MA.)

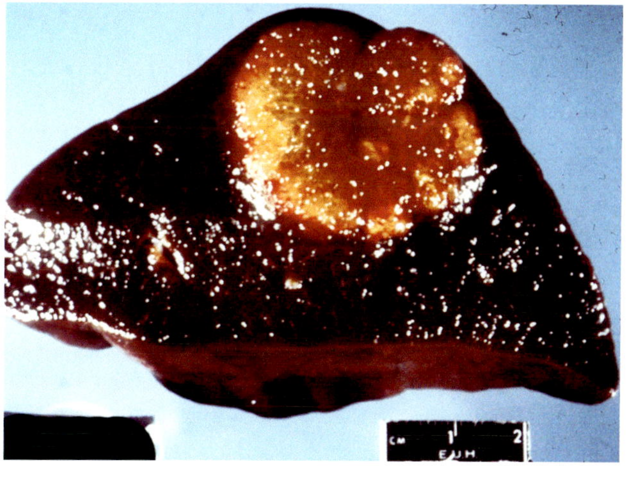

Fig. 5.107 Metastatic mucinous cystadenocarcinoma. This gross image (unfortunately poorly focused) shows a glistening, slimy-appearing nodule of tumor. This is a mucinous cystadenocarcinoma

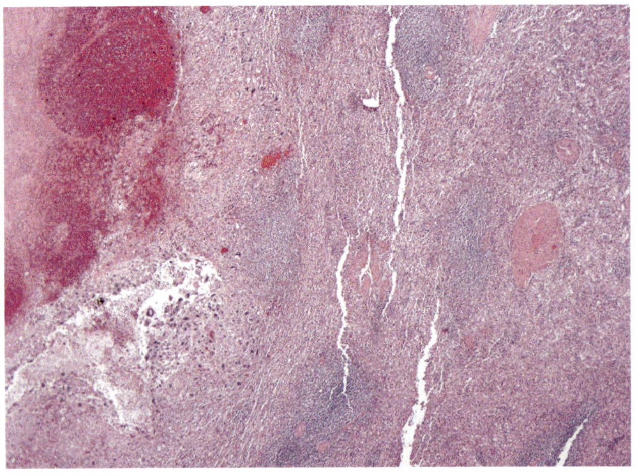

Fig. 5.108 Metastatic choriocarcinoma. Metastatic choriocarcinoma has a propensity for spleen metastases. This is likely due to the vascular nature of the spleen, which is amenable to growth and proliferation of the tumor cells. In this low-magnification image, normal spleen (*right*) is adjacent to an area of hemorrhage and necrosis (*left*). Some large tumor cells are notable (*lower middle*)

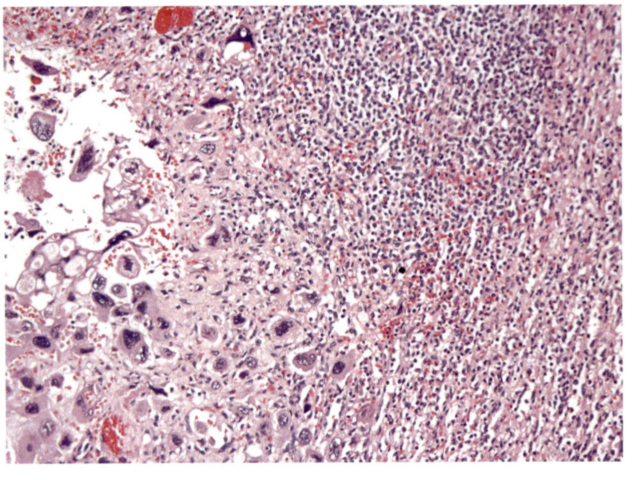

Fig. 5.109 Metastatic choriocarcinoma. The large and pleomorphic tumor cells of choriocarcinoma are obvious adjacent to the normal red pulp and lymphoid cells of spleen

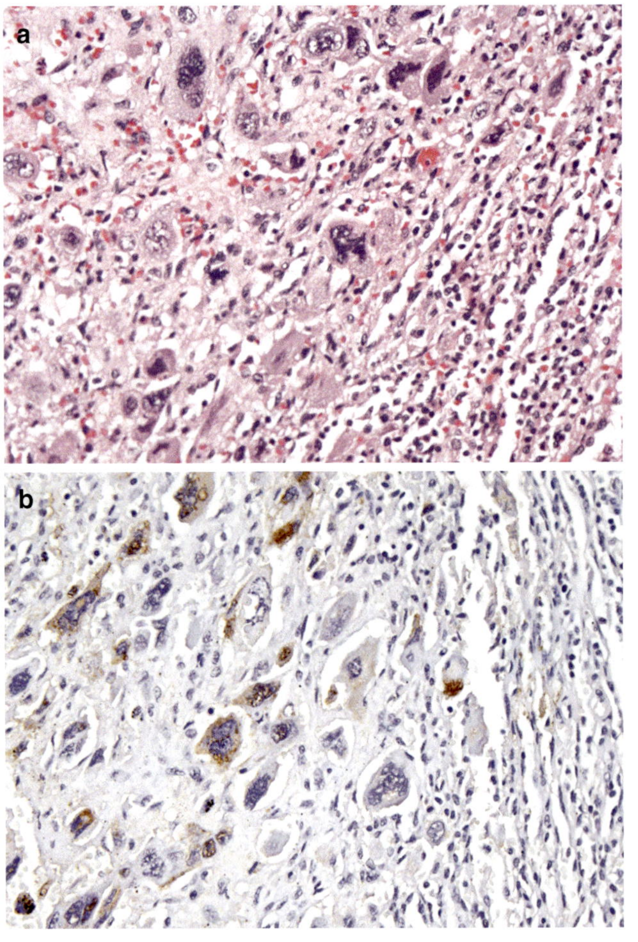

Fig. 5.110 Metastatic choriocarcinoma, H&E (**a**), beta human chorionic gonadotropin (HCG) stain. Bizarre, pleomorphic tumor cells of choriocarcinoma are seen (**a**) that stain positively for beta HCG (**b**)

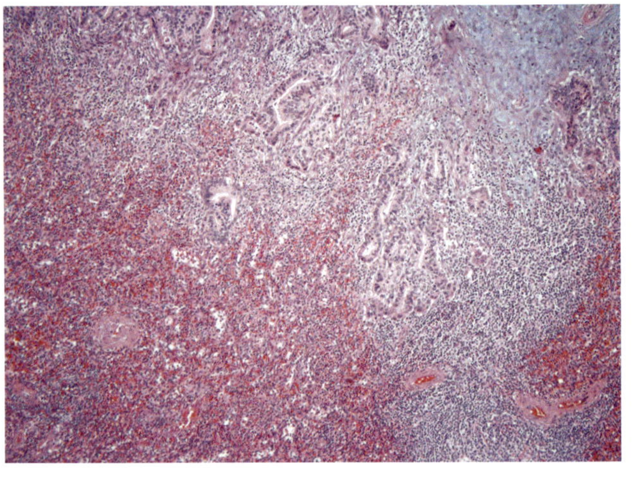

Fig. 5.111 Metastatic cholangiocarcinoma. An example of metastatic cholangiocarcinoma in the spleen. Normal parenchyma is present (*lower left*) with abnormal tumor glands present and some myxoid degeneration (*upper right*)

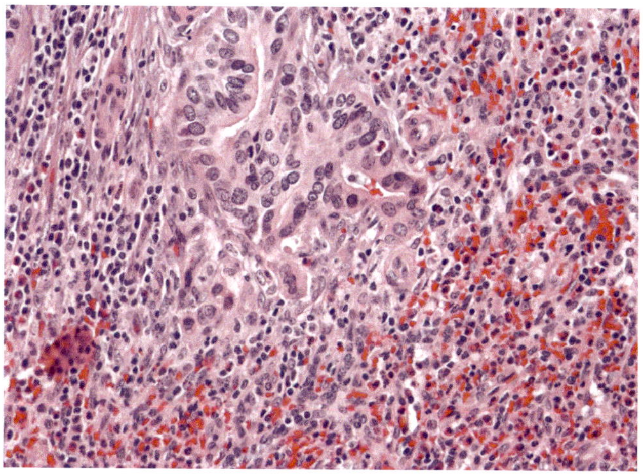

Fig. 5.112 Metastatic cholangiocarcinoma. The abnormal glands of cholangiocarcinoma in this case are present in red pulp

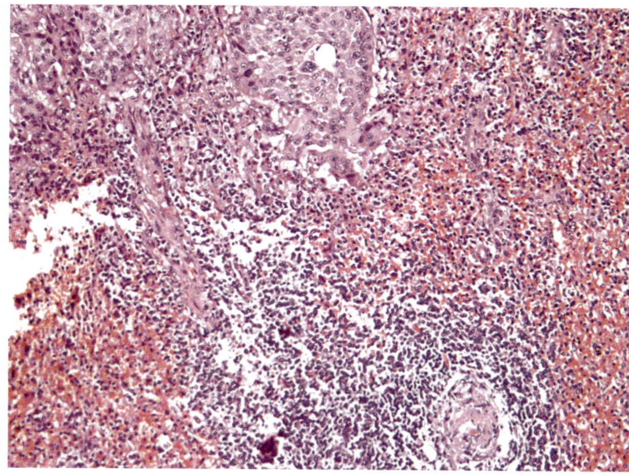

Fig. 5.114 Metastatic squamous cell carcinoma. An example of metastatic squamous cell carcinoma. In this example, the nests of tumor cells are adjacent to white pulp structures

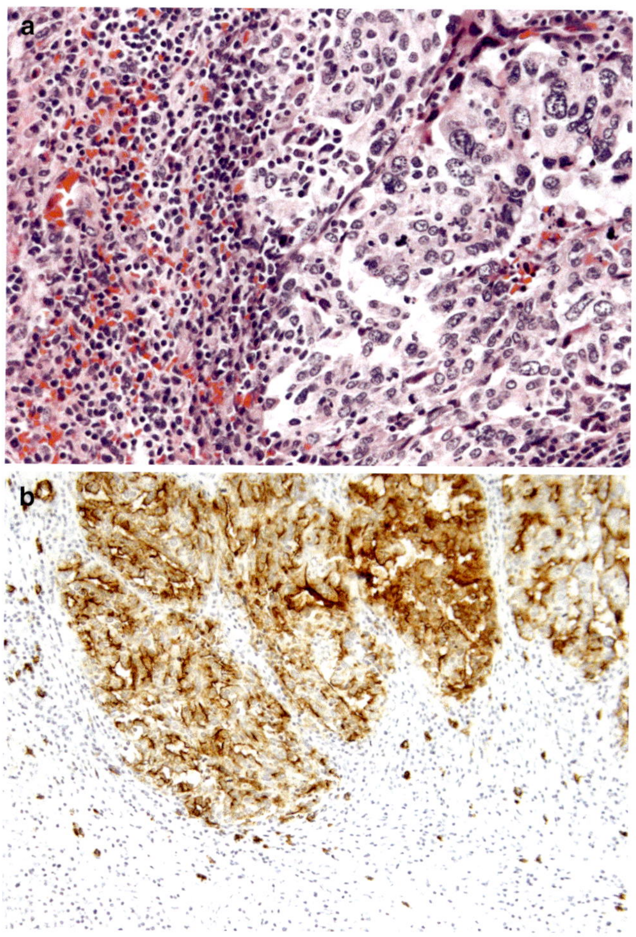

Fig. 5.113 Metastatic papillary serous carcinoma, H&E (**a**), CA-125 stain. The highly abnormal cells of the carcinoma (*right*) form vague papillary structures (**a**). CA-125 is strongly expressed in this tumor (**b**)

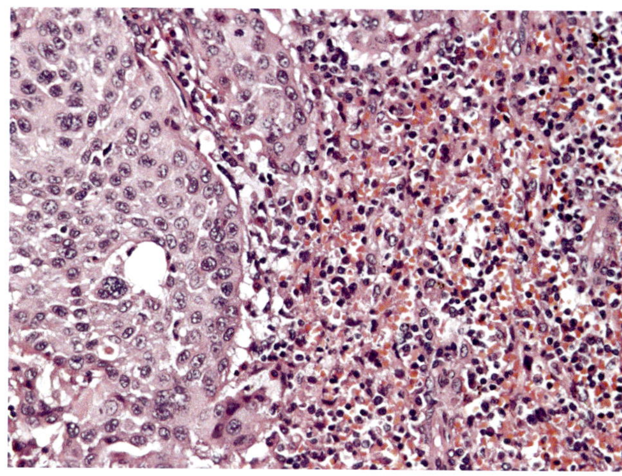

Fig. 5.115 Metastatic squamous cell carcinoma. High-magnification image of squamous cell carcinoma in spleen. Note the pleomorphism and large amounts of pink cytoplasm in the tumor cells. Metastases to the spleen are relatively rare, despite the many examples illustrated here

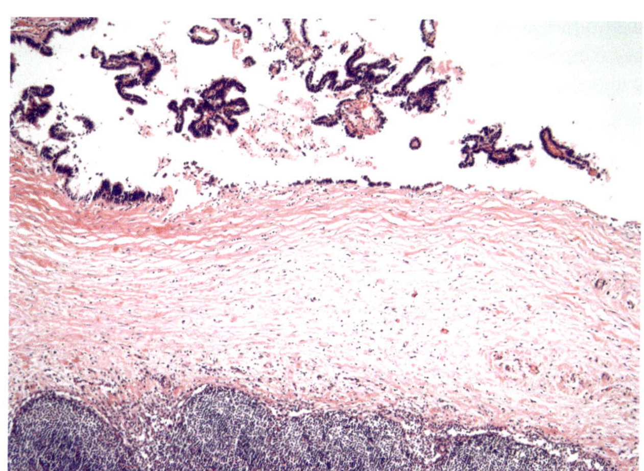

Fig. 5.116 Metastatic cystadenocarcinoma. In the upper portion, there is a cystic space with abnormal papillary fronds of tumor. This is surrounded by a thick fibrous capsule, with normal splenic parenchyma in the lower portion of the image

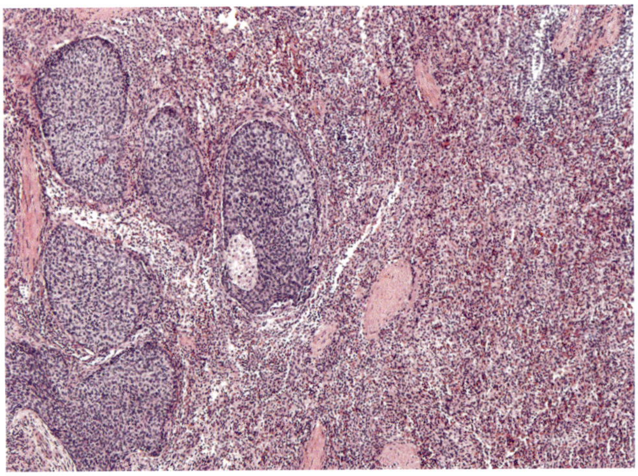

Fig. 5.117 Metastatic transitional cell carcinoma (TCC). An example of TCC metastatic to spleen. Spleen metastases in most tumors are a late-stage event. This TCC shows nests of tumor cells with an appearance of peripheral palisading, which is seen in relatively well-differentiated tumors

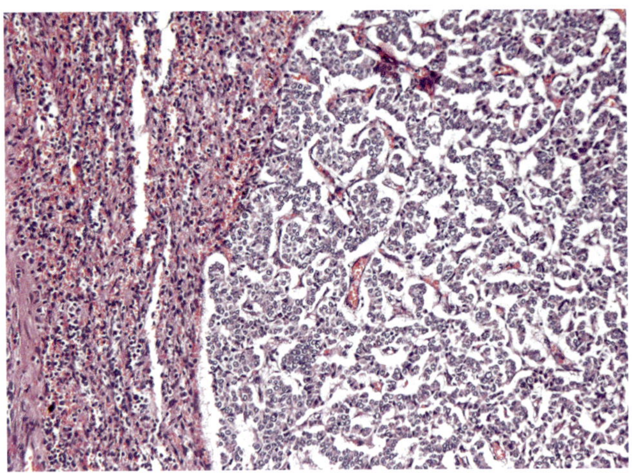

Fig. 5.118 Metastatic neuroendocrine carcinoma. An example of a well-differentiated neuroendocrine tumor metastatic to spleen. The trabecular pattern of the tumor cells is suggestive of this diagnosis but not pathognomonic

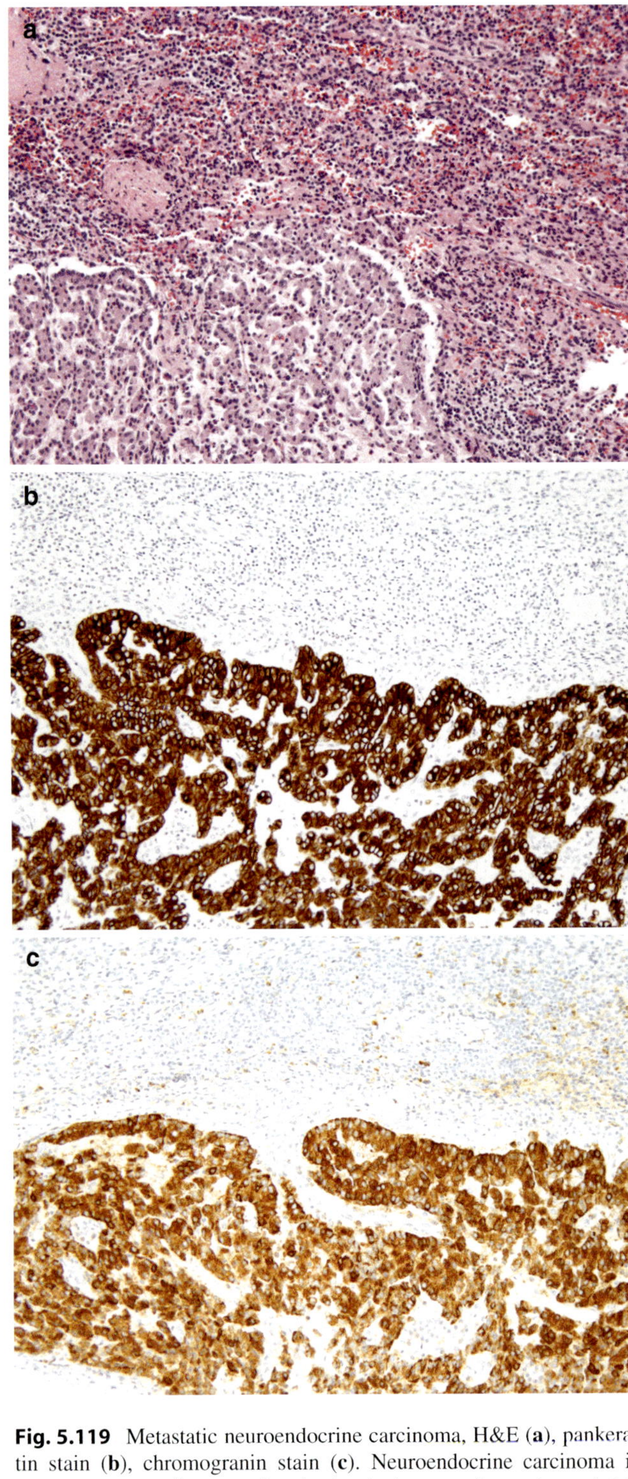

Fig. 5.119 Metastatic neuroendocrine carcinoma, H&E (**a**), pankeratin stain (**b**), chromogranin stain (**c**). Neuroendocrine carcinoma is present next to adjacent uninvolved splenic parenchyma (*upper*) (**a**), which is positive for pankeratin (**b**) and chromogranin (**c**)

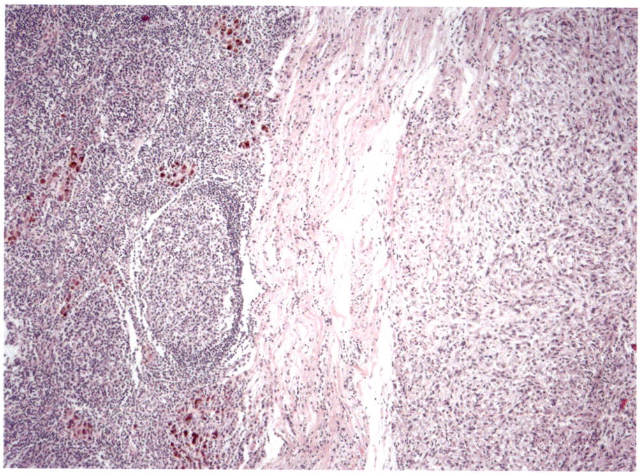

Fig. 5.120 Metastatic sarcomatoid carcinoma. In this example, the sarcomatoid carcinoma (*right*) has a spindled appearance, mimicking a sarcoma. There is some hemosiderin deposition (brown pigment in macrophages) in the normal splenic parenchyma. This may be due to previous hemorrhages in these areas

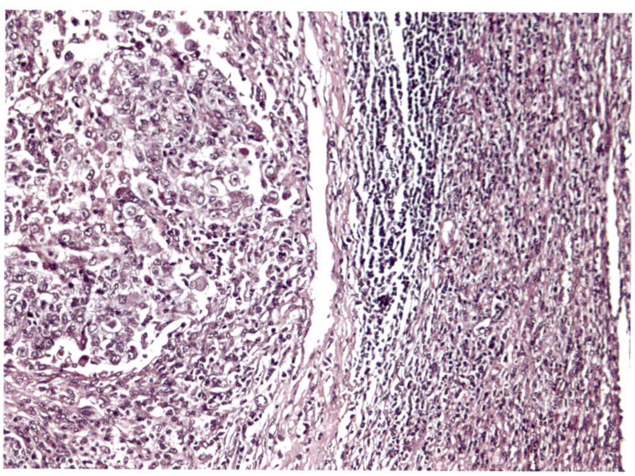

Fig. 5.121 Metastatic melanoma. Melanoma (*left*) can be seen in the spleen but is a rare, late-stage event. The tumor cells are highly pleomorphic. Immunohistochemical confirmation with stains such as S-100, HMB-45, and Melan A should be performed if there is not a clear history of widespread metastatic disease

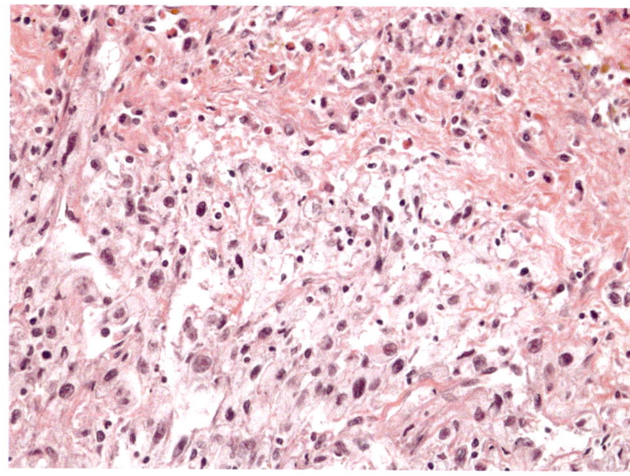

Fig. 5.122 Metastatic malignant fibrous histiocytoma (MFH). Sarcomas are only rarely metastatic and even more rarely involve the spleen. This example represents MFH (or high-grade sarcoma) and has marked cellular pleomorphism

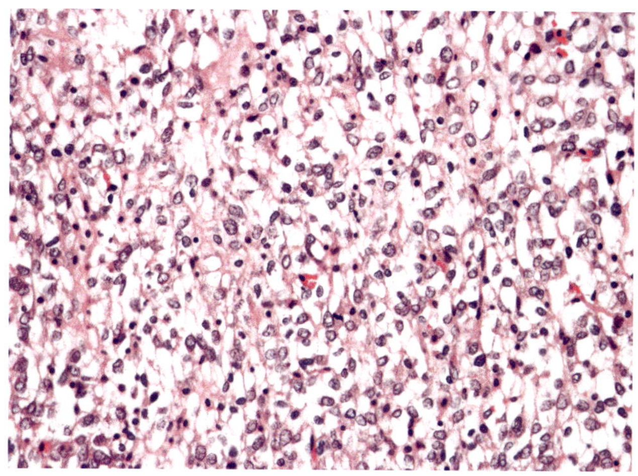

Fig. 5.123 Metastatic myxoid liposarcoma. An example of myxoid liposarcoma metastatic to the spleen. The tumor cells are highly atypical and easily distinguishable from normal splenic parenchyma. (*Image courtesy of* L. Weiss, Aliso Viejo, CA.)

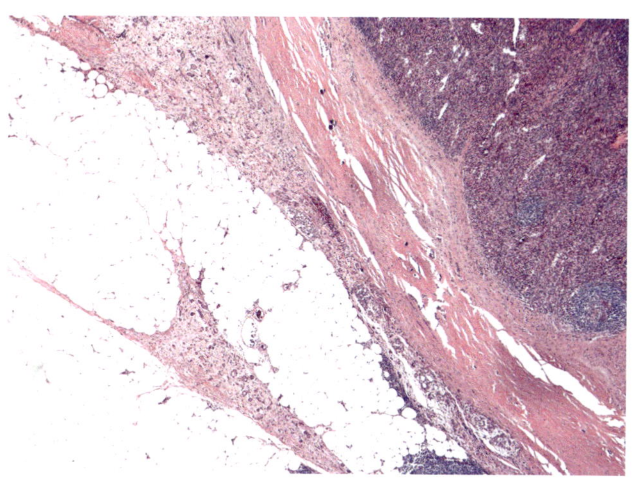

Fig. 5.124 Mesothelioma. Although extremely rare, mesothelioma can encase the spleen when there is abdominal involvement, as in this example. Mesothelioma is present (*left*) attached to the splenic capsule, which is thickened. Normal splenic parenchyma is present at right. (*Image courtesy of* D. Hudnall, New Haven, CT)

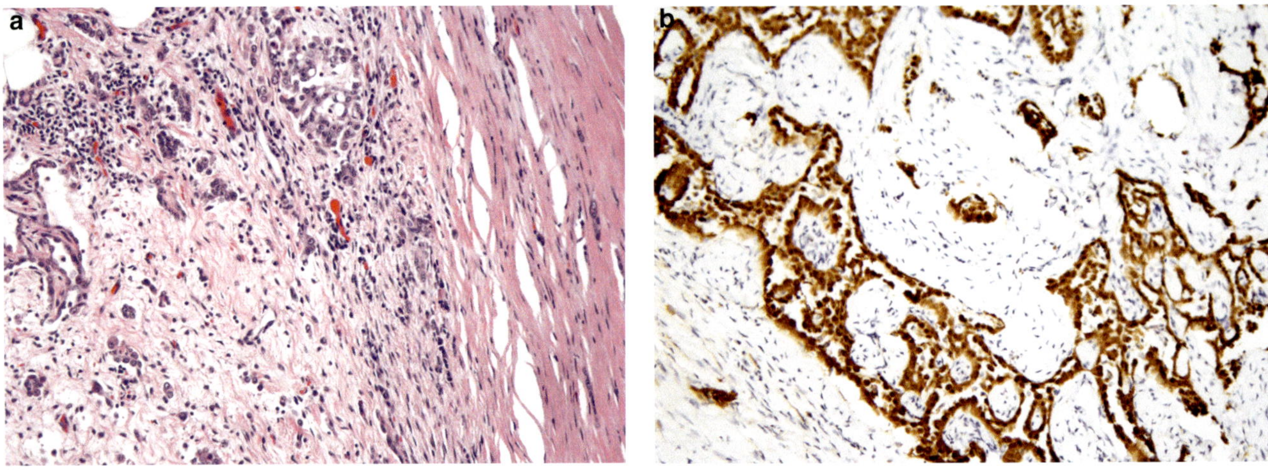

Fig. 5.125 Mesothelioma, H&E (**a**), calretinin stain (**b**). Higher magnification of mesothelioma from above (**a**), which is strongly positive for calretinin (**b**)

Reference

1. O'Malley DP, George TI, Orazi A, Abbondanzo S. Armed Forces Institute of Pathology: Benign and reactive conditions of lymph node and spleen [book]. Silver Spring, MD: American Registry of Pathology Press; 2009.

6 Reactive and Systemic Conditions

A broad range of reactive and systemic disorders may be manifest in the spleen. Trauma and physical damage to the spleen are discussed, as well as their histologic appearances. Perisplenitis, or sugar-coated spleen, is illustrated. Fibrocongestive splenomegaly, a final common pathway of chronic vascular compromise, is shown. The splenic effects of red blood cell disorders, including thalassemia, sickle cell disease, and hereditary spherocytosis, are illustrated. Granulomas of a variety of causes are shown. The splenic manifestations of the rare histiocyte-derived disorder, Langerhans cell histiocytosis, is shown. A variety of unusual disorders, including Castleman disease, and effects of various therapies are illustrated. A broad range of autoimmune disorders will have significant findings in the spleen and these are discussed. Finally, this chapter shows a range of congenital immunodeficiency disorders that involve the spleen.

D.P. O'Malley, *Atlas of Spleen Pathology*, Atlas of Anatomic Pathology,
DOI 10.1007/978-1-4614-4672-9_6, © Springer Science+Business Media New York 2013

6.1 Rupture and Trauma

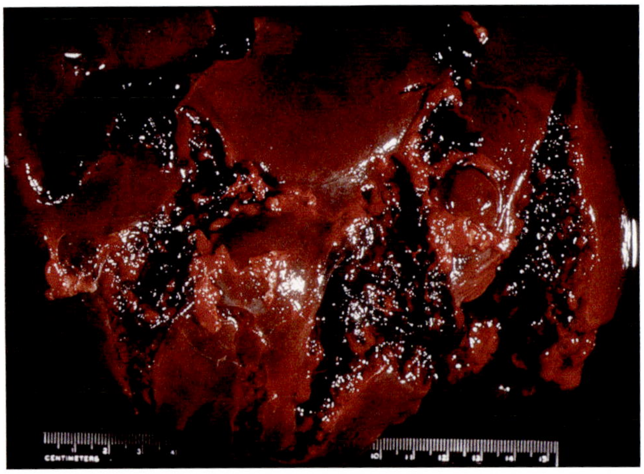

Fig. 6.1 Splenic rupture. Gross image of splenic rupture. Although the etiology is not apparent, the presence of numerous jagged, connected ruptures suggests the possibility of significant abdominal trauma. In this case, there is adherent clotted blood within the jagged edges of ruptured capsular material. (*Image courtesy of* D. Farhi, Atlanta, GA.)

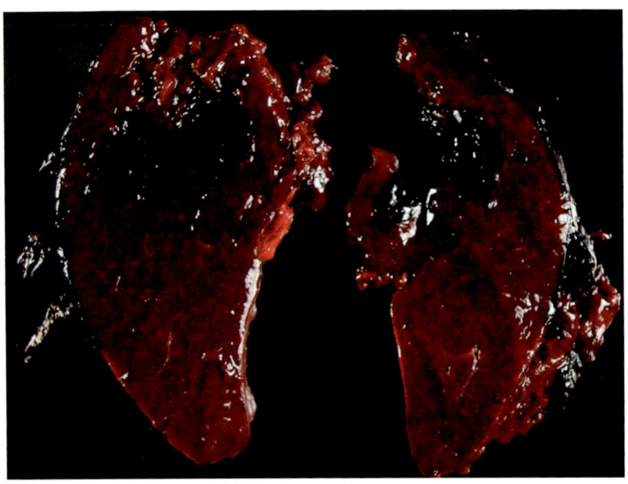

Fig. 6.2 Splenic rupture. Cut section of a spleen with "spontaneous" rupture. Large amounts of freshly clotted blood (dark red-black) are within the parenchyma of the spleen. Most cases of "spontaneous" splenic rupture are better referred to as *pathologic rupture*; they have an identifiable etiology and are not truly spontaneous. (*Image courtesy of* A. Sohani, Boston, MA.)

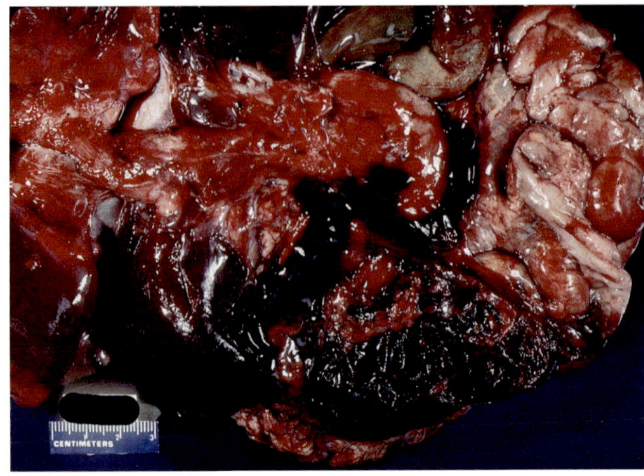

Fig 6.3 Splenic rupture. En bloc resection of spleen (*center*) intestine (*upper right*) and pancreas (*center and extending to left*). There is clotted blood adherent to the splenic hilum and the splenic capsule appears torn (*lower*)

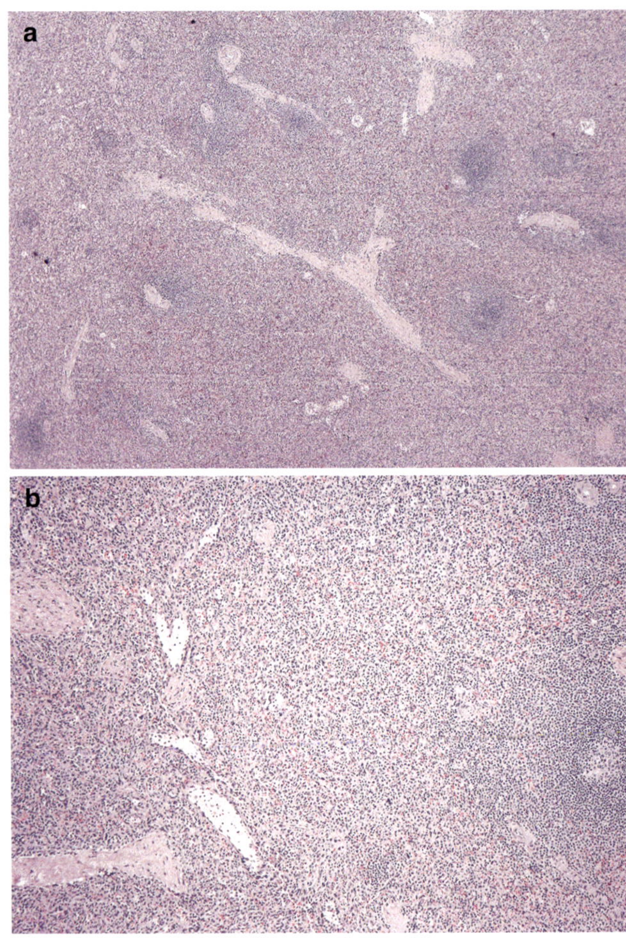

Fig. 6.4 Splenic rupture, incidental trauma. This patient experienced pathologic rupture of the spleen due to coughing. The only morphologic findings are an increase in red pulp and some fibrocongestive splenomegaly. Coughing causes an increase in intraperitoneal pressure and under the appropriate circumstances, a weakened splenic capsule could rupture due to coughing. (**a**, low magnification; **b**, intermediate magnification)

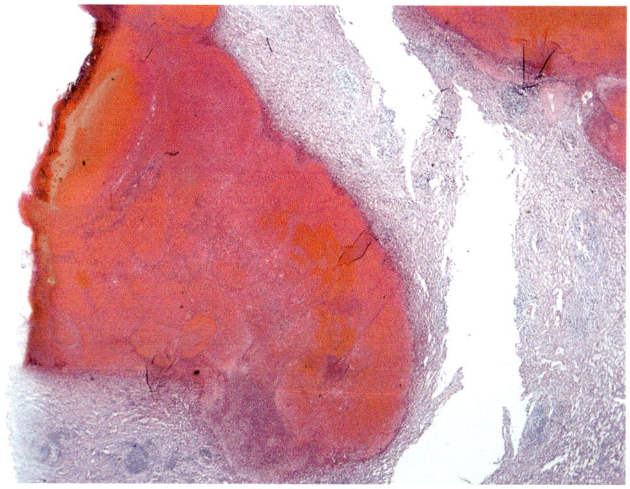

Fig. 6.5 "Spontaneous" splenic rupture. Low-magnification image of a spleen after spontaneous rupture. There is recent hemorrhage and no etiology is apparent. However, true spontaneous rupture is extraordinarily rare. In most cases, some cause—whether traumatic, systemic, or otherwise—can be discovered and indicates "pathologic" rupture. Spontaneous rupture should be reserved for cases where other causes are excluded both by history and histologic examination

6.2 Infarction and Embolization

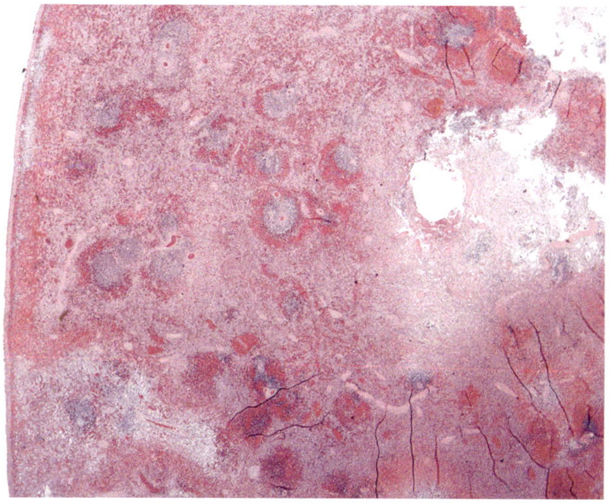

Fig. 6.7 Recent infarction. Low-magnification image of recent infarction of the spleen. Note the overall preservation of the splenic architecture. However, there is extensive hemorrhage in the interstitial spaces, both in red pulp and surrounding white pulp nodules

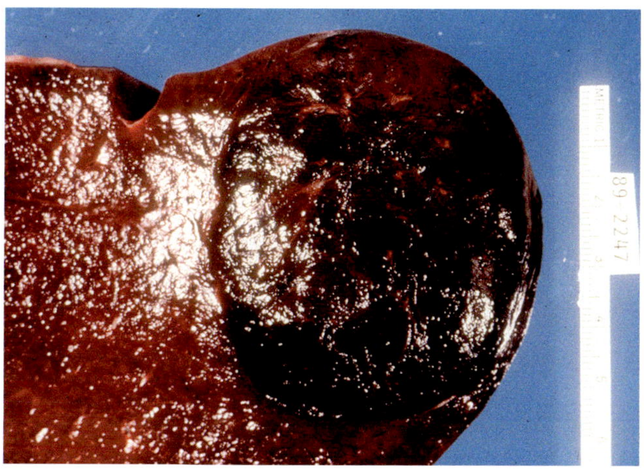

Fig. 6.6 Recent injury/hematoma. Gross image of recent injury to spleen with hematoma formation. This cannot be reliably distinguished from early infarction. As with most traumatic injuries, they occur at the convex surface, which is adjacent to the ribs and the outside environment. (*Image courtesy of* D. Farhi, Atlanta, GA.)

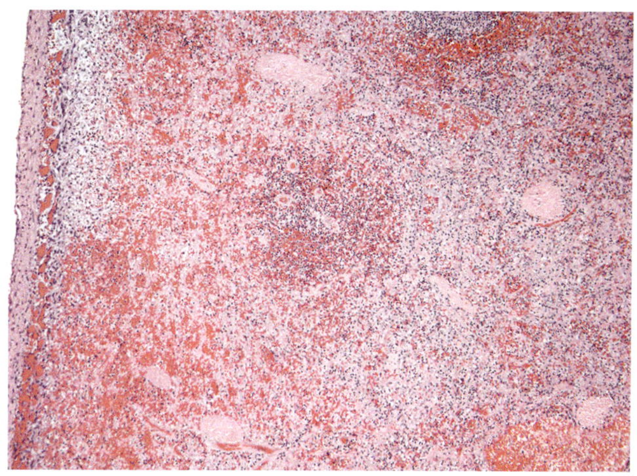

Fig. 6.8 Recent infarction. Higher-magnification image of recent infarction of the spleen. Besides extensive hemorrhage, there is a loss of the definition of individual cell features, suggesting early necrosis

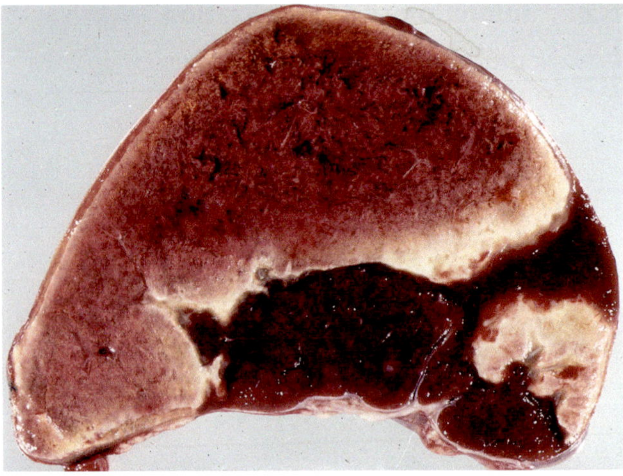

Fig. 6.9 Remote infarction. Gross image of remote infarct of spleen. In the lower portions there is relatively preserved splenic parenchyma with its typical dark-red appearance. The upper portion shows a rim of yellow-tan fibrous tissue, representing organizing infarcted material. The central portion appears less cohesive and would eventually resolve by liquefaction or fibrosis. In this case, the patient had chronic myelogenous leukemia. (*Image courtesy of* D. Farhi, Atlanta, GA.)

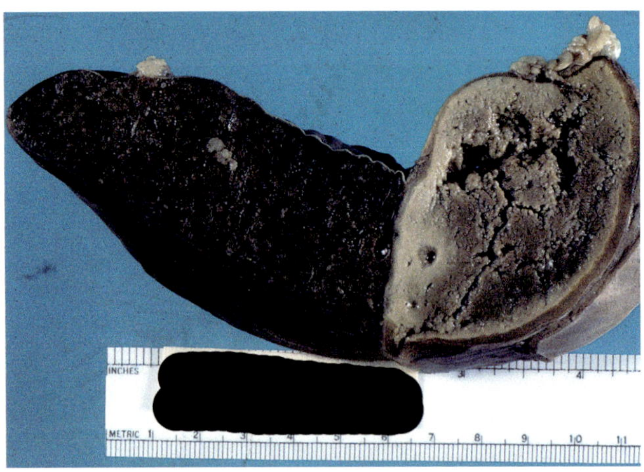

Fig. 6.10 Remote infarction. Gross image of sectioned, fixed spleen. The left portion is uninvolved whereas the right portion has clear evidence of infarction. The infarct shows central collapse. In addition, there is prominent thickening of the capsule overlying the area of infarct, compatible with perisplenitis. (*Image courtesy of* D. Farhi, Atlanta, GA.)

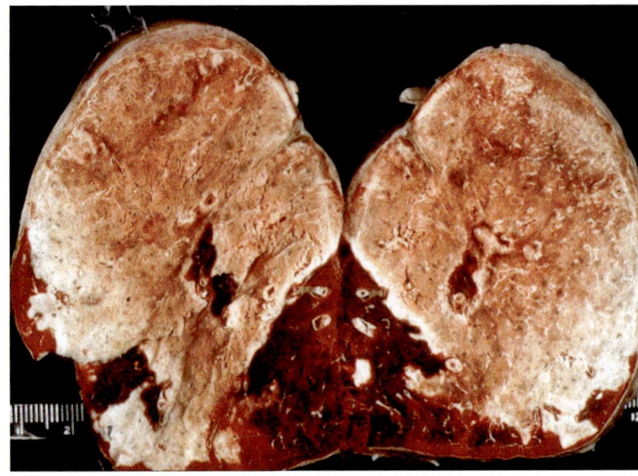

Fig. 6.11 Remote infarction. Massive infarction in this gross example of a bisected spleen. The pale, yellow-tan areas represent infarcted tissue. No specific etiology is apparent from gross examination. (*Image courtesy of* D. Farhi, Atlanta, GA.)

Fig. 6.12 Remote infarction. A large nodular infarcted area of spleen. In this case, the remote nature of the infarct is apparent from the dense fibrotic gross appearance of the affected area. (*Image courtesy of* D. Farhi, Atlanta, GA.)

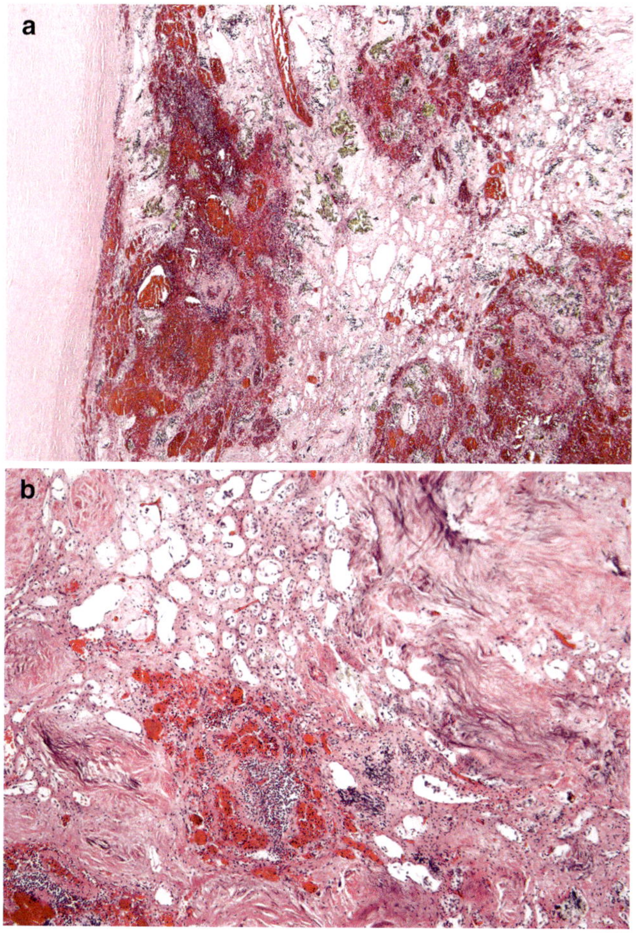

Fig. 6.13 Remote infarction. In this example, a remote infarction is present. There is considerable hemorrhage, which may be due to localized vascular damage, or possibly an infarction of a vascular lesion such as a hemangioma (**a**). Note the deposition of large amounts of hemosiderin, due to remote hemorrhage (**b**)

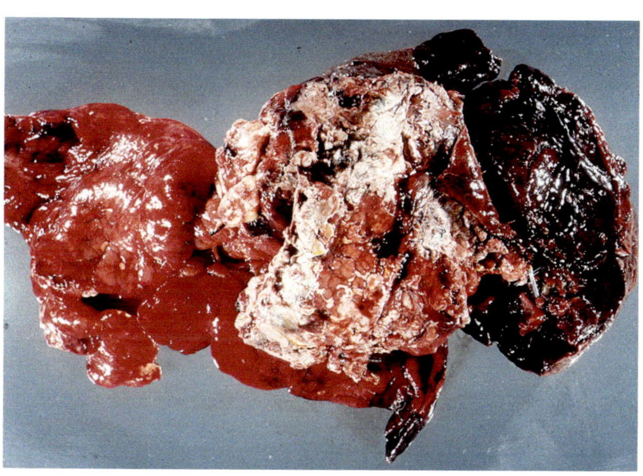

Fig. 6.14 Pancreatitis and necrosis. Composite resection of the spleen and pancreas. Extensive pancreatic necrosis and infarction has occurred adjacent to, and partly involving, the spleen

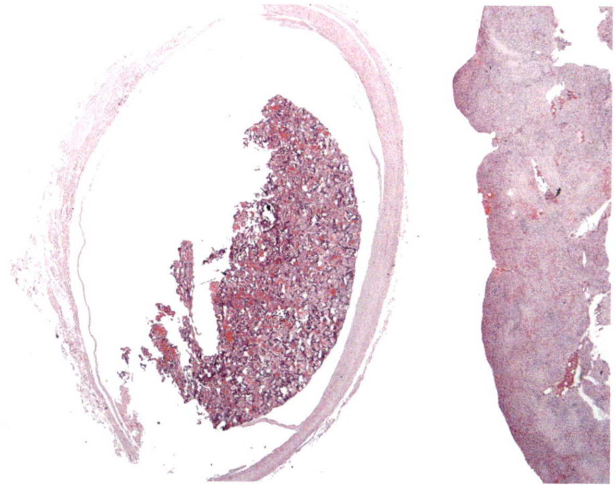

Fig. 6.15 6.c-1 Embolization. A splenic artery showing some fragments of gelatin foam material used to embolize the spleen. The material occludes the vessel, with aggregation and clotting of blood, depriving the spleen of oxygenated blood

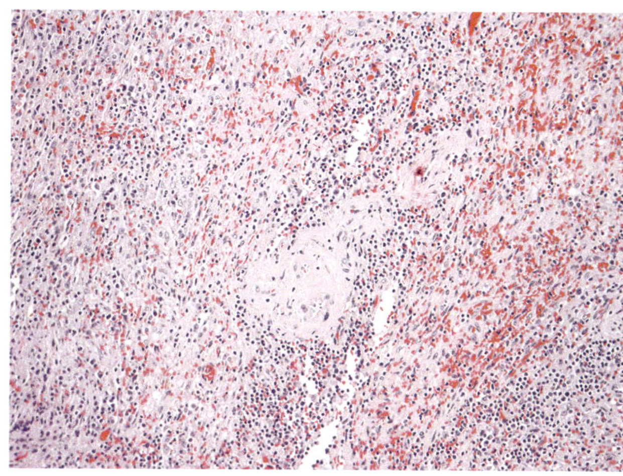

Fig. 6.16 Embolization. A high-magnification image of the spleen after recent embolization (approximately 2 h before splenectomy). In this case, there is prominent vascular congestion, with areas of early necrosis spreading radially from arterioles. The progression from ischemia to infarction to necrosis is inevitable, unless there is reoxygenation. The morphologic signs of early ischemia may be subtle

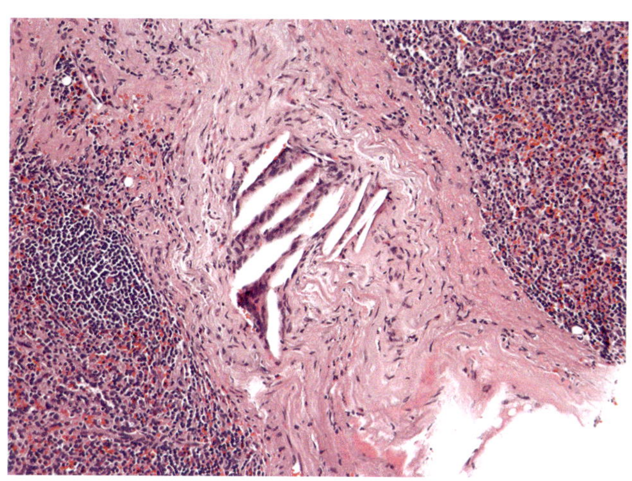

Fig. 6.17 Thrombus. Splenic artery showing evidence of remote thrombus formation, with cholesterol clefts within the thrombus

6.3 Perisplenitis (Sugar-Coated Spleen)

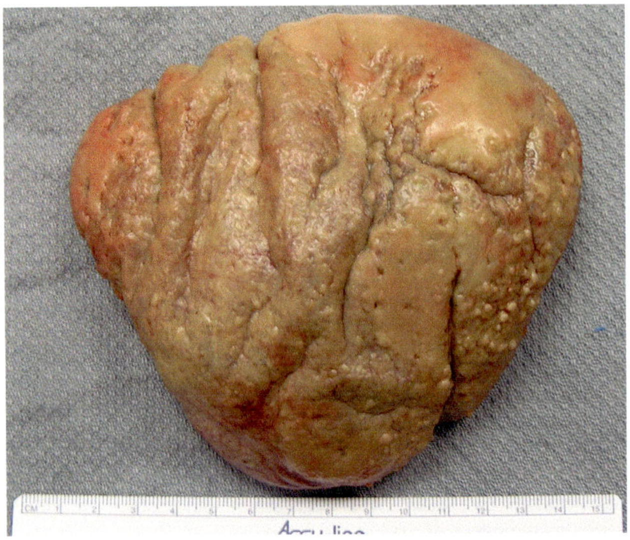

Fig. 6.18 Perisplenitis. An example of dense, severe perisplenitis. The yellow fibrous coating on this spleen completely hides the normal appearance and color of the organ. (*Image courtesy of* D. Coon, Las Vegas, NV.)

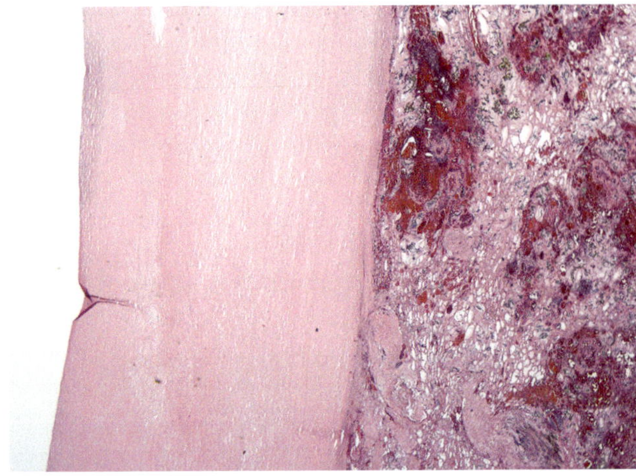

Fig. 6.20 Perisplenitis. Histologic image of perisplenitis. Inflammation of the surface of the spleen (*eg*, the peritoneal surface) leads to eventual fibrous thickening of the splenic capsule. This capsule is 5–10× thicker than is expected

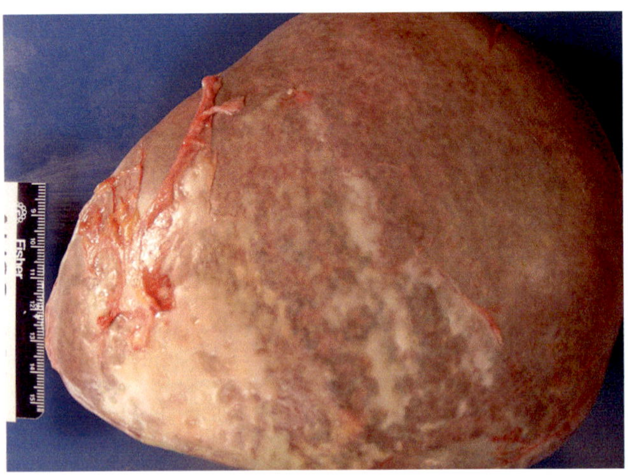

Fig. 6.19 Perisplenitis. Another example of perisplenitis, also called "sugar-coated" spleen because of the dense fibrous capsule. (*Image courtesy of* R. Cooke, Brisbane, Australia.)

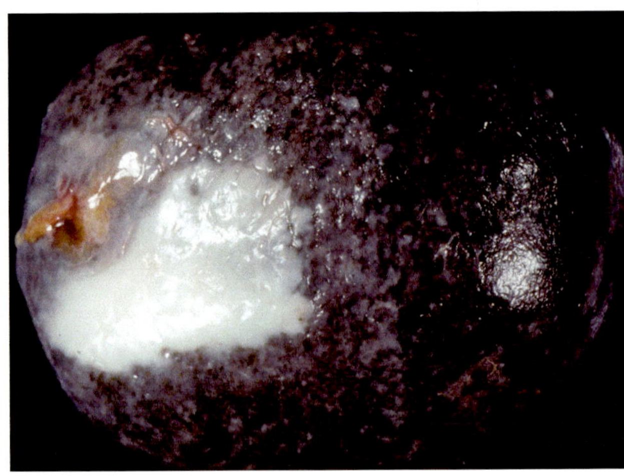

Fig. 6.21 Focal perisplenitis. Focal capsular thickening and fibrosis, possibly due to localized trauma. This area appears shiny and white and is on the convex surface of the spleen, as expected

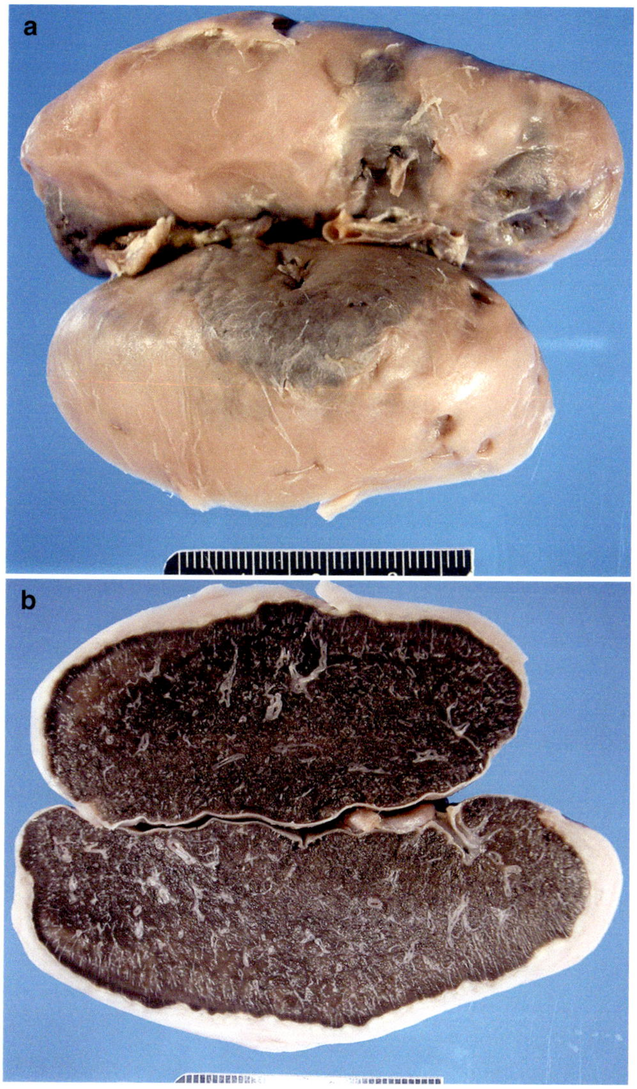

Fig. 6.22 Perisplenitis. The surface of the spleen is coated with a thick hyaline material (**a**), which on cut surface (**b**) is thick, pale fibrotic material. (*From* O'Malley et al. [1]; with permission.)

6.4 Congestive Splenomegaly

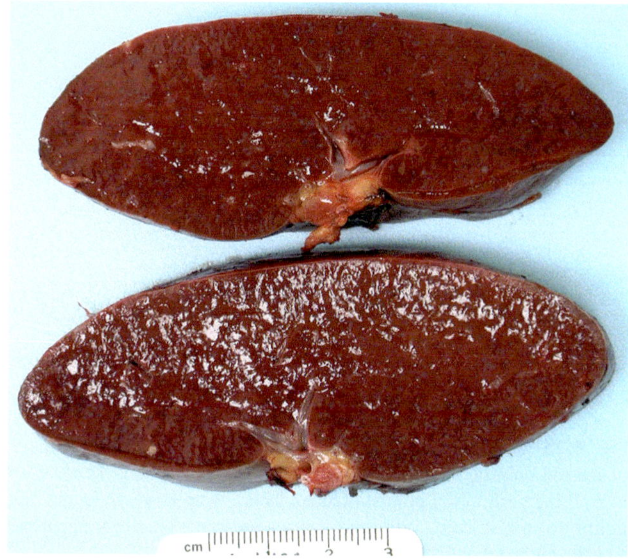

Fig. 6.23 Fibrocongestive splenomegaly. No focal abnormalities are seen. There is more prominent red pulp, with less apparent white pulp, but no specific abnormalities are apparent on gross examination. (*From* O'Malley et al. [1]; with permission.)

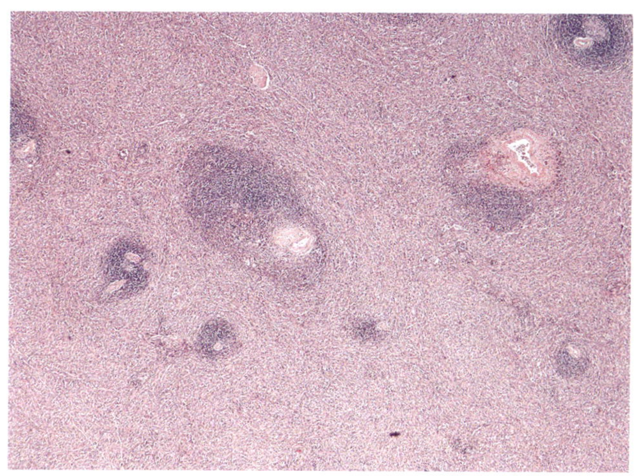

Fig. 6.24 Fibrocongestive splenomegaly. Low-magnification image of fibrocongestive splenomegaly. In most cases, there is a marked expansion of the red pulp compartment, with increased stromal elements and a generalized increase in entrapped red blood cells

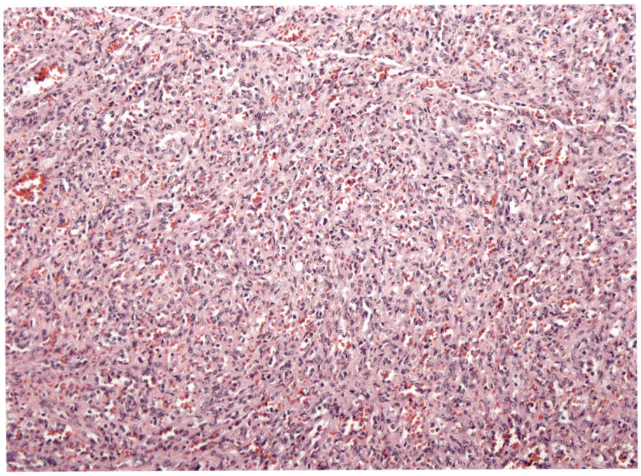

Fig. 6.25 Fibrocongestive splenomegaly. High-magnification image of the red pulp in a case of fibrocongestive splenomegaly. There is an overall increase in red pulp structures. There is an increase in the cellularity of the splenic cords, and sinuses are somewhat less apparent. This is due to an overall increase in stromal elements

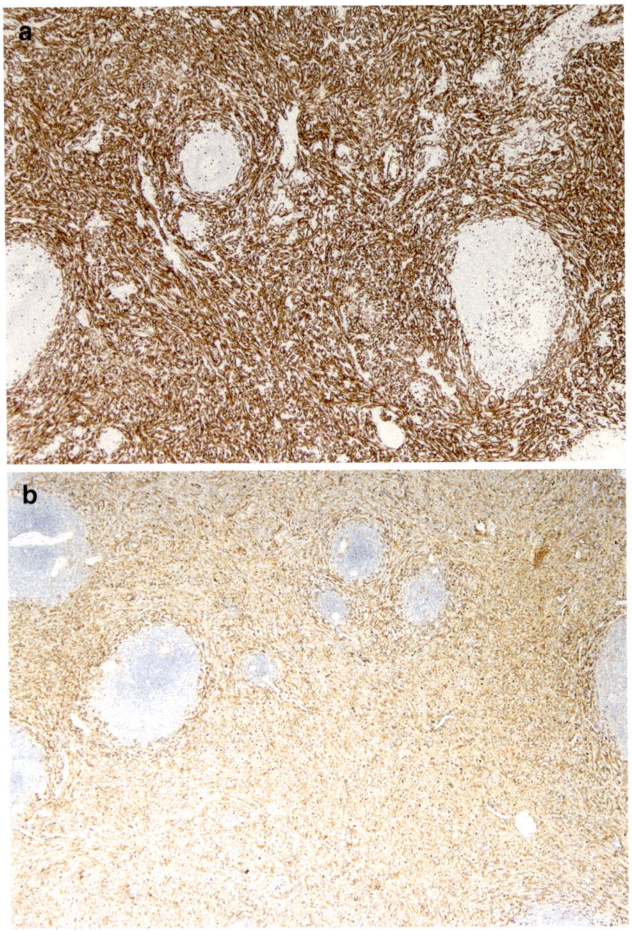

Fig. 6.26 Fibrocongestive splenomegaly, CD8 stain, CD163 stain. A CD8 stain of fibrocongestive splenomegaly shows a marked increase in the number of splenic sinuses in the red pulp (**a**), and CD163 shows a concurrent increase in macrophages in the splenic cords (**b**)

6.5 Red Blood Cell Disorders

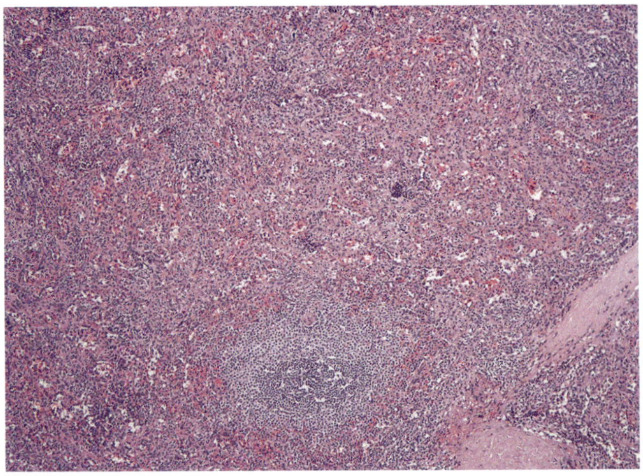

Fig. 6.27 Thalassemia. In this case of severe thalassemia, there is expansion of red pulp with significant extramedullary hematopoiesis in the red pulp. The hematopoietic components are predominantly erythroid elements

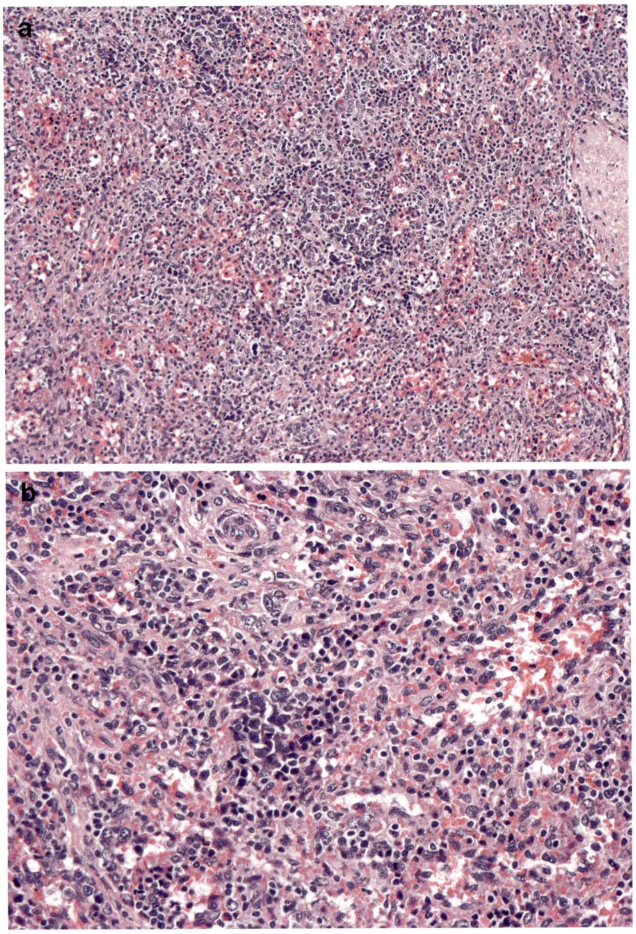

Fig. 6.28 Thalassemia. Intermediate- (**a**) and high-magnification (**b**) images of red pulp in severe thalassemia. Both show a combination of immature and mature erythroid precursors within the red pulp

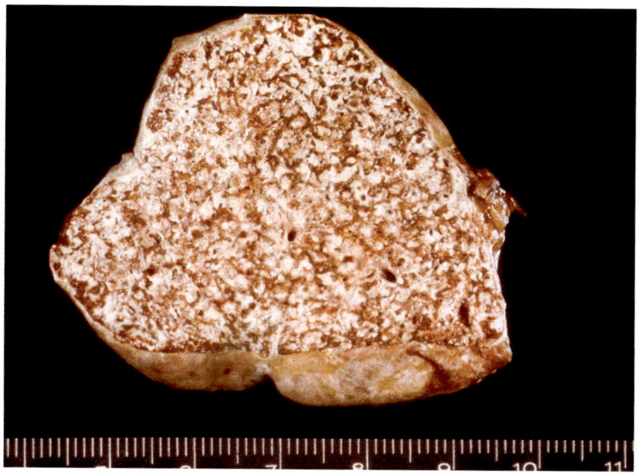

Fig. 6.29 Sickle cell disease, infarcted spleen. Gross image of a "late" effect of sickle cell disease. After many infarcts, the spleen is ultimately left as a small, scarred, fibrotic tissue fragment. No viable normal splenic parenchyma is present. (*Image courtesy of* D. Farhi, Atlanta, GA.)

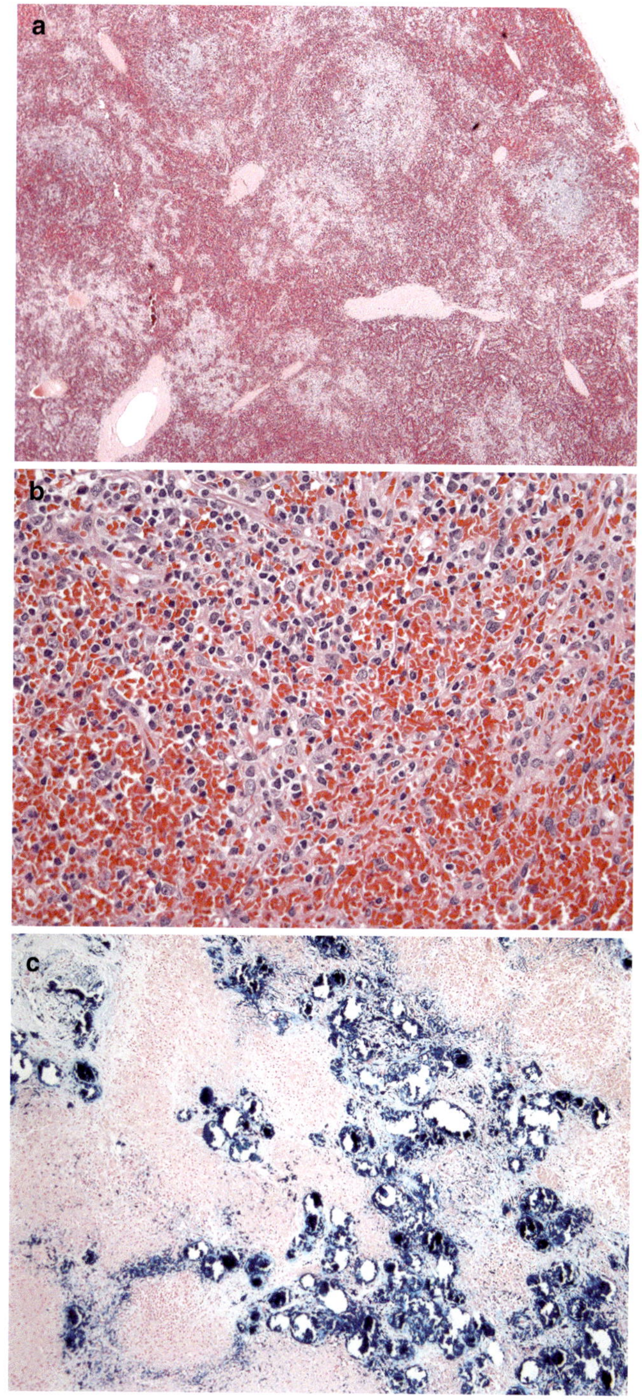

Fig. 6.30 Sickle cell disease, splenic sequestration crisis. In sickle cell disease (hemoglobin SS), patients occasionally have severe abdominal pain associated with acute infarction of the spleen. Termed "splenic sequestration crisis," proper management often leads to splenectomy. (**a**) Massive red pulp congestion in an architecturally distorted spleen. (**b**) High magnification shows red blood cell congestion of white pulp with some associated fibrosis in white pulp regions. Careful examination reveals the abnormal pointed shape of many of the red blood cells, characteristic of sickle cell anemia. **c**, An iron stain in a case of splenic sequestration, indicating long-term deposition of iron in the spleen due to repeated infarction and hemolysis of abnormal red blood cells

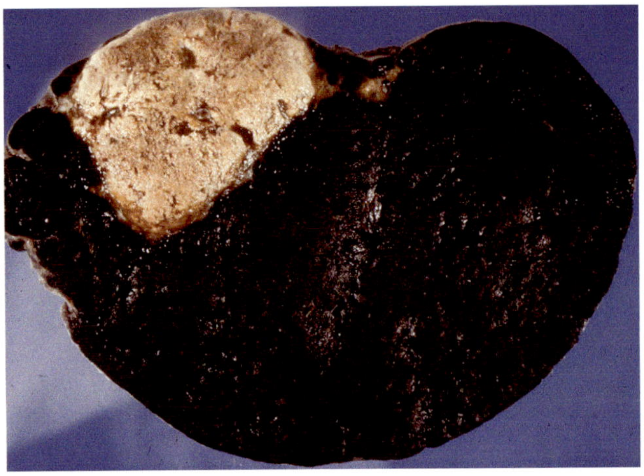

Fig. 6.31 Sickle cell disease, focal infarction. Gross image of a spleen with sickle cell disease. Infarction begins focally, and may be precipitated by a sickling crisis or other causes of hypoxia. As infarcts resolve, they become firm and yellow, replacing normal parenchyma with macrophages and fibrosis

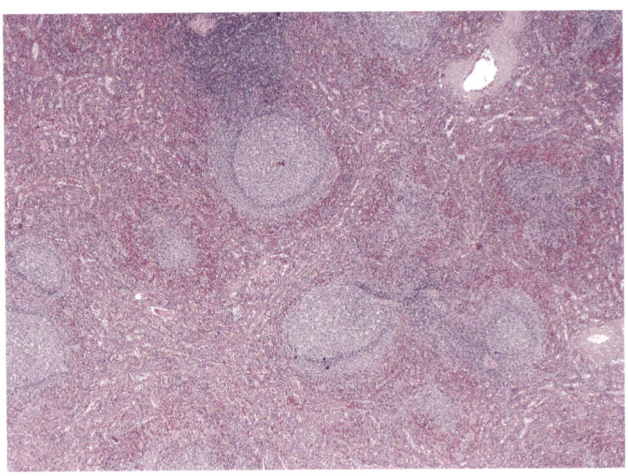

Fig. 6.32 Sickle cell disease. In this example of sickle cell disease, there is follicular hyperplasia. This is likely from a young patient, as there is little evidence of fibrosis or previous infarction

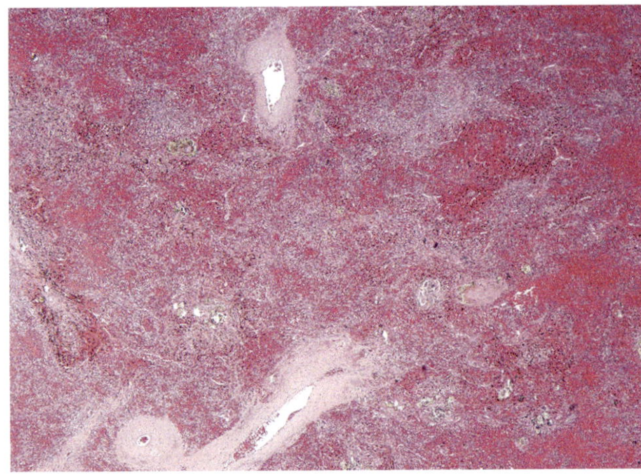

Fig. 6.33 Sickle cell disease. Low-magnification image of a spleen with "late" changes of sickle cell disease. There is depletion of white pulp, marked red blood cell congestion of red pulp, and numerous foci of hemosiderin deposition from previous hemolytic events

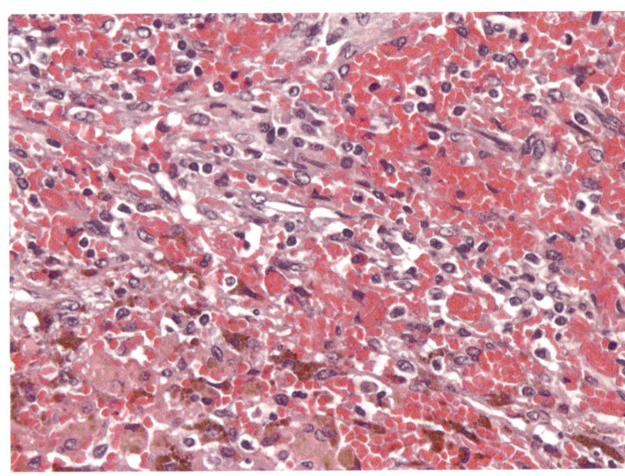

Fig. 6.34 Sickle cell disease. High-magnification image of sickle cell disease in spleen. There is hemosiderin deposition, red blood cell congestion, and focal erythrophagocytosis. Although not immediately apparent, careful inspection of red blood cells will reveal some with irregular pointed ends (sickled cells)

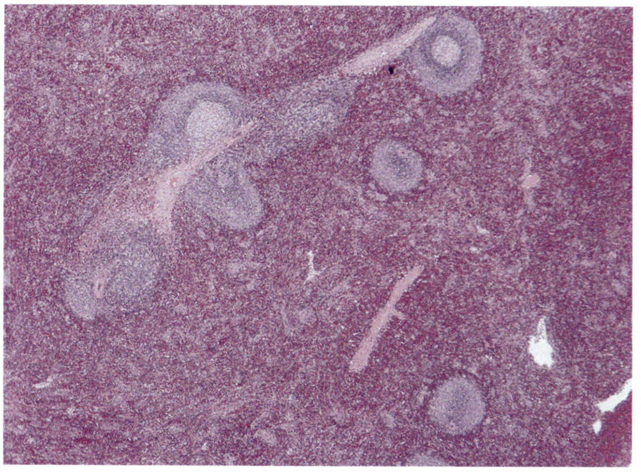

Fig. 6.35 Hemoglobin SC disease. Hemoglobin SC disease will often have less severe symptoms than sickle cell disease. In this case, there is preservation of white pulp structures with congestion of red pulp

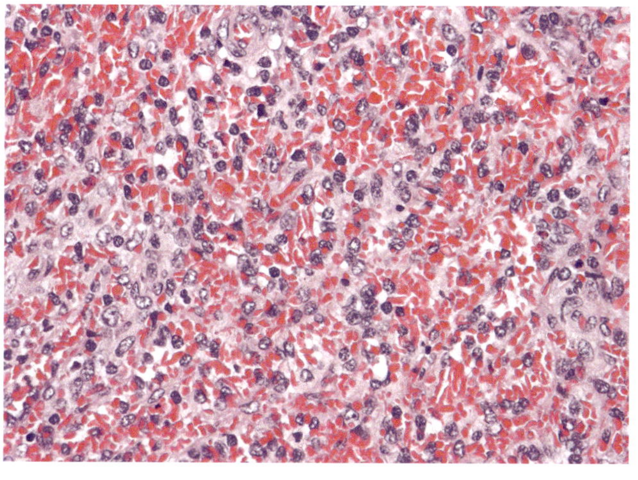

Fig. 6.36 Hemoglobin SC disease. High-magnification image of hemoglobin SC disease. As in typical sickle cell disease, there is congestion of red pulp structures with abnormal red blood cells. The irregular and pointed nature of the red blood cells can be appreciated within the sinuses

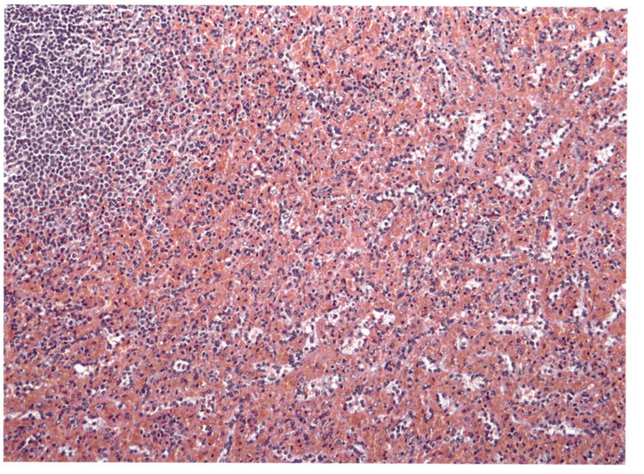

Fig. 6.37 Hereditary spherocytosis (HS). HS often leads to splenomegaly. In HS, the red pulp is congested with red blood cells. The spherocytic nature is not immediately apparent. However, although subtle, careful inspection reveals only round cross-sections of cells, with none showing the normal "biconcave disc" appearance

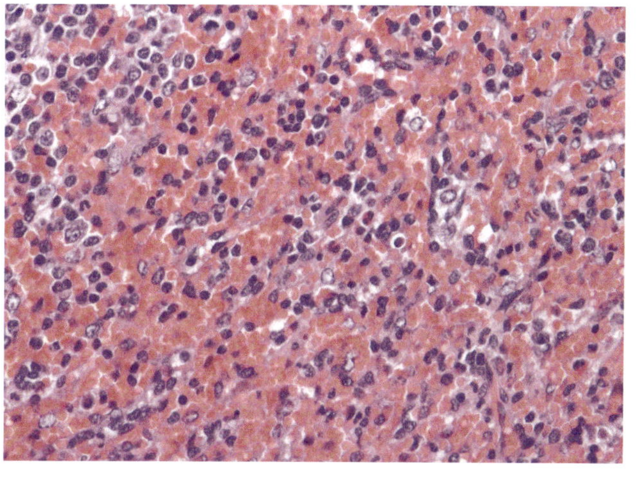

Fig. 6.38 Hereditary spherocytosis. High-magnification image of HS in the spleen. The red cells fill the sinuses and cords of the red pulp

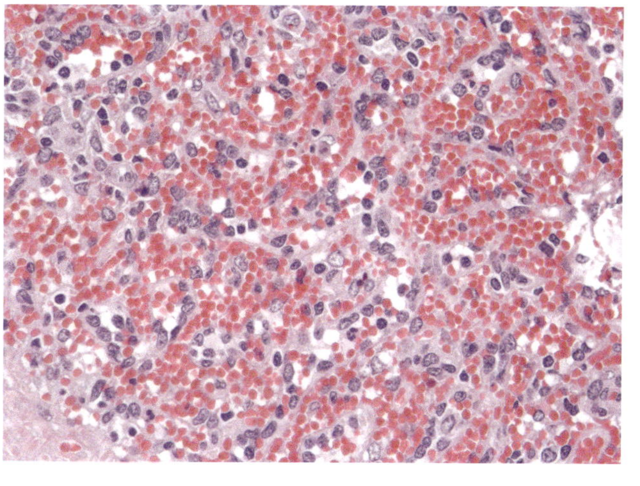

Fig. 6.39 Hereditary spherocytosis. Another high-magnification image of HS red blood cells in the spleen. In this image, there is congestion of sinuses and virtually all cells have a round (*eg*, spherical) aspect

6.6 Granulomas

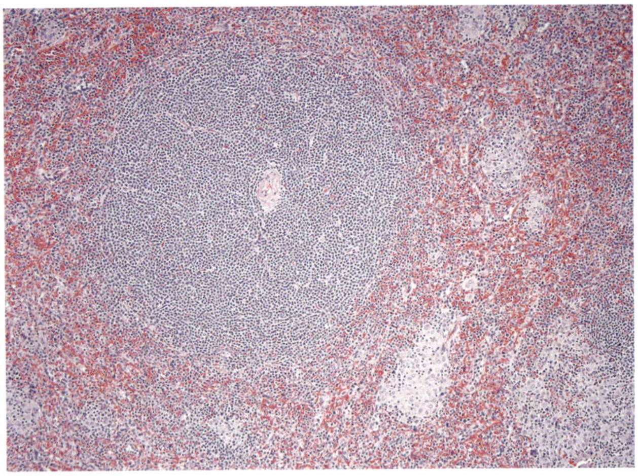

Fig. 6.40 Granuloma. Small epithelioid granulomas adjacent to a white pulp nodule. No organisms or foreign materials were identified. Although the possibility of organisms should be pursued in almost all cases, in most circumstances they are not found and the cause of the granulomas remains unknown

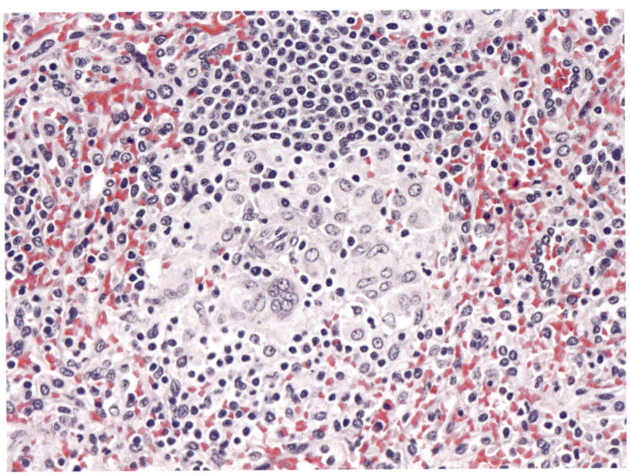

Fig. 6.41 Granuloma. High-magnification image of an epithelioid granuloma showing a combination of plump epithelioid macrophages, occasional giant cells, and variable numbers of admixed lymphocytes

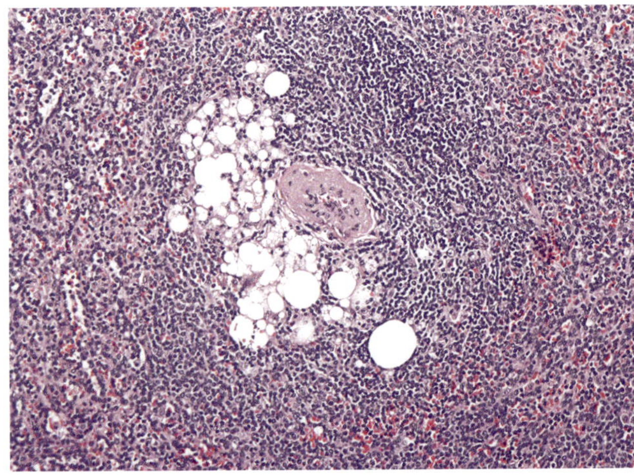

Fig. 6.42 Lipoid granuloma. High-magnification image of a lipoid granuloma of spleen. Lipoid granulomas represent small depositions of lipid material in the spleen, with subsequent inflammation. The lipid elements are likely deposited from the vascular system, as there is no intrasplenic adipose tissue under normal circumstances. Lipoid granulomas are fairly common and nonspecific. They are not typically associated with infection

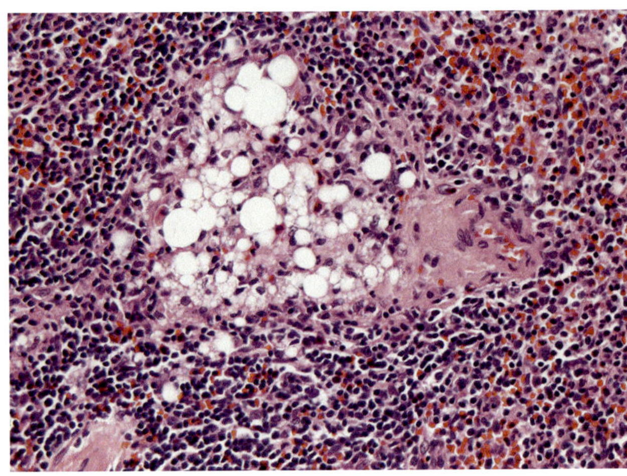

Fig. 6.43 Lipoid granuloma. Another example of a lipoid granuloma. Epithelioid histiocytes are not seen in contrast to the findings in many infection-associated granulomas

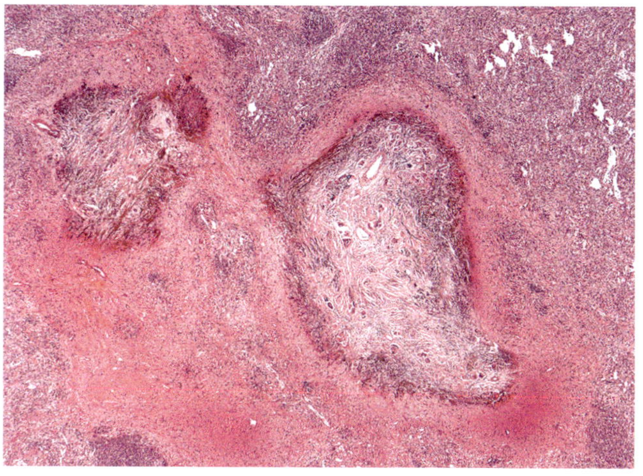

Fig. 6.44 Granuloma, primary biliary cirrhosis. Examples of granulomas in the spleen associated with primary biliary cirrhosis (PBC). Although not entirely specific, the presence of granulomas in the spleen is common in PBC. In this case, there is also extensive fibrosis and hemorrhage with deposition of hemosiderin

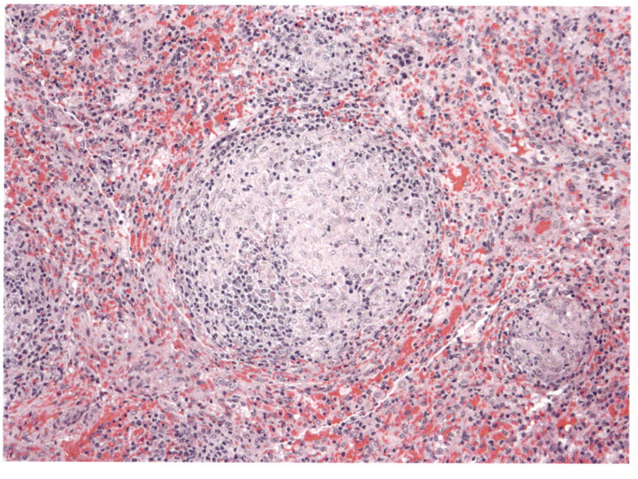

Fig. 6.45 Granuloma associated with follicular lymphoma. The germinal center in this example is nearly completely replaced with an epithelioid granuloma. Granulomas can occur in association with any lymphoma type

6.7 Langerhans Cell Histiocytosis

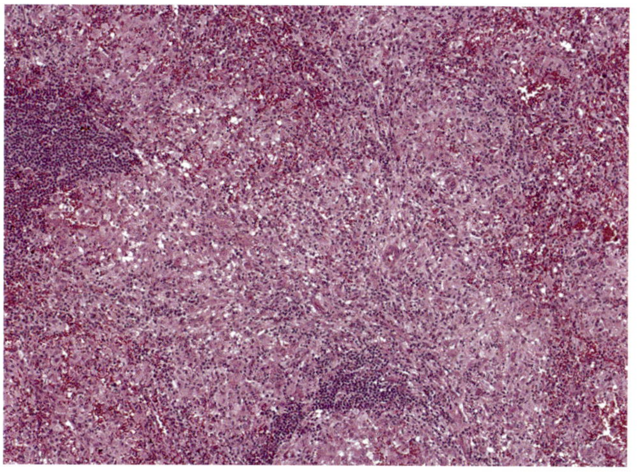

Fig. 6.46 Langerhans cell histiocytosis. Involvement of the spleen by Langerhans cell histiocytosis is relatively rare, and most common in disseminated or systemic disease. In this case, there are nodules of abnormal Langerhans cells in the red pulp, adjacent to white pulp structures. (*Image courtesy of* S. Abbondanzo, Washington, DC.)

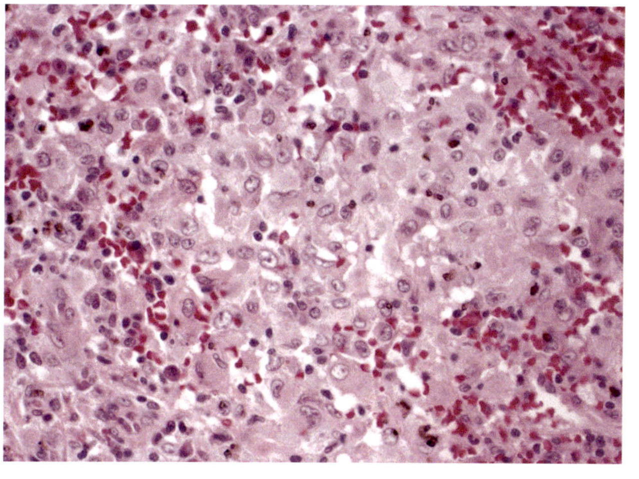

Fig. 6.47 Langerhans cell histiocytosis. High-magnification image of abnormal Langerhans cells in the spleen. These cells have ovoid nuclei, some with nuclear grooves, and copious amounts of pink cytoplasm. (*Image courtesy of* S. Abbondanzo, Washington, DC.)

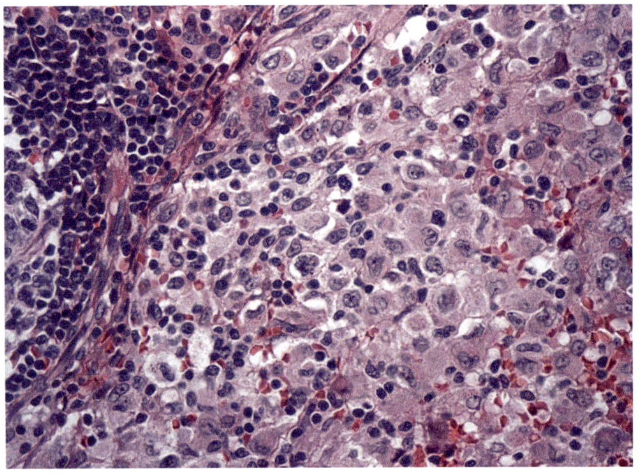

Fig. 6.48 Langerhans cell histiocytosis. Another high-magnification image of Langerhans cell histiocytosis in the spleen. In contrast to other tissues, eosinophilic infiltrates are not as prominent in the spleen. (*Image courtesy of* P. Banks, Charlotte, NC.)

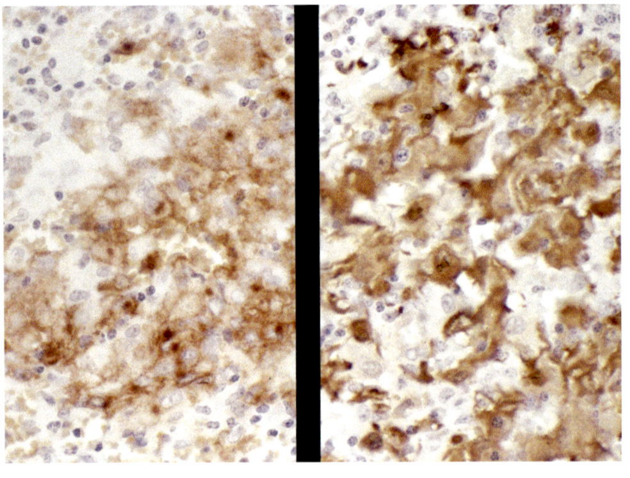

Fig. 6.49 Langerhans cell histiocytosis, CD1a stain, S-100 stain. Stains for CD1a (*left*) and S-100 (*right*) are positive in the Langerhans cells. (*Image courtesy of* S. Abbondanzo, Washington, DC.)

6.8 Storage Diseases

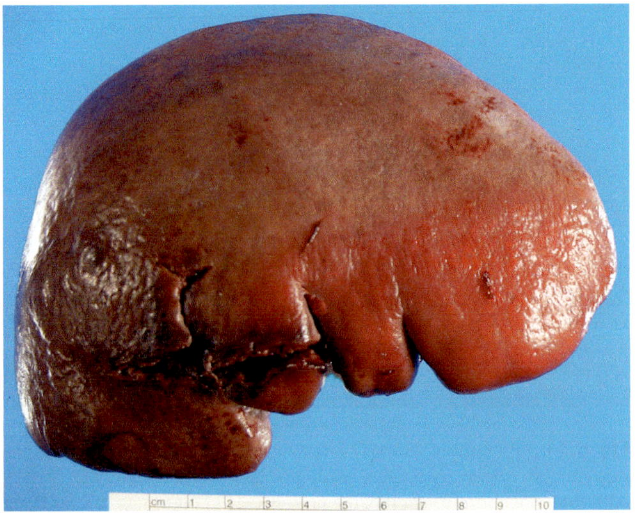

Fig. 6.50 Amyloidosis. Gross image of unsectioned spleen with diffuse involvement by amyloidosis. Note the overall enlargement, pale color, and prominence of the grooves at the edges. (*Image courtesy of* M. Kamionek, Boston, MA.)

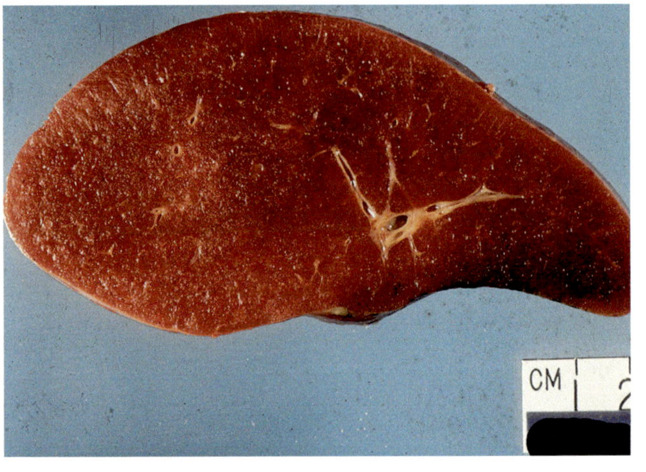

Fig. 6.51 Amyloidosis. Low-magnification gross image of spleen with amyloid deposition. Note that there is some abnormal color (red-orange compared to deep red-purple) and a vague waxy appearance. This spleen has extensive amyloid deposition but some sparing of normal structures. (*Image courtesy of* D. Farhi, Atlanta, GA.)

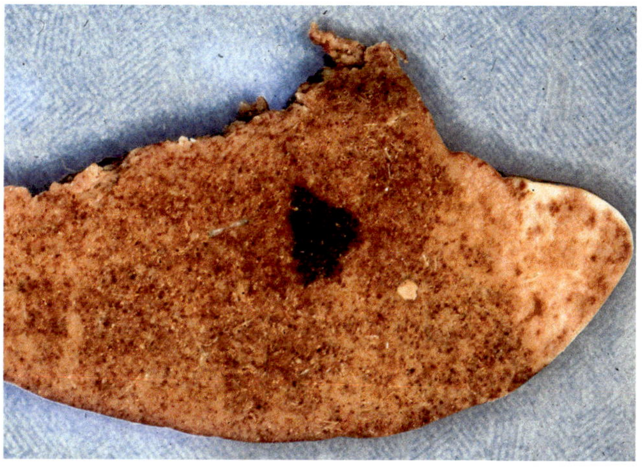

Fig. 6.52 Amyloidosis. Gross image of a sectioned spleen with extensive amyloid deposition. The yellow firm amyloid is present with only small spots of red representing residual vascular or lymphoid elements. This appearance is grossly referred to as "lardaceous," referring to the waxy, "greasy" appearance of the cut section. (*Image courtesy of* D. Farhi, Atlanta, GA.)

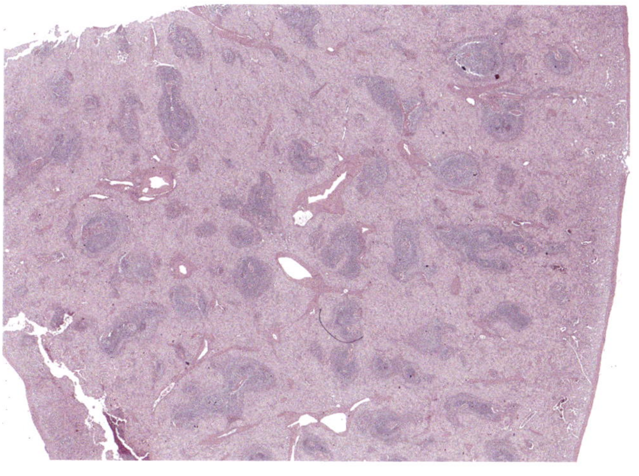

Fig. 6.53 Amyloidosis. Low-magnification image of early deposition of amyloid in spleen. In its earliest stages, amyloid will be deposited primarily in the walls of blood vessels. It is usually most often seen in the walls of smaller caliber vessels such as arterioles. There is no prominent increase in plasma cells

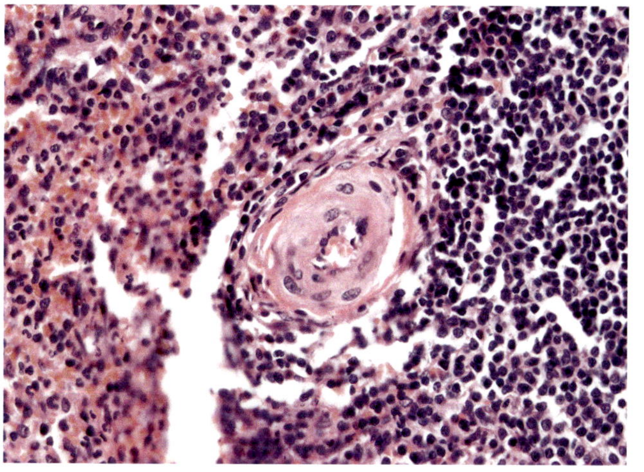

Fig. 6.54 Amyloidosis. High-magnification image of early amyloid deposition in splenic vessels. In this case, the wall of an arteriole shows the characteristic glassy, hyaline material

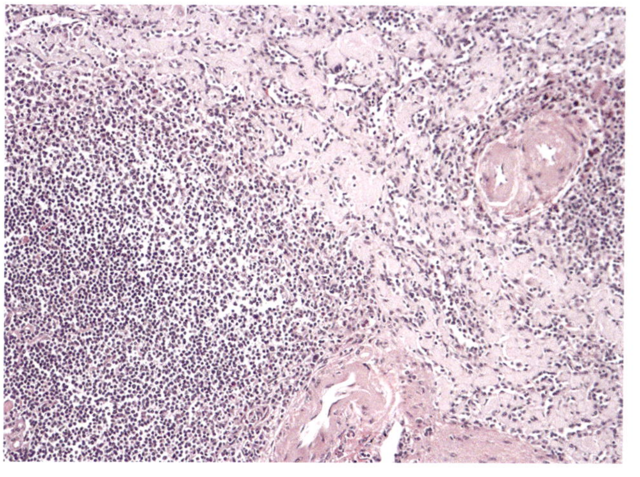

Fig. 6.55 Amyloidosis. As the extent of amyloid deposition increased, there is uniform involvement of the blood vessel walls, and then deposition in cordal macrophages

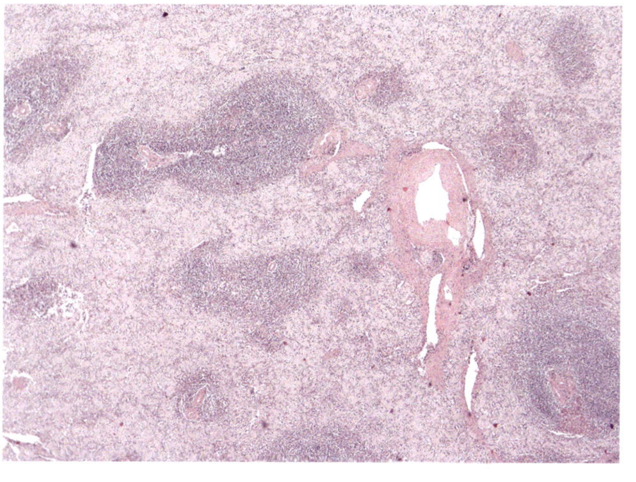

Fig. 6.56 Amyloidosis. In later stages of amyloidosis, the expansion of cordal macrophages is massive, replacing much of the functional area of the red pulp. There is sparing of the white pulp

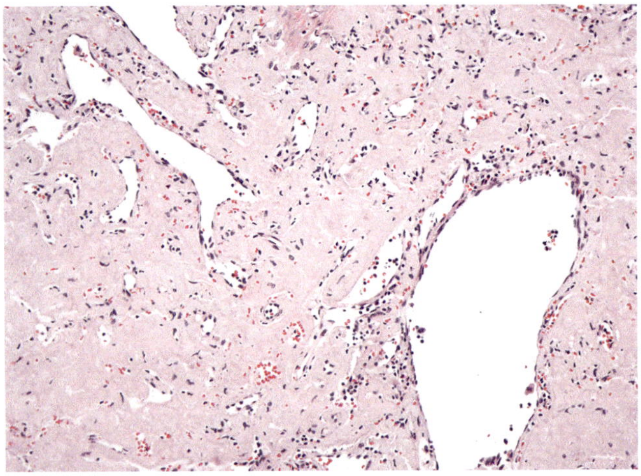

Fig. 6.57 Amyloidosis. In the end stage, nearly all splenic structures are replaced by diffuse sheets of deposited amyloid. There is a combination of extra- and intracellular deposition of the amyloid protein

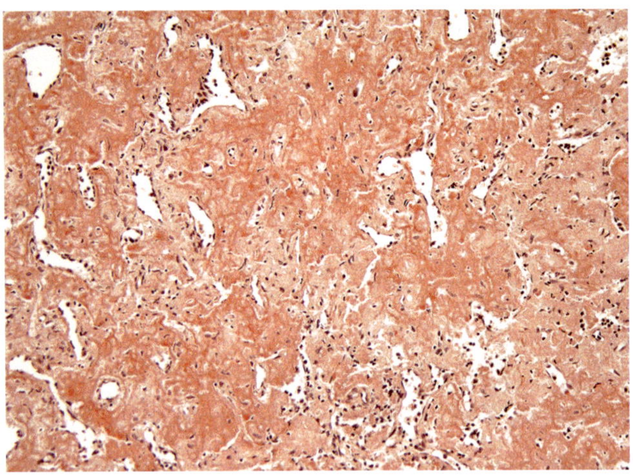

Fig. 6.58 Amyloidosis. A Congo red stain of amyloid shows the deep red staining expected from amyloid. All types of amyloid (*eg*, primary, secondary, inherited) are positive for Congo red staining

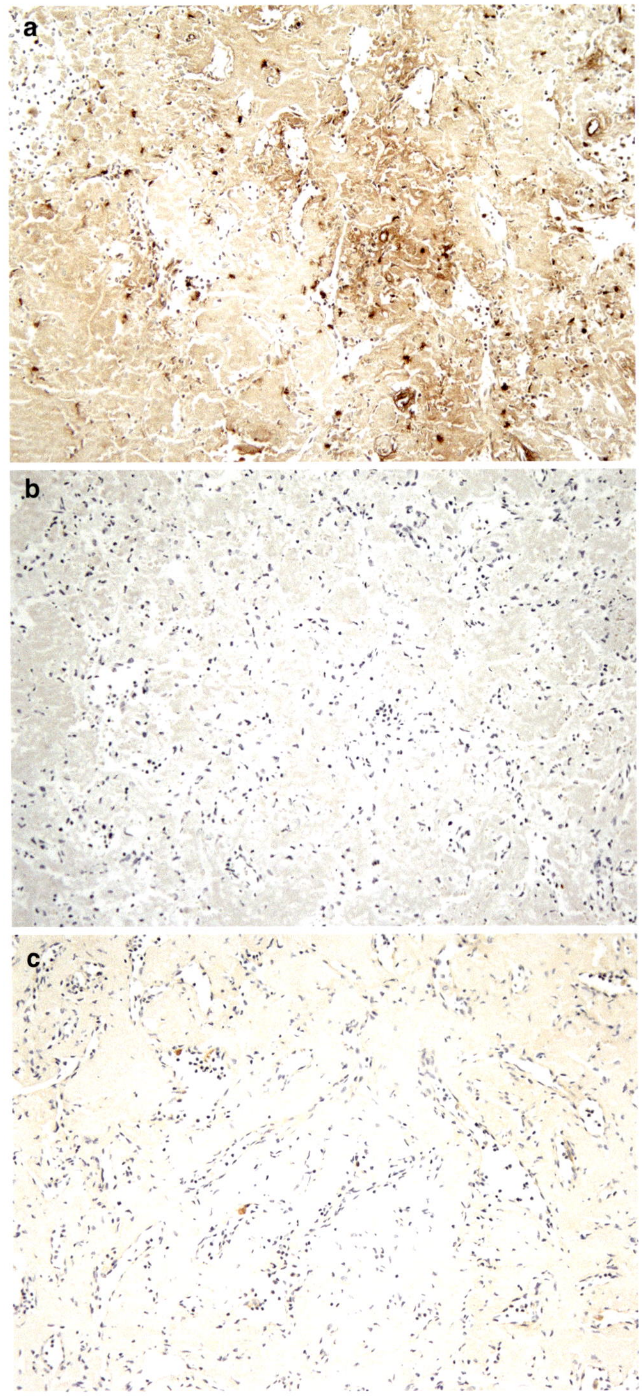

Fig. 6.59 Amyloidosis. Immunohistochemical staining of amyloid; in this case, there is staining for Amyloid P (**a**), with no staining for Amyloid A (**b**) or prealbumin (**c**). This is most compatible with a primary amyloidosis as is seen with plasma cell dyscrasia

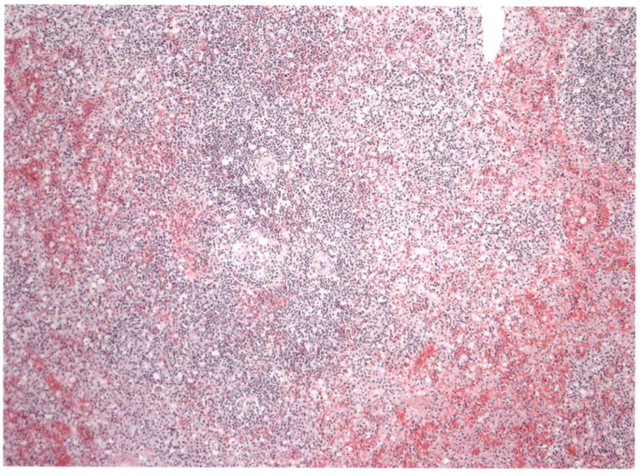

Fig. 6.60 Sialidosis. Low-magnification image of spleen in a patient with sialic acid storage disease. As with other metabolic disorders, the abnormal product is often stored in splenic macrophages

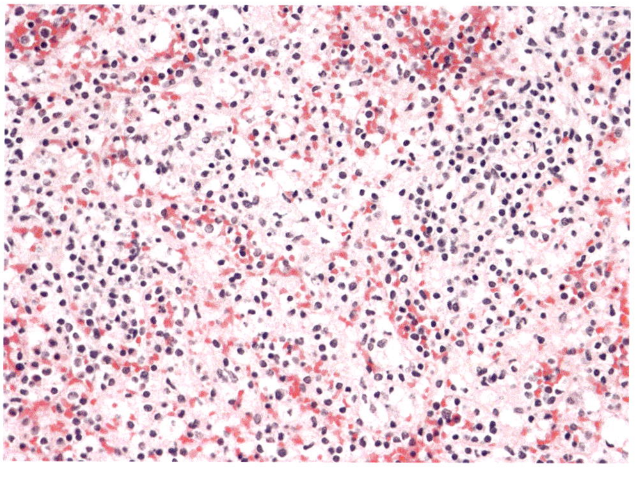

Fig. 6.61 Sialidosis. High-magnification image of red pulp showing an overall increase in the size of splenic cords. The cords are expanded by macrophages with pale or pink, slightly granular cytoplasm filled with abnormal metabolic product

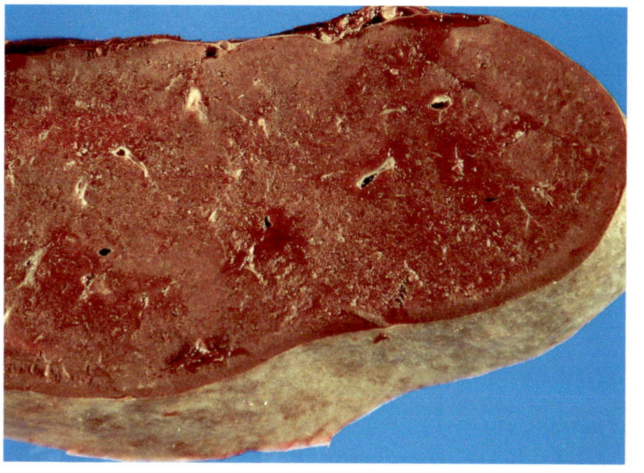

Fig. 6.62 Niemann-pick disease. Gross photo of spleen diffusely involved with an adult form of Neimann-Pick disease. As with other metabolic diseases, spleen is frequently a site of deposition of the abnormal accumulated metabolic product. In this case, there is an expansion of the entire spleen, giving it an overall pale appearance. This is due to the massive proliferation of histiocytes filled with the abnormal metabolic product. (*Image courtesy of* J. Sidhu, Johnson City, NY.)

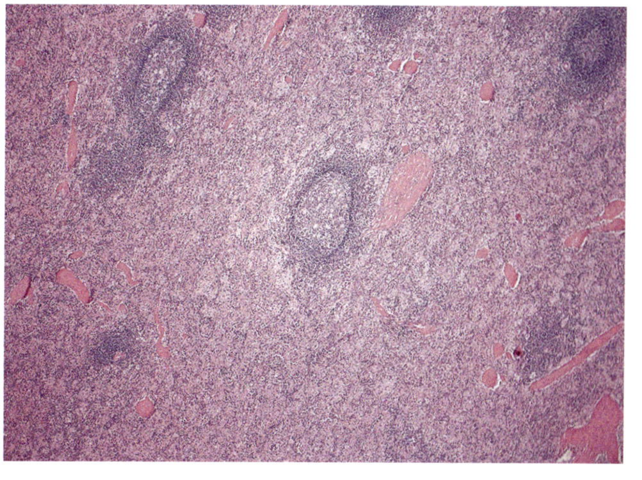

Fig. 6.63 Niemann-Pick disease. Low-magnification image of the spleen with extensive involvement by Niemann-Pick disease. White pulp structures appear normal, with massive expansion of the red pulp

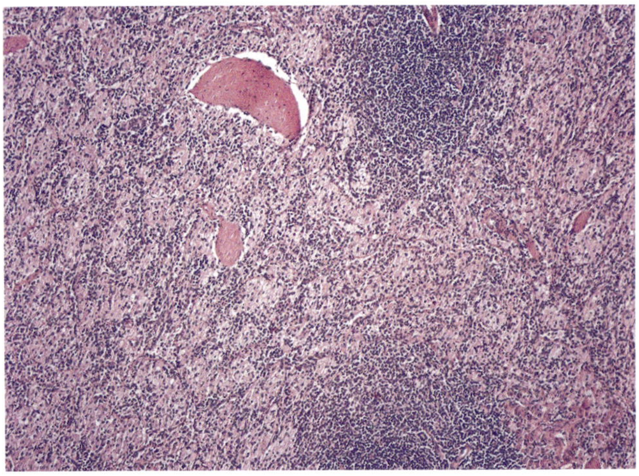

Fig. 6.64 Niemann-Pick disease. Intermediate-magnification image of Niemann-Pick disease in spleen. The splenic cords are massively expanded with macrophages filled with abnormal metabolic product

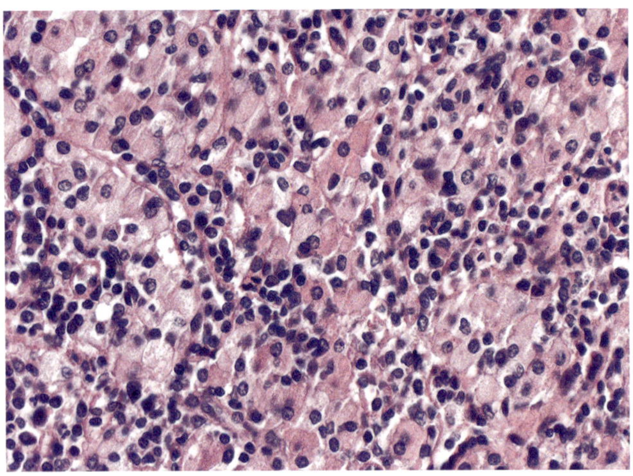

Fig. 6.65 Niemann-Pick disease. High-magnification image of splenic red pulp with massive expansion by the macrophages in Niemann-Pick disease. The cytoplasm of the macrophages is distended with large amounts of granular cytoplasm

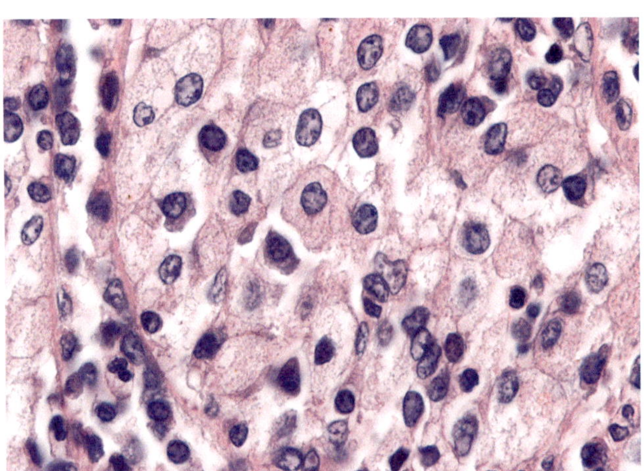

Fig. 6.66 Niemann-Pick disease. Very high magnification of the macrophages of Niemann-Pick disease. The cytoplasm has an appearance of many small vacuoles with small granular debris

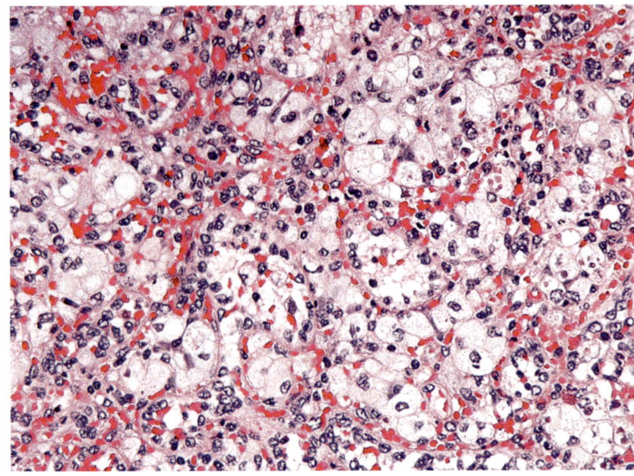

Fig. 6.67 Niemann-Pick disease. An example of a late adult-onset case of Niemann-Pick disease. In this case, the macrophages have a clearer, foamy appearance. (*Image courtesy of* A. Orazi, New York, NY.)

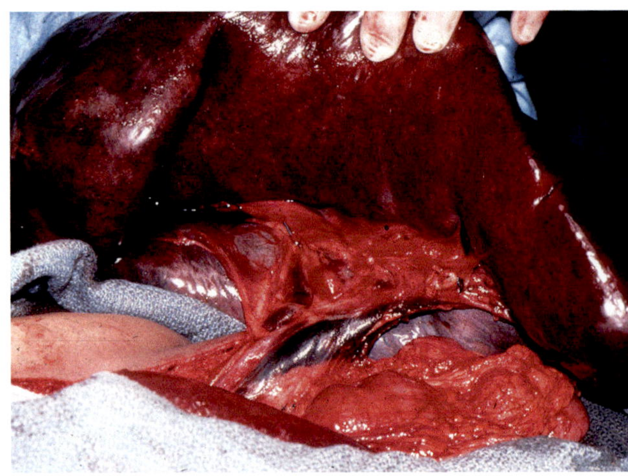

Fig. 6.68 Gaucher disease. Intrasurgical image of a massively enlarged spleen in Gaucher disease. The size of the spleen compared to the hand of the surgeon illustrates the massive enlargement. (*Image courtesy of* L. Morgenstern, Los Angeles, CA.)

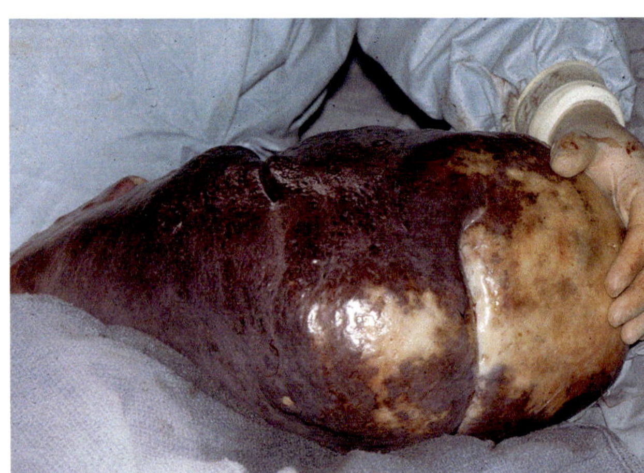

Fig. 6.69 Gaucher disease. Another intrasurgical view of a massively enlarged spleen in Gaucher disease. The variegated yellow areas are likely areas of remote infarct with resolution and fibrosis. (*Image courtesy of* L. Morgenstern, Los Angeles, CA.)

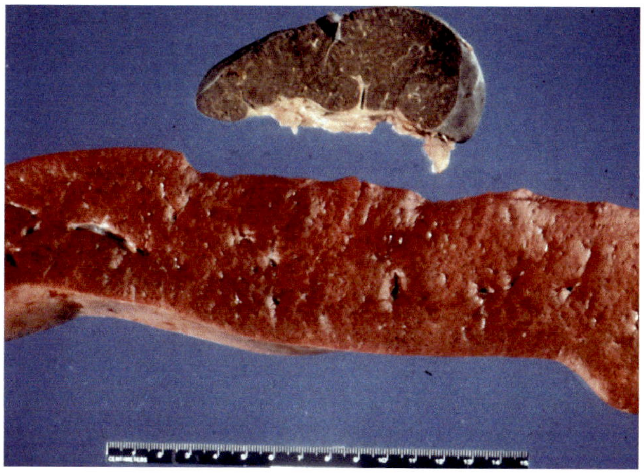

Fig. 6.70 Gaucher disease. Gross image of sectioned spleen with Gaucher disease. A section of a normal (fixed) spleen is included in the upper portion for scale. The Gaucher spleen (*lower*) is massively enlarged. There is a loss of normal red and white pulp architecture, due to the massive expansion of histiocytes filled with abnormal metabolic product. (*Image courtesy of* D. Farhi, Atlanta, GA.)

Fig. 6.71 Gaucher disease. Massively enlarged spleen with extensive involvement by Gaucher disease. The cut surface is shiny, suggesting deposition in the red pulp, rather than the typical "beefy" appearance of typical red pulp disease

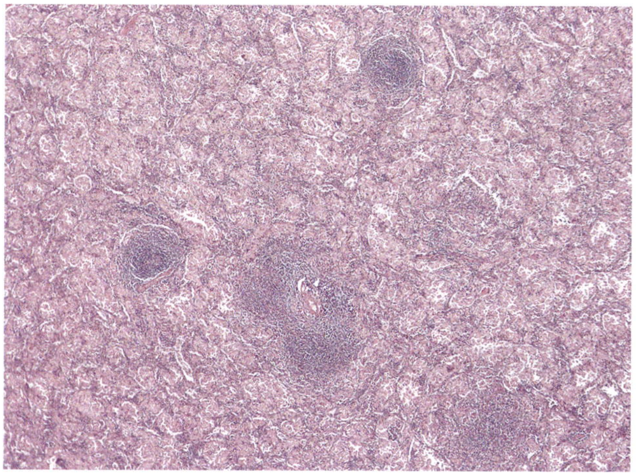

Fig. 6.72 Gaucher disease. Low-magnification image of Gaucher disease in the spleen. White pulp structures are maintained but there is expansion and distortion of the red pulp. This is due to a marked increase in macrophages in the splenic cords

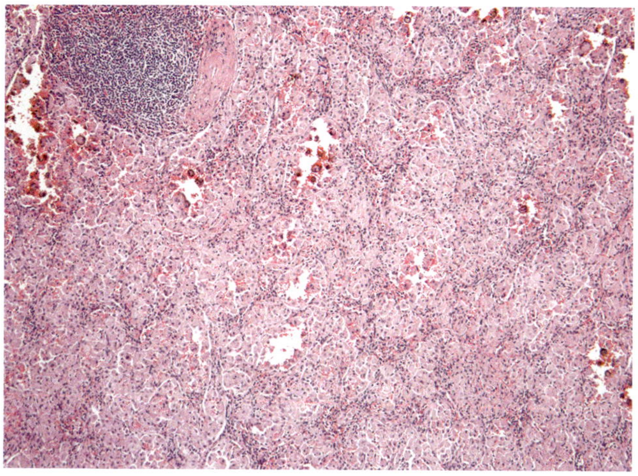

Fig. 6.73 Gaucher disease. Another example of Gaucher disease in the spleen. The enlarged macrophages are pink, with a massive increase in cytoplasm

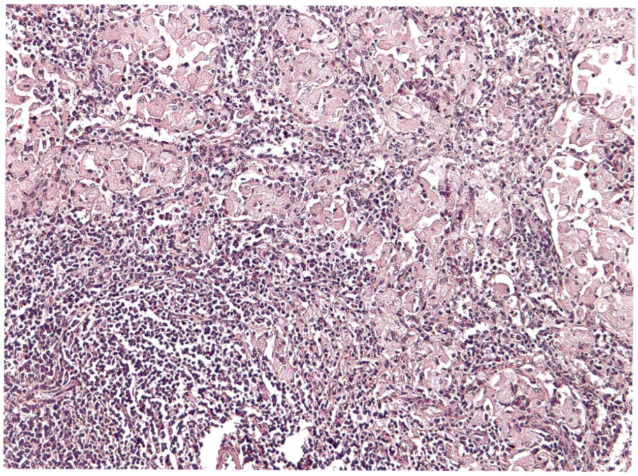

Fig. 6.74 Gaucher disease. Splenic sinuses are present but are often compressed by the increase in splenic cordal macrophages in Gaucher disease

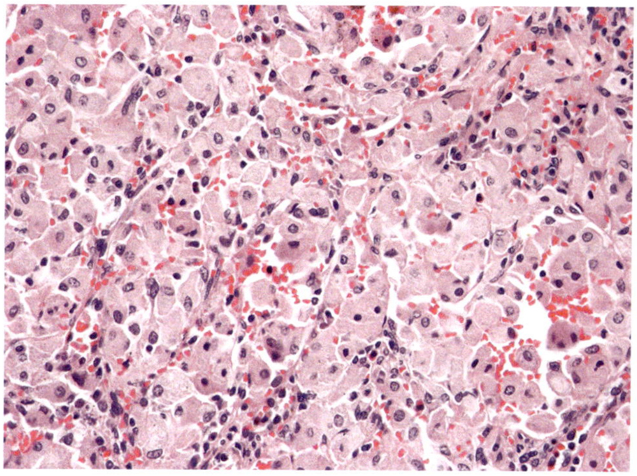

Fig. 6.75 Gaucher disease. High-magnification image of the macrophages of Gaucher disease. The cytoplasm is traditionally described as appearing like "wrinkled tissue paper"

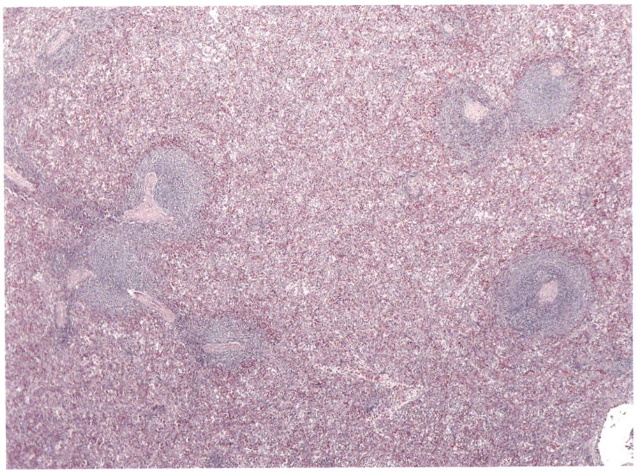

Fig. 6.76 Sea-blue histiocytosis. Most cases of sea-blue histiocytosis are thought to represent a variant of Neimann-Pick disease. As in other storage diseases, there is an increase in red pulp macrophages

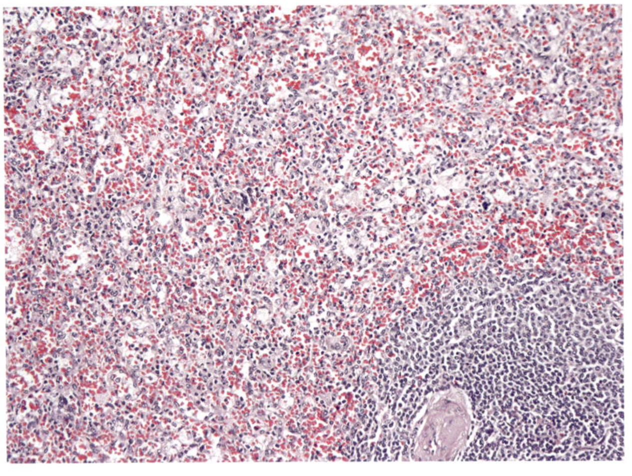

Fig. 6.77 Sea-blue histiocytosis. Red pulp histiocytes are increased, have a foamy appearance, and, on hematoxylin and eosin (H&E) stain, a faint brown pigmentation

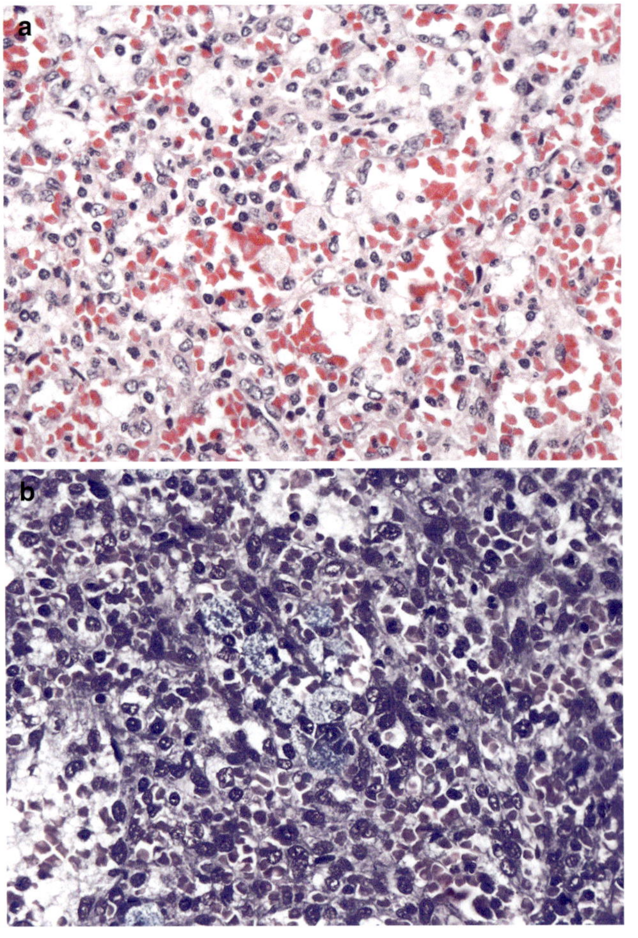

Fig. 6.78 Sea-blue histiocytosis, H&E, Giemsa stain. High-magnification image of the foamy histiocytes of sea-blue histiocytosis (**a**) and a Giemsa stain (**b**), which is the origin of the "sea-blue" color

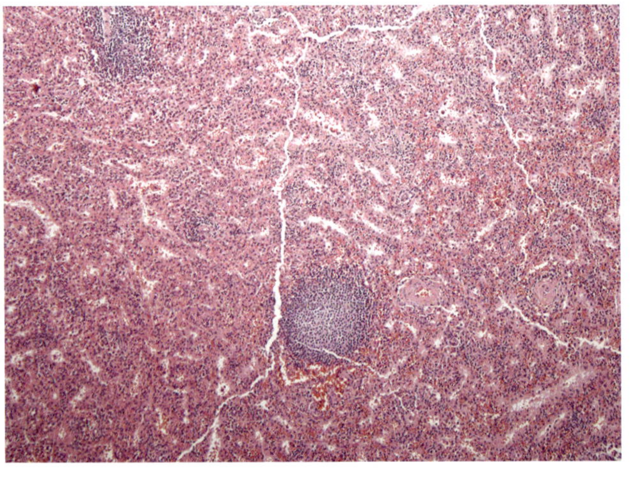

Fig. 6.79 Crystal-storing histiocytosis. Crystal-storing histiocytosis is a rare complication of plasma cell neoplasms. There is deposition of immunoglobulin crystals in macrophages of the spleen and rarely other tissue. The histiocytes of the red pulp have pink cytoplasmic granules

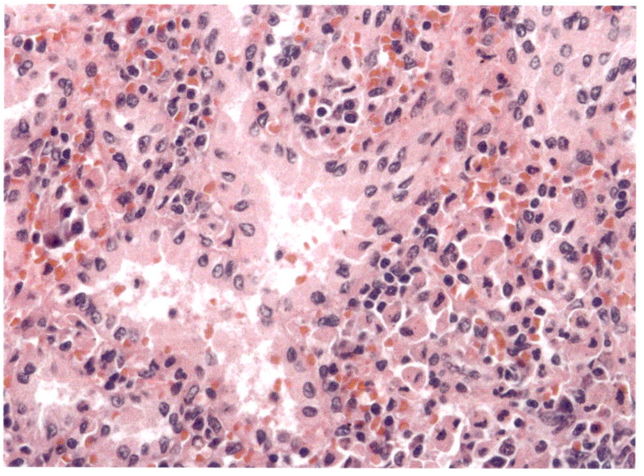

Fig. 6.80 Crystal-storing histiocytosis. High-magnification image of crystal-storing histiocytosis. The macrophages of the red pulp have deposition of pink material. There is also expansion of the sinus lining cells with similar material

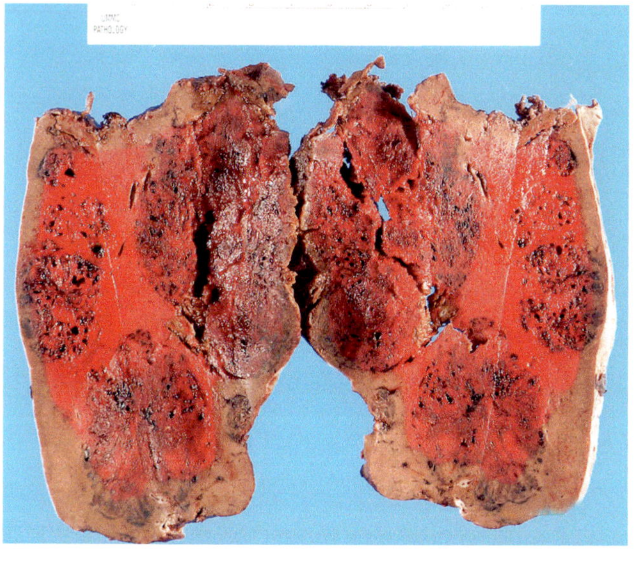

Fig. 6.81 Storage disease. Spleen is massively enlarged with foamy histiocytes and has an abnormal spongy appearance. Although this was likely a metabolic storage disease, it was not well-characterized. (*Image courtesy of* M. Kamionek, Boston, MA.)

6.9 Castleman Disease

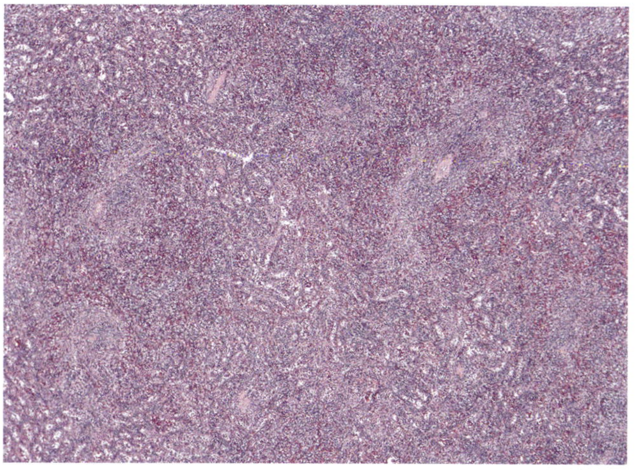

Fig. 6.82 Plasma cell Castleman disease. Low-magnification image of spleen with involvement by plasma cell Castleman disease. There is an increase in cellularity in the red pulp, with retention of the PALS regions

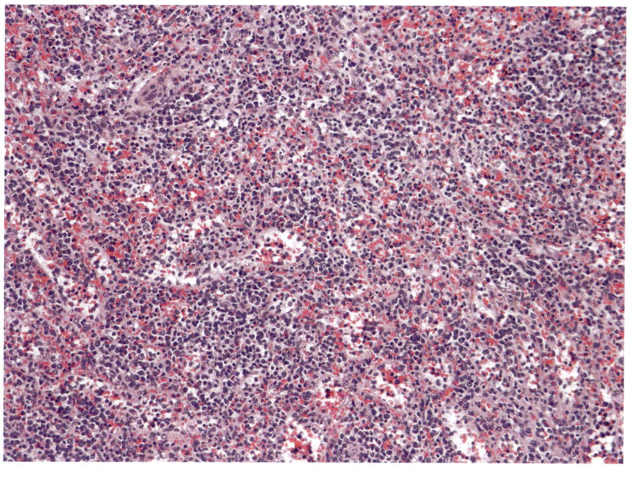

Fig. 6.83 Plasma cell Castleman disease. Plasma cell Castleman disease can be seen in the spleen, most often in patients with HIV/AIDS. Hyaline-vascular Castleman disease is exceedingly rare in the spleen. Increased plasma cells are seen in the red pulp in this case

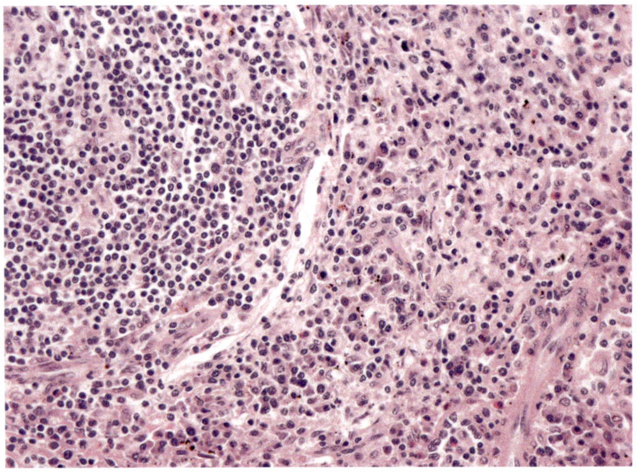

Fig. 6.84 Plasma cell Castleman disease. Follicle (*left*) with increased plasma cells in the red pulp in plasma cell Castleman disease. The plasma cells are typically polyclonal. Many cases show expression of human herpesvirus (HHV)-8

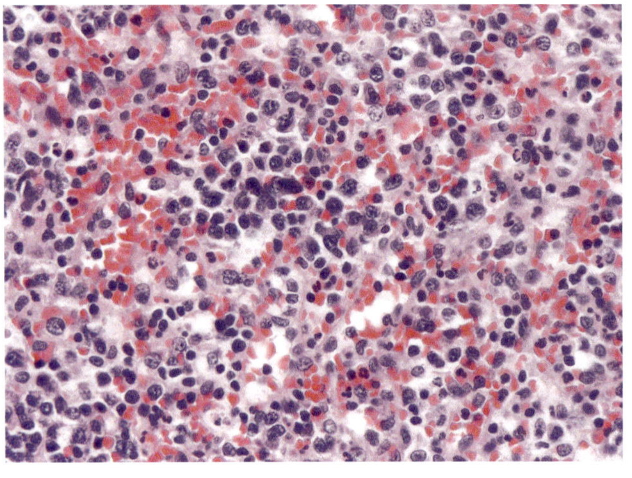

Fig. 6.85 Plasma cell Castleman disease. High magnification of the red pulp of plasma cell Castleman disease. Plasma cells are present mostly in cords but some are present in sinuses. Some mild cytologic atypia can be seen in the plasma cells

6.10 Therapy Effects

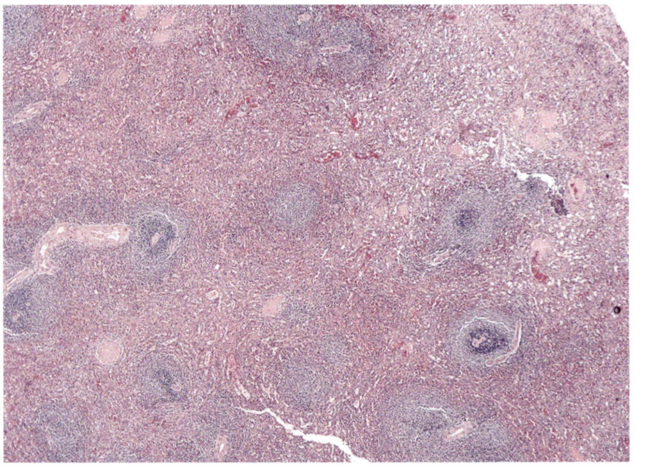

Fig. 6.86 Corticosteroid therapy. In this spleen, there is an apparently normal distribution of red and white pulp structures. The most frequent change associated with steroid therapy is the presence of many primary follicles but few secondary follicles (*eg*, germinal centers)

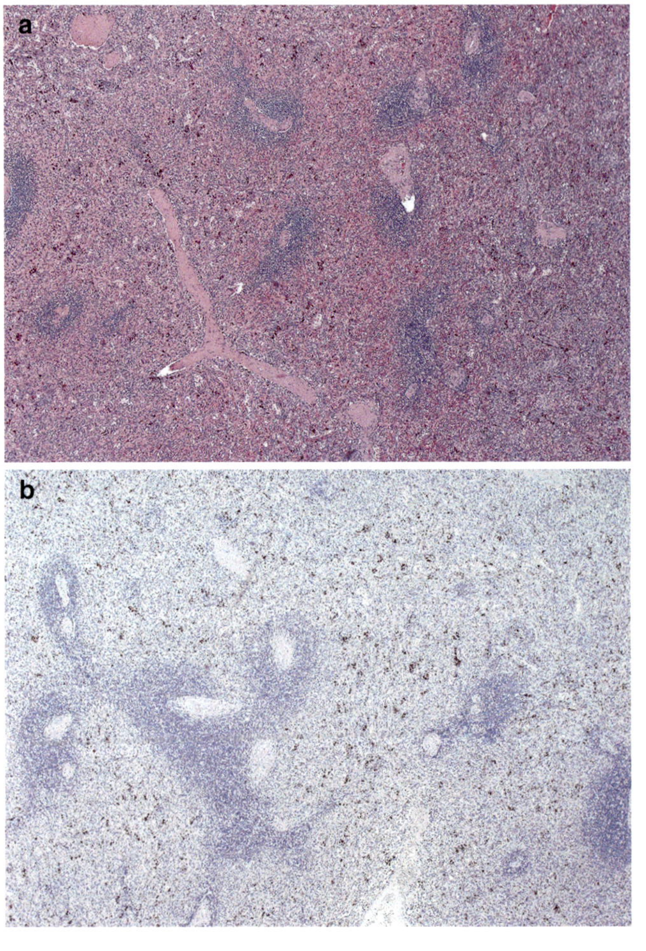

Fig. 6.87 Rituximab therapy, immune thrombocytopenic purpura, H&E, CD20 stain. Rituximab, an anti-CD20 monoclonal antibody, can be used as a therapy for immune thrombocytopenic purpura. In this case, H&E (**a**) shows preservation of PALS regions but diminution of white pulp nodules. CD20 staining shows a complete absence of CD20+ B-cells (**b**)

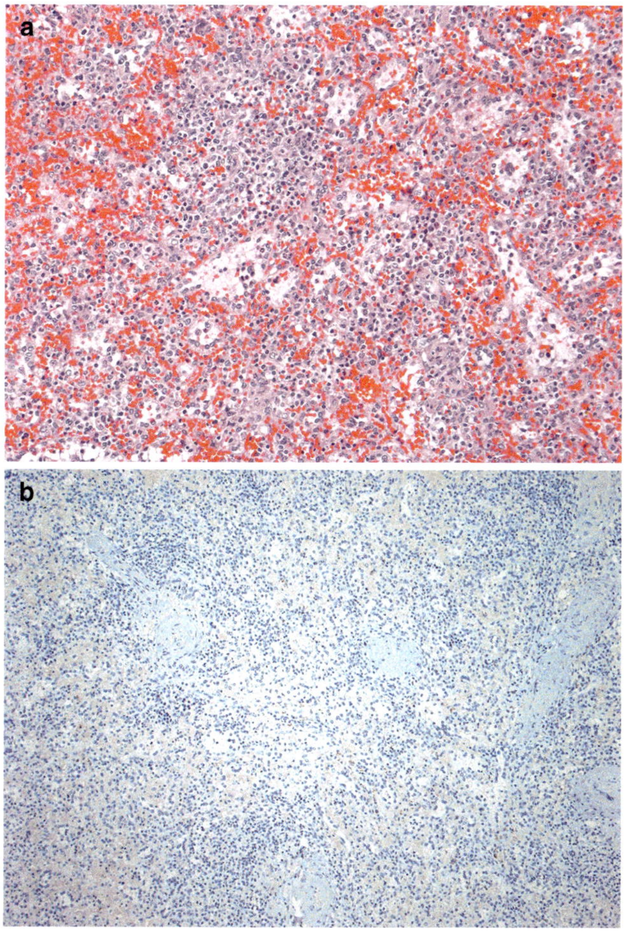

Fig. 6.88 Rituximab therapy, immune thrombocytopenic purpura, H&E, CD20 stain. Another example of rituximab therapy in spleen. H&E staining of red pulp structures (**a**) and CD20 staining of similar areas (**b**)

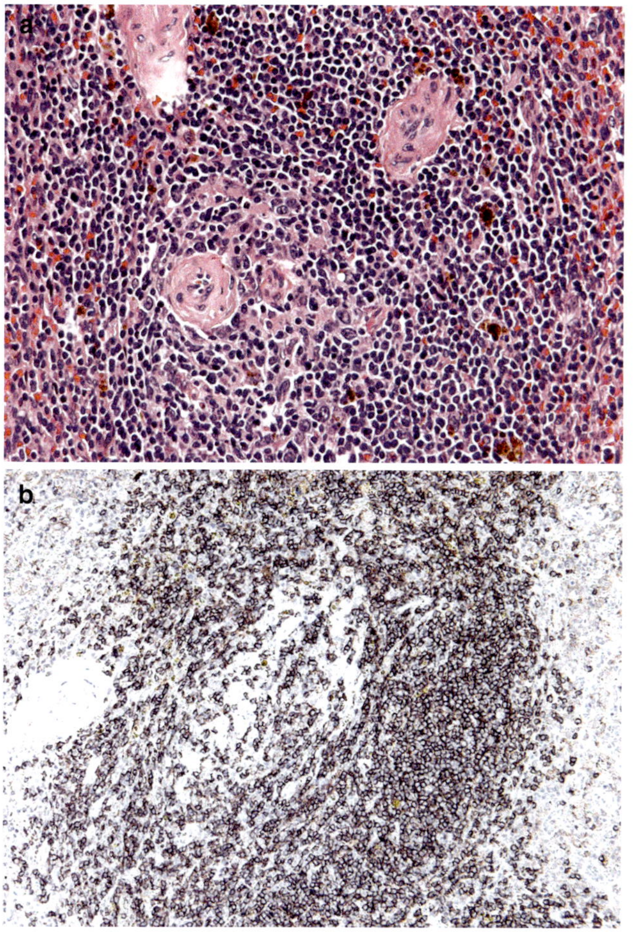

Fig. 6.89 Phenytoin therapy, H&E, CD3 stain. Treatment with phenytoin in this case was associated with an expansion of the PALS regions (**a**) and increase in polyclonal T cells, highlighted by CD3 staining (**b**). (*Images courtesy of* A. Orazi, New York, NY.)

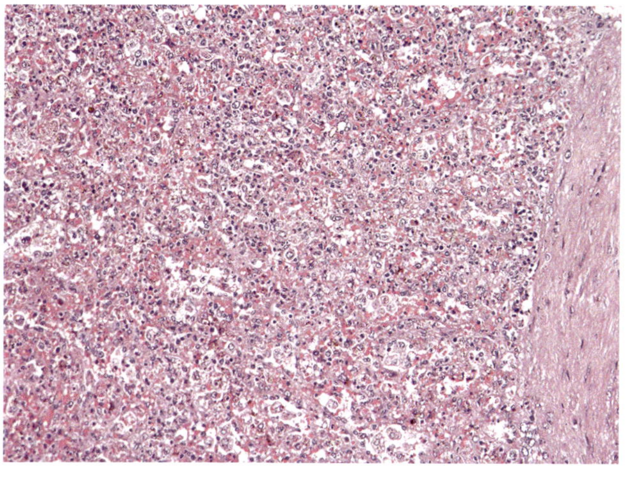

Fig. 6.90 Status post-bone marrow transplantation (BMT). After BMT, many changes can be seen in the splenic architecture. There is evidence of extramedullary hematopoiesis, which is not an uncommon finding after BMT. In this case, both erythroid and immature myeloid elements are present. Also, increased histiocytes, many with prominent hemosiderin in their cytoplasm, are seen

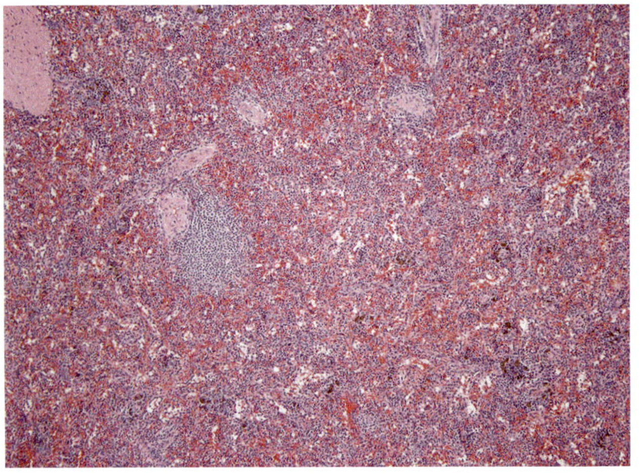

Fig. 6.91 Status post-bone marrow transplantation. Low-magnification image showing a general preservation of white pulp structures. The histologic changes in post-BMT are variable, depending on the type of underlying disorder and pre-transplant therapy

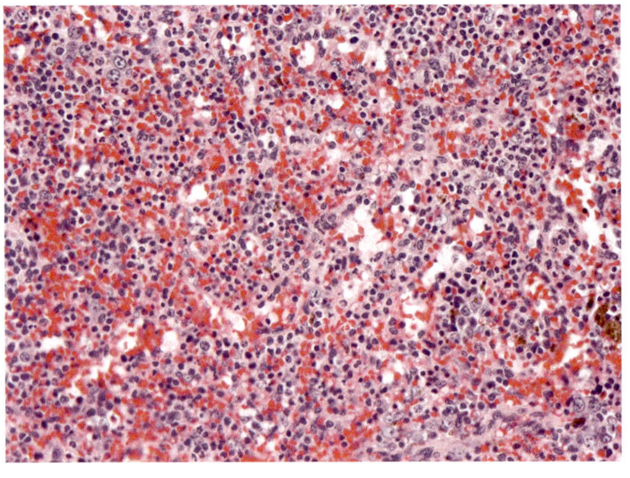

Fig. 6.92 Status post-bone marrow transplantation. This example of post-BMT spleen shows a combination of vascular congestion, numerous erythroid precursors, and some small lipogranulomas. Red pulp architecture is mostly inapparent

6.11 Other Disorders

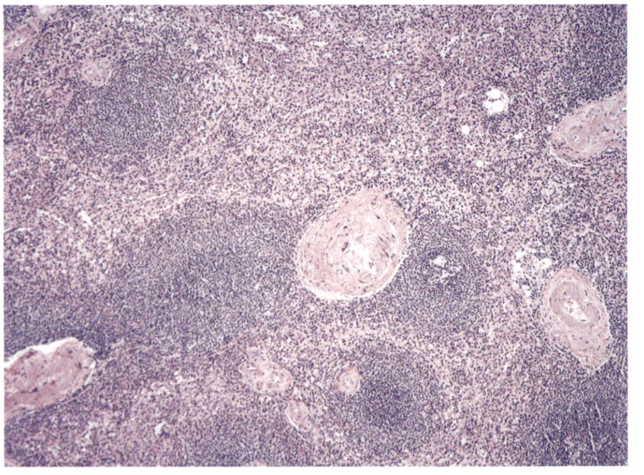

Fig. 6.93 Diabetes mellitus. An intermediate magnification of a spleen in a patient with diabetes mellitus. Although the changes are typically nonspecific, in splenic samples there may be abnormalities noted in vascular elements, as in this case. Note the thickening of the blood vessel walls. In rare cases, calcification may be seen within walls of splenic vessels

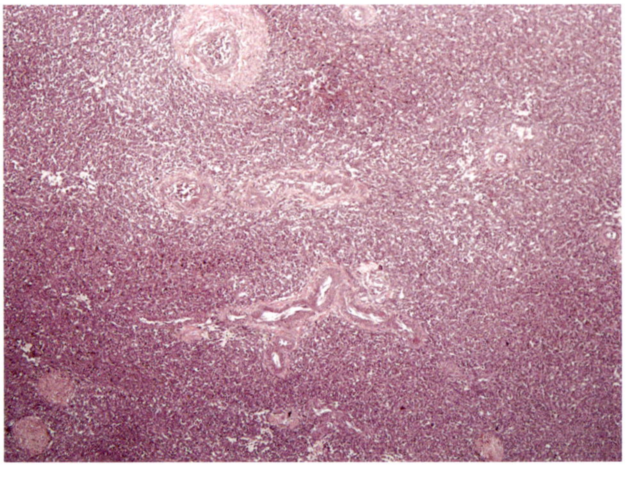

Fig. 6.94 Radiation effect. Spleen approximately 2 days after patient experienced significant accidental exposure to large amounts of radiation (8800 rad). Most lymphocytes are necrotic. There is some evidence of hemosiderin, likely due to lysis of red blood cells. The vessels appear abnormal, likely due to endothelial cell damage

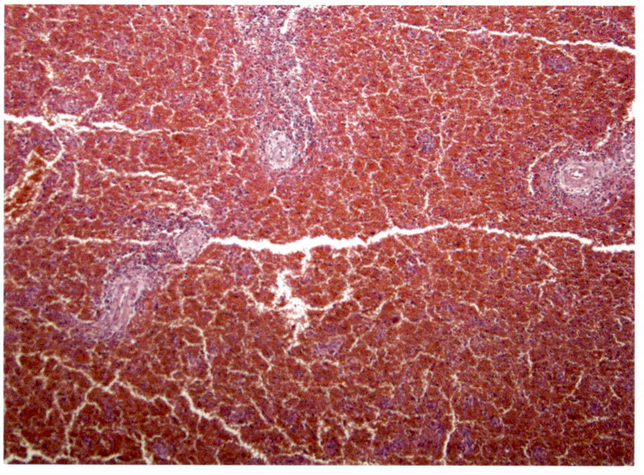

Fig. 6.95 Radiation effect. Accidental exposure to radiation (220 rad), 9 days after exposure. There is some retention of lymphoid cells but red pulp structures appear obliterated with only red blood cell congestion

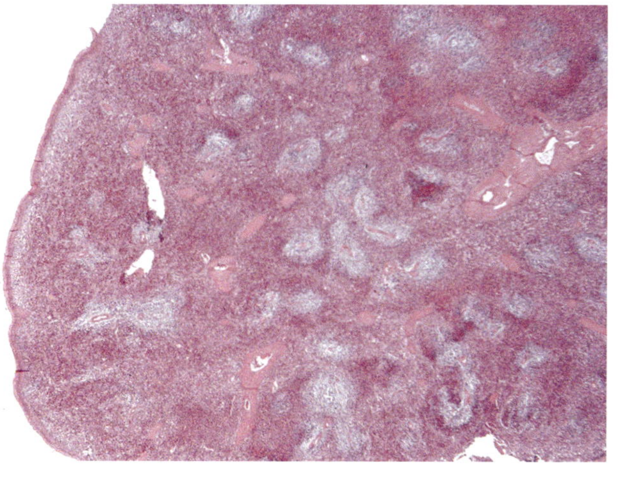

Fig. 6.96 Radiation effect. Effect of atomic bomb exposure (Nagasaki, Japan, 1945). Note the depletion of lymphocytes and loss of clear architectural definition of red and white pulp structures

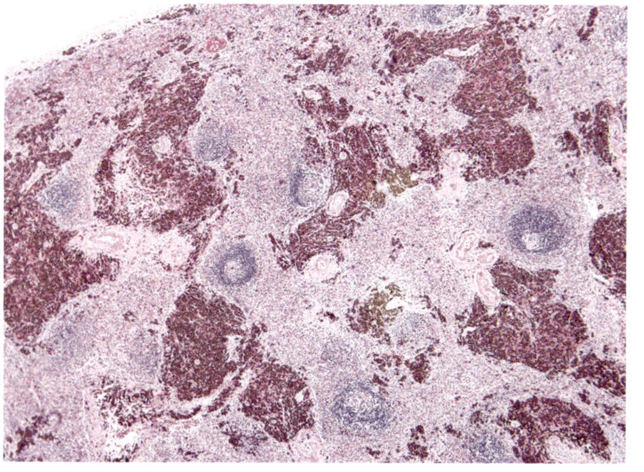

Fig. 6.97 Thorotrast deposition. The spleen shows deposition of abnormal brown pigmented material in red pulp macrophages. The material is Thorotrast, a radiologic contrast material used in the past, that has an association with development of angiosarcoma. (*Image courtesy of* L. Weiss, Aliso Viejo, CA.)

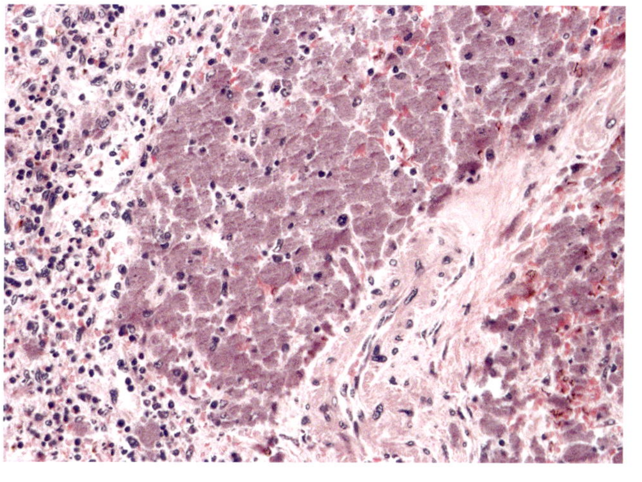

Fig. 6.98 Thorotrast deposition. High-magnification image of macrophages with Thorotrast material deposition in clusters of large macrophages in the red pulp. (*Image courtesy of* L. Weiss, Aliso Viejo, CA.)

6.12 Hemophagocytic Syndrome

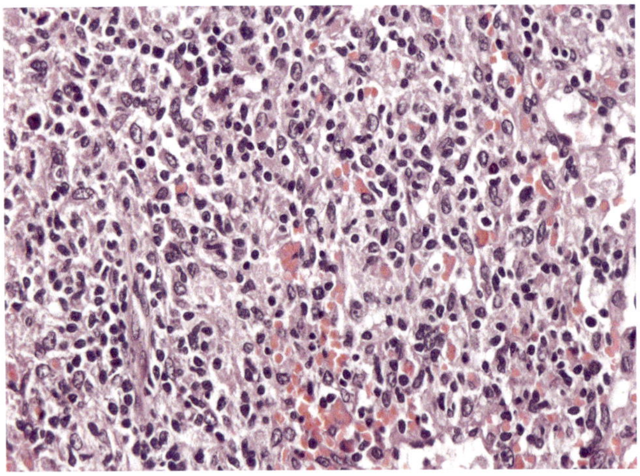

Fig. 6.99 Hemophagocytic syndrome. Prominent hemophagocytosis is often seen in the spleen in cases of hemophagocytic syndrome (HPS). Many etiologies may be associated with HPS, including infections and neoplasms. In this case, there is an increase in lymphoid cells and histiocytes in the splenic red pulp. In one large histiocyte (*center*), ingestion of red blood cells (erythrophagocytosis) is seen

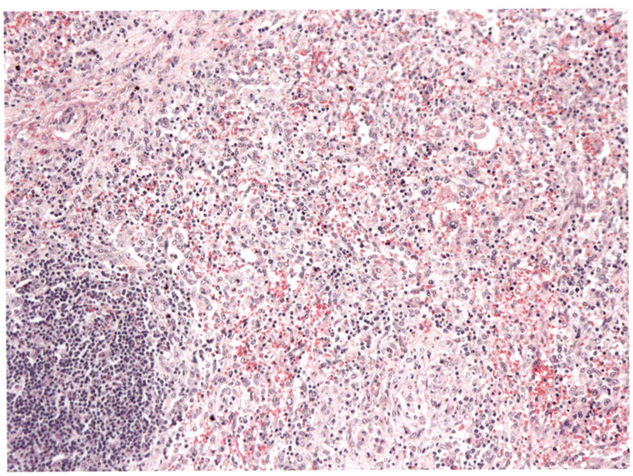

Fig. 6.100 Hemophagocytic syndrome. Intermediate magnification of spleen with a massive increase in red pulp histiocytes. There is also evidence of extramedullary hematopoiesis in the red pulp. Although subtle, this spleen also shows evidence of hemophagocytosis

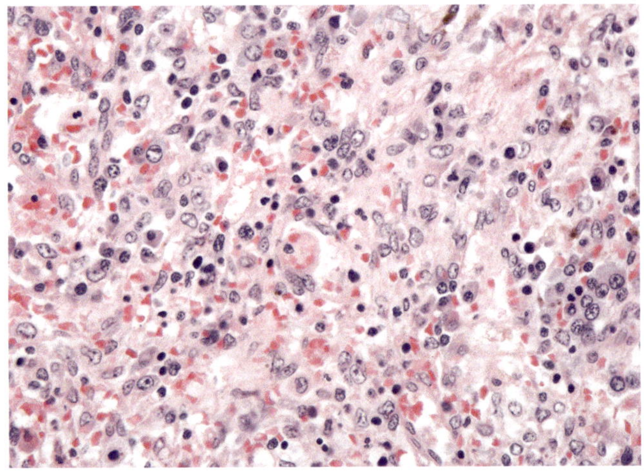

Fig. 6.101 Hemophagocytic syndrome. In a higher-magnification image, there is evidence of hemophagocytosis. No etiology is apparent. In cases of infection-associated hemophagocytic syndrome, Epstein-Barr virus (EBV) is often the underlying etiology

6.13 Autoimmune Disorders

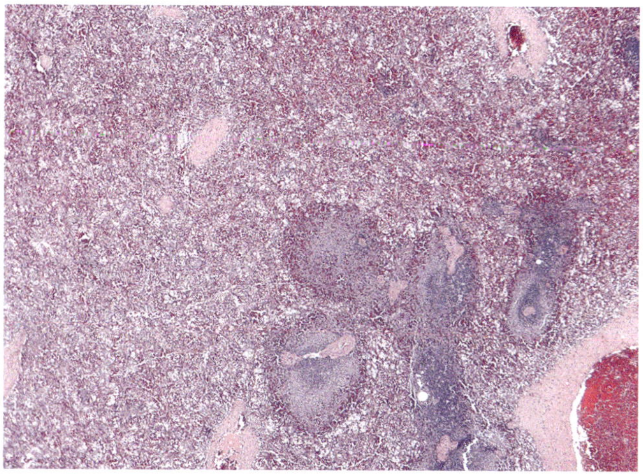

Fig. 6.102 Immune thrombocytopenic purpura (ITP). Low-magnification image of the spleen removed for ITP. In most cases, splenectomy is performed after several attempts at nonsurgical therapy. In this case, the disease is unmodified by therapy. There is a hyperplasia of white pulp structures (typical of an autoimmune disorder), with an increase in foamy macrophages in the white pulp, due to immune-mediated ingestion of platelets

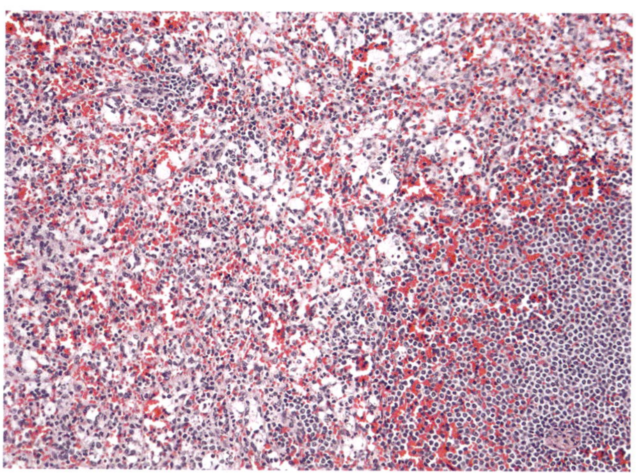

Fig. 6.103 Immune thrombocytopenic purpura. Intermediate magnification of ITP without previous therapy shows white pulp (*lower right*) with an increase in foamy macrophages in the red pulp. Touch preparations on a fresh specimen would show platelets inside of macrophages. It should be noted that this finding is only typical in cases of untreated disease

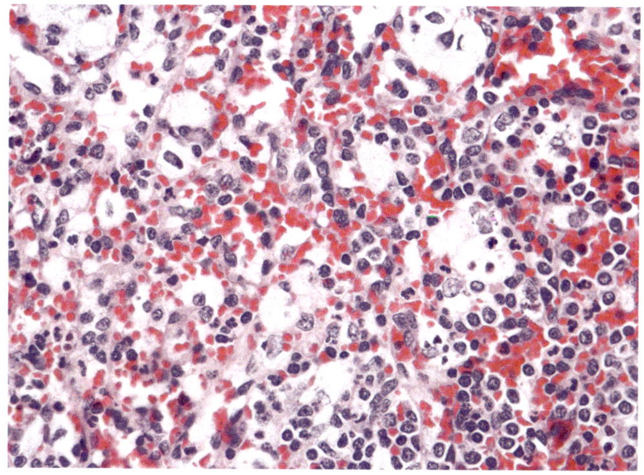

Fig. 6.104 Immune thrombocytopenic purpura. High-magnification image of foamy macrophages of untreated ITP. Note the macrophages are present in the splenic cords

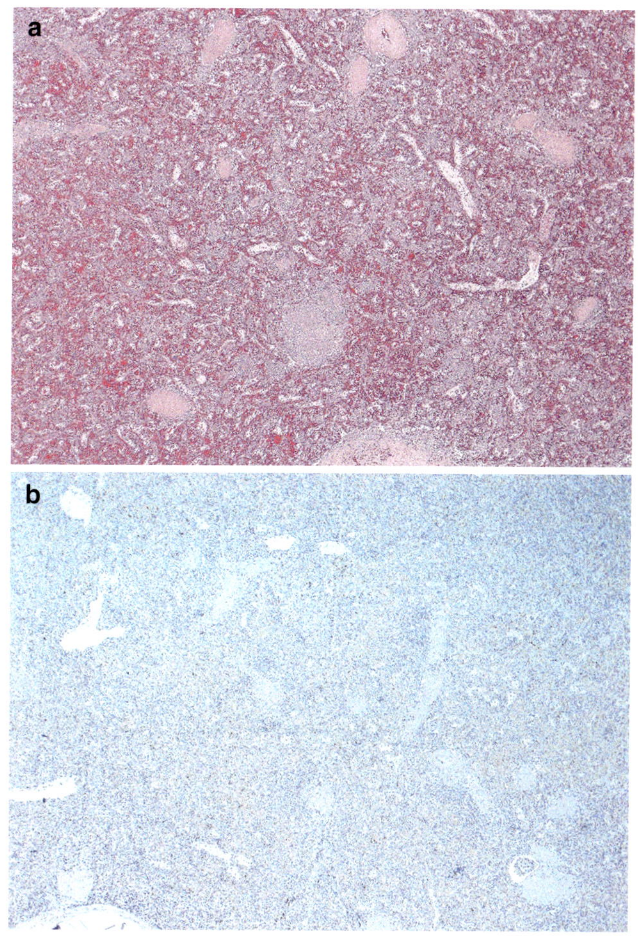

Fig. 6.105 Refractory immune thrombocytopenic purpura, H&E (**a**), CD20 immunohistochemical stain (**b**). Histologic and immunohistochemical staining for refractory ITP. Refractory ITP is resistant to several therapies and usually results in splenectomy as a therapeutic measure. In this case, the patient was treated with rituximab, an anti-CD20 therapy, which depleted the normal numbers of B-cells present in the spleen

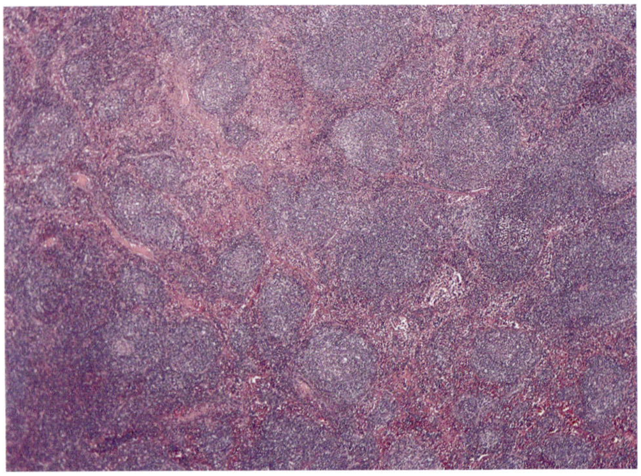

Fig. 6.106 Immune thrombocytopenic purpura. An example of white pulp lymphoid hyperplasia associated with ITP. As mentioned previously, multiple therapies are often attempted before splenectomy, so the "unmodified" appearance of ITP pathology is rare

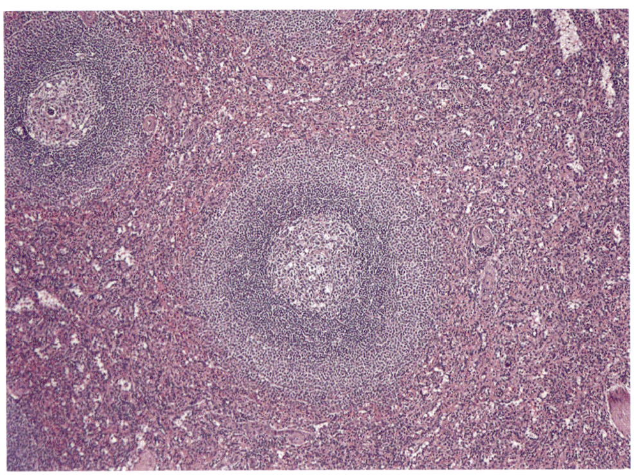

Fig. 6.107 Immune thrombocytopenic purpura. Another common finding in treated ITP is marginal zone hyperplasia. This may be seen with or without hyperplasia of germinal centers

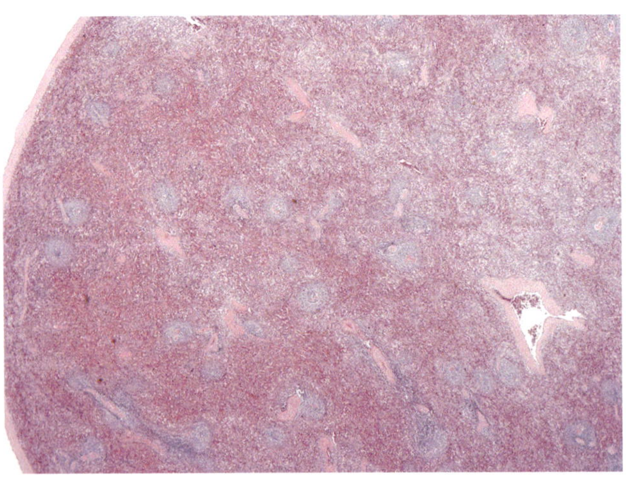

Fig. 6.108 Immune thrombocytopenic purpura. Low-magnification image of treated ITP. There is an overall increase in the amount of white pulp present but reactive germinal centers are not seen. This finding can be seen in ITP treated with corticosteroids

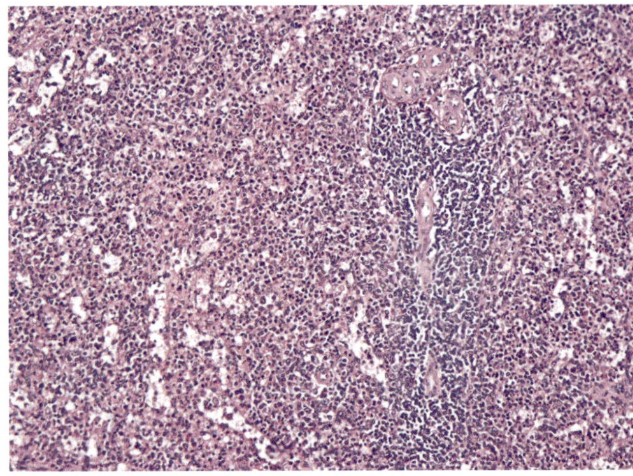

Fig. 6.109 Immune thrombocytopenic purpura with rheumatoid arthritis. In this case, two autoimmune disorders, rheumatoid arthritis and ITP, are seen in the same patient. The spleen shows features of treated ITP, including diminution of B-cell structures. A finding typical of rheumatoid arthritis is the presence of increased plasma cells in the red pulp

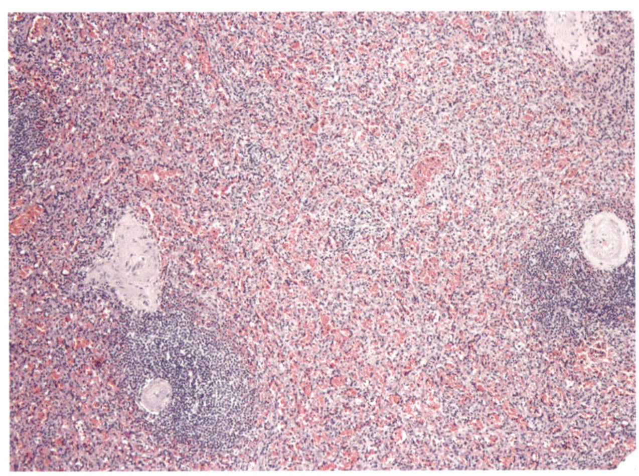

Fig. 6.110 Autoimmune hemolytic anemia. Low-magnification image of spleen in a patient with autoimmune hemolytic anemia. White pulp structures appear normal but there is an expansion of red pulp, with some vascular congestion and an increase in nucleated elements within red pulp structures

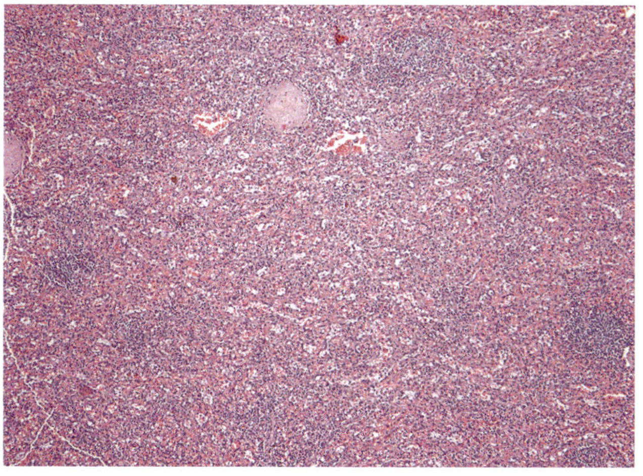

Fig. 6.111 Autoimmune hemolytic anemia. Another low-magnification image of spleen in a patient with autoimmune hemolytic anemia. In this case, white pulp structures are diminished in size, suggesting the possibility of previous immunosuppressive therapies such as steroids. In this case, there is also red pulp congestion and increased nucleated elements in the red pulp

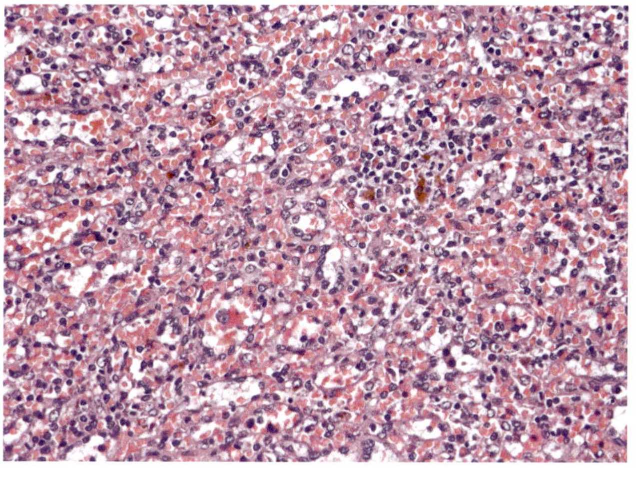

Fig. 6.112 Autoimmune hemolytic anemia. Intermediate-magnification image of the spleen in a patient with autoimmune hemolytic anemia. Red pulp shows congestion of cords and increased number of red blood cells. There is hemosiderin deposition, indicative of red blood cell destruction

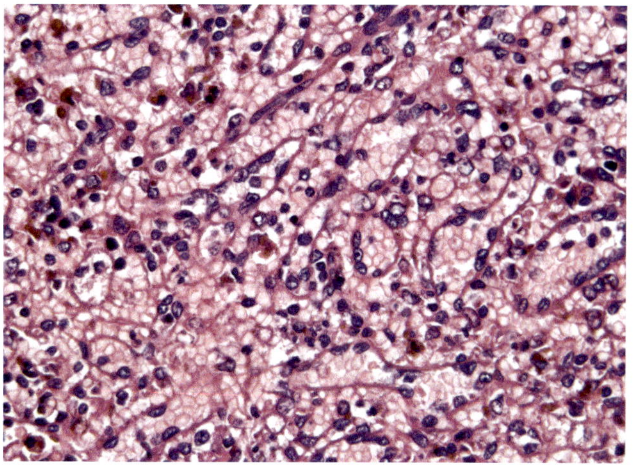

Fig. 6.113 Autoimmune hemolytic anemia. High-magnification image of spleen with autoimmune hemolytic anemia. In this image, the individual red blood cells (pale cells with pink cytoplasmic borders) are seen filling red pulp sinuses and cords

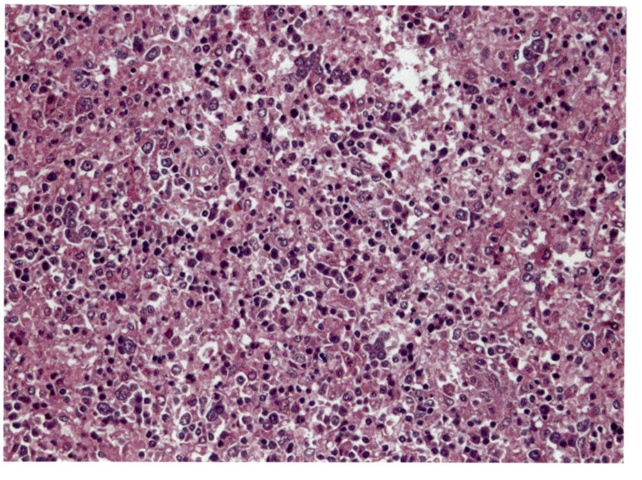

Fig. 6.114 Autoimmune hemolytic anemia. Spleen in a patient with autoimmune hemolytic anemia. In this patient, besides the presence of red pulp congestion, there is significant extramedullary hematopoiesis consisting of erythroid precursors and some myeloid elements. This may be due to compensatory hyperplasia of erythroid elements and crowding of normal hematopoietic elements in the marrow space

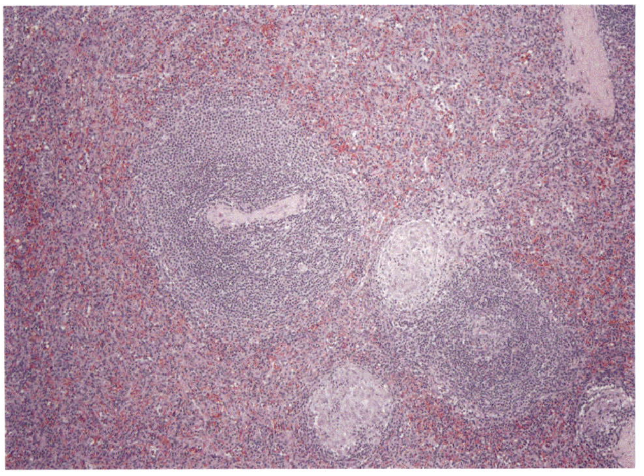

Fig. 6.115 Sarcoidosis. Low-magnification of spleen in a patient with sarcoidosis. Sarcoidal granulomas can be distributed throughout the spleen. They are formed from epithelioid histiocytes and often have sharp, well-defined borders

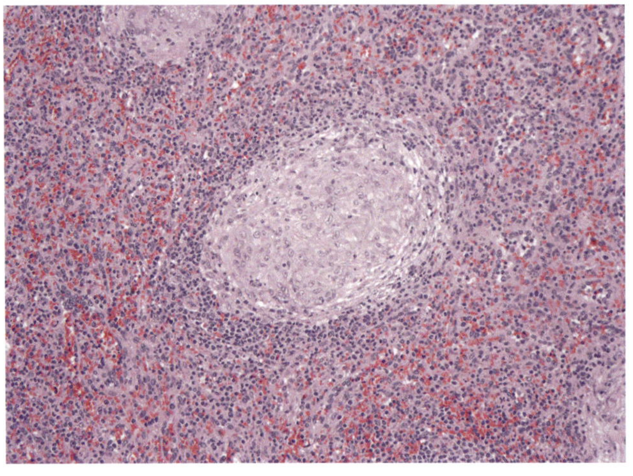

Fig. 6.116 Sarcoidosis. Granulomas of sarcoid can often replace germinal centers, as in this example

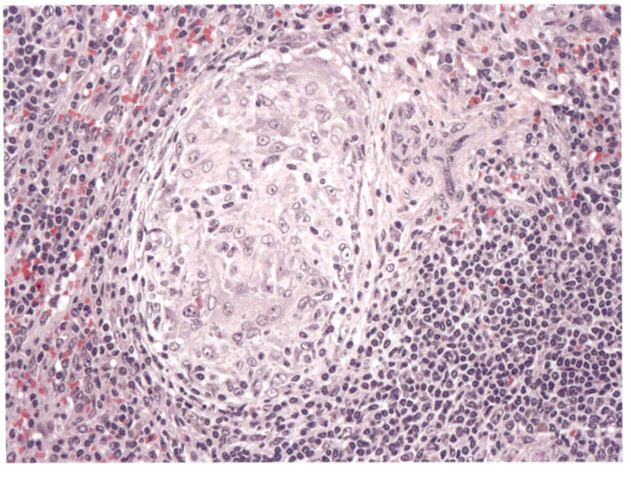

Fig. 6.117 Sarcoidosis. The epithelioid histiocytes of sarcoidal granulomas have bland nuclei and voluminous pink cytoplasm. Occasional infiltrating lymphocytes (T cells) are seen

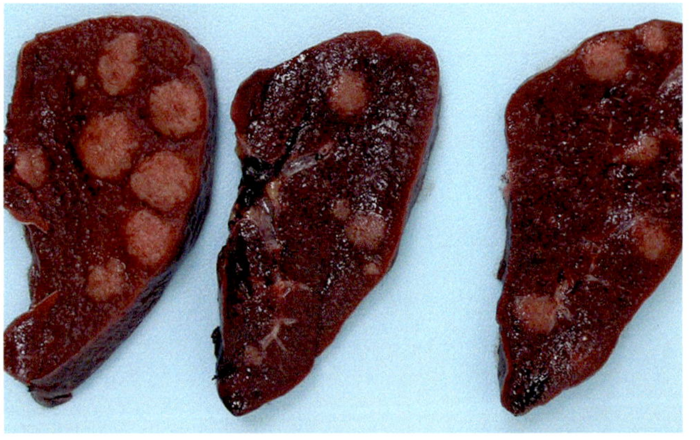

Fig. 6.118 Sarcoidosis. Gross image of sectioned spleen with numerous sarcoid-associated granulomas. These can vary widely in size and the extent of splenic involvement can be highly variable. (*From* O'Malley et al. [1]; with permission.)

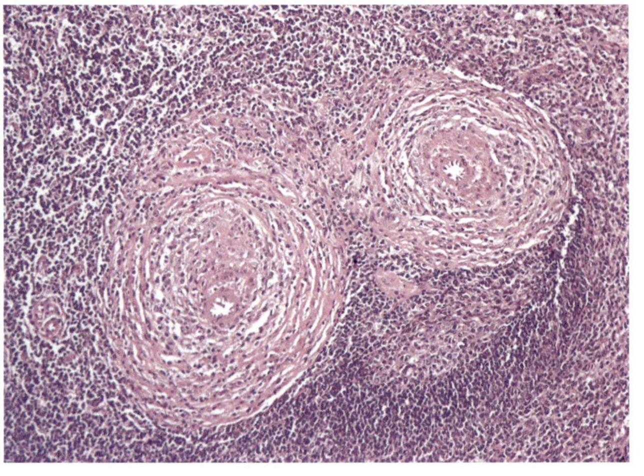

Fig. 6.119 Systemic lupus. Vascular change is the spleen in a case of systemic lupus. These changes are termed "onion-skinning" of the vessels, describing the laminated thickening of the vascular walls. This finding is fairly common in the spleen of systemic lupus but is not entirely specific

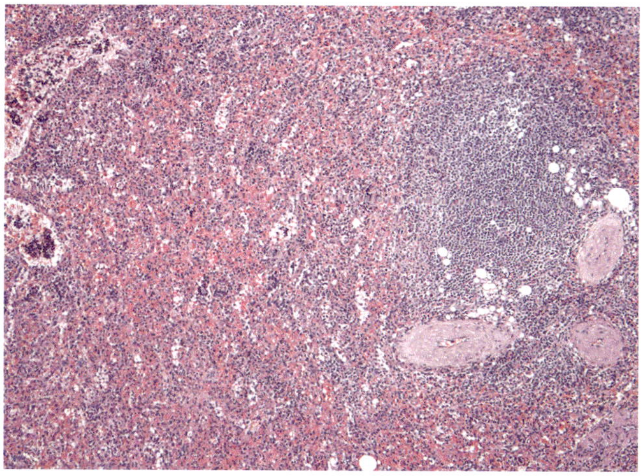

Fig. 6.120 Hemolytic-uremic syndrome. Hemolytic-uremic syndrome is a systemic activation of the coagulation system due to a variety of etiologies. Splenic changes are common but are nonspecific. These include infarcts, vascular congestion, and extramedullary hematopoiesis. In this case, some dilated sinuses are seen, as well as clusters of erythroid precursors and some megakaryocytes. A few unrelated lipoid granulomas are also seen in this case

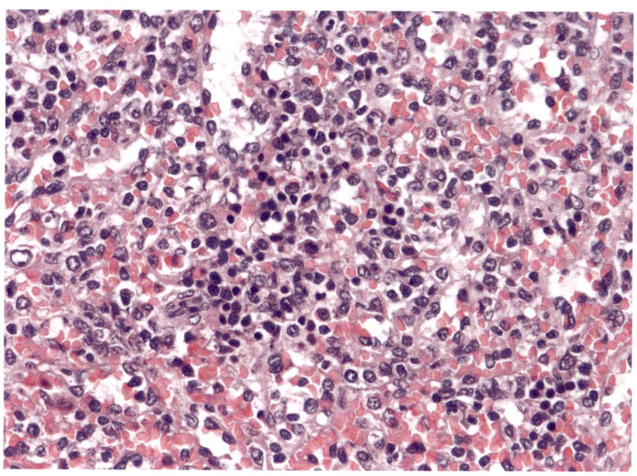

Fig. 6.121 Hemolytic-uremic syndrome. In this high-magnification image of red pulp in hemolytic-uremic syndrome, numerous red blood cell precursors are seen

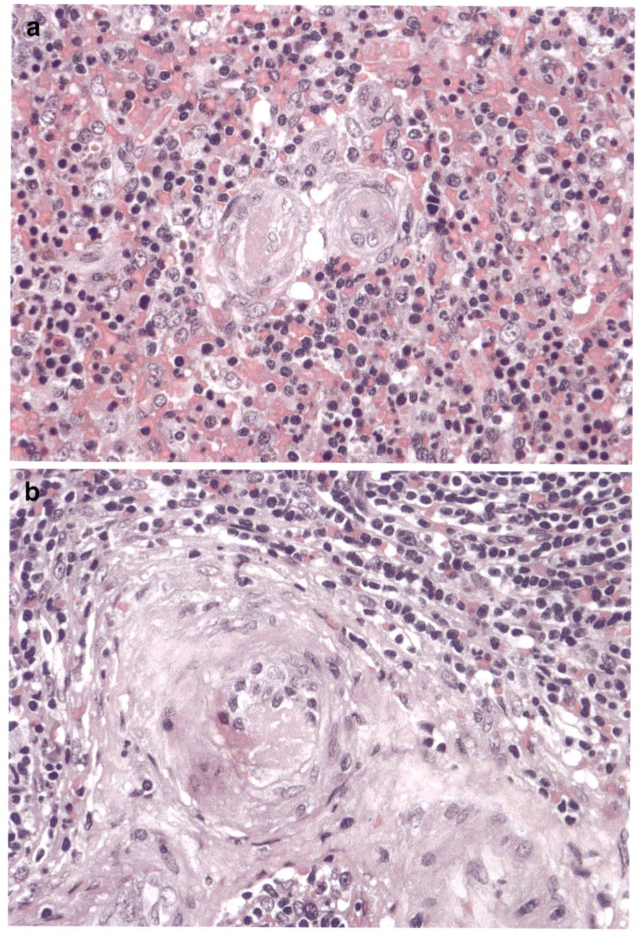

Fig. 6.122 Thrombotic-thrombocytopenic purpura (TTP). **a** and **b**, Examples of small thrombi formed in splenic arterioles in TTP

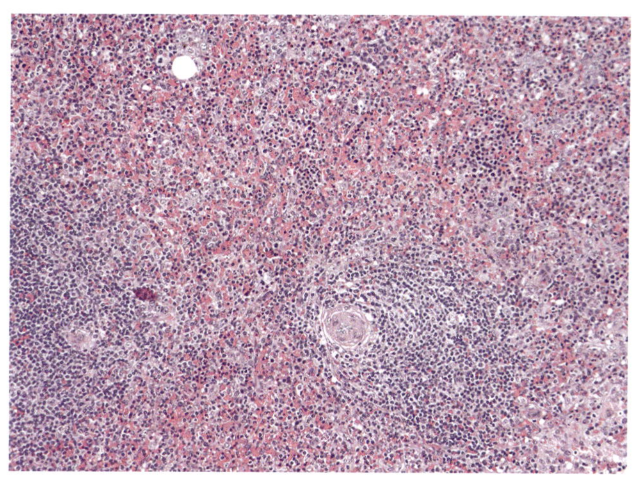

Fig. 6.123 Thrombotic-thrombocytopenic purpura. A low-magnification image of spleen showing white and red pulp components. In this case, there is evidence of nucleated red blood cell precursors in the red pulp. In addition, there are scattered immature appearing mononuclear cells. These are immature myeloid elements

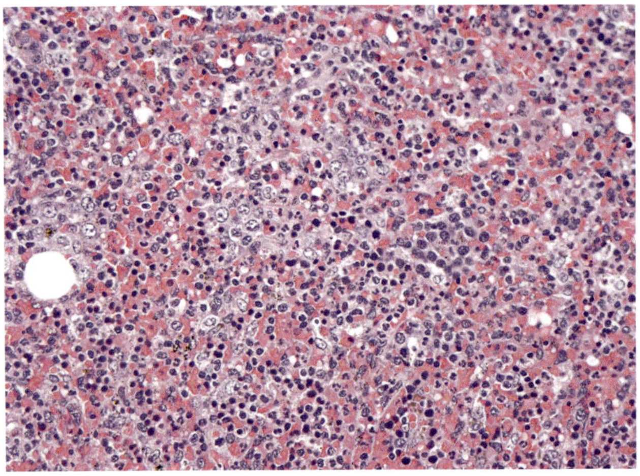

Fig. 6.124 Thrombotic-thrombocytopenic purpura. Higher-magnification image of a spleen with TTP. In the red pulp, there is an admixture of nucleated red blood cell precursors and immature myeloid elements with open chromatin, irregular nuclei, and moderate amounts of pale to pink cytoplasm. These myeloid elements are promyelocytes and myelocytes, not myeloblasts

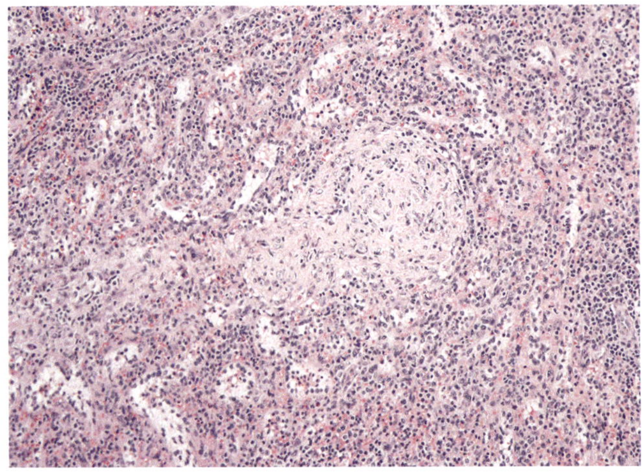

Fig. 6.126 Polyarteritis nodosa. Polyarteritis nodosa with thickening of splenic arterioles. Note that the adjacent red pulp is unaffected

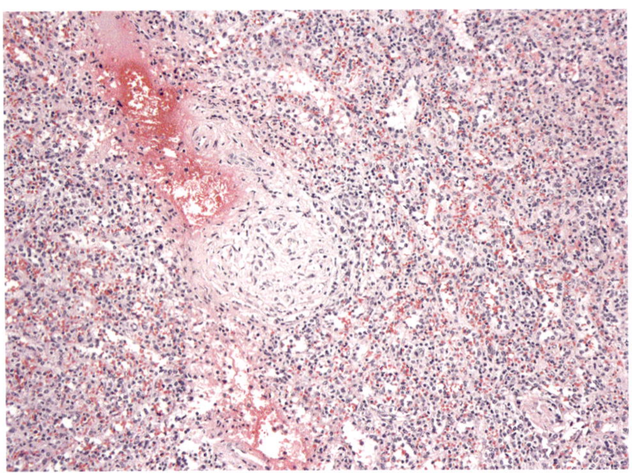

Fig. 6.125 Polyarteritis nodosa. Polyarteritis nodosa involving the spleen. The medial portions of splenic arteries/arterioles are markedly expanded

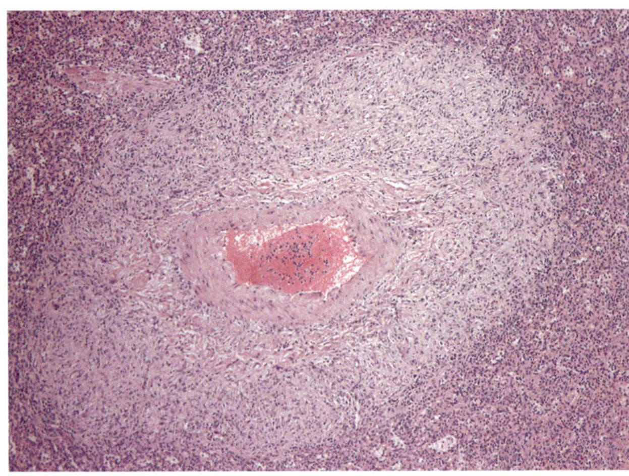

Fig. 6.127 Polyarteritis nodosa. A massively enlarged artery in a case of polyarteritis nodosa in the spleen. All components of the vessel are enlarged, with the adventitial layer showing significant expansion

6.14 Immunodeficiency (Inherited and Acquired)

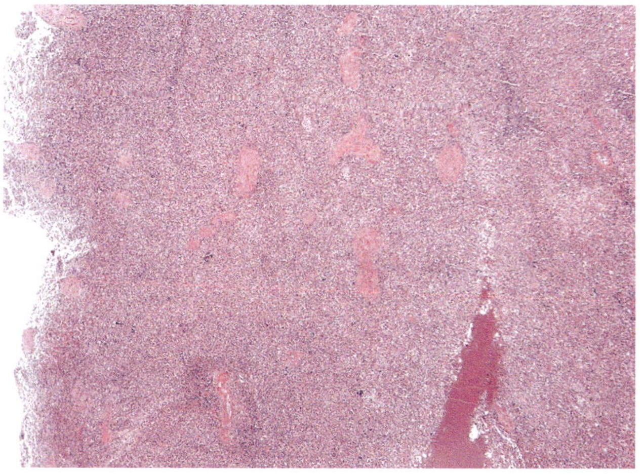

Fig. 6.128 Nezelof syndrome. Part of the spectrum of severe combined immunodeficiency (SCID), the spleen in Nezelof syndrome shows significant depletion of white pulp elements. B-cell nodules are absent, and only minimal T-cells areas (PALS regions) are seen. (*Slide courtesy of* T. George, Palo Alto, CA.)

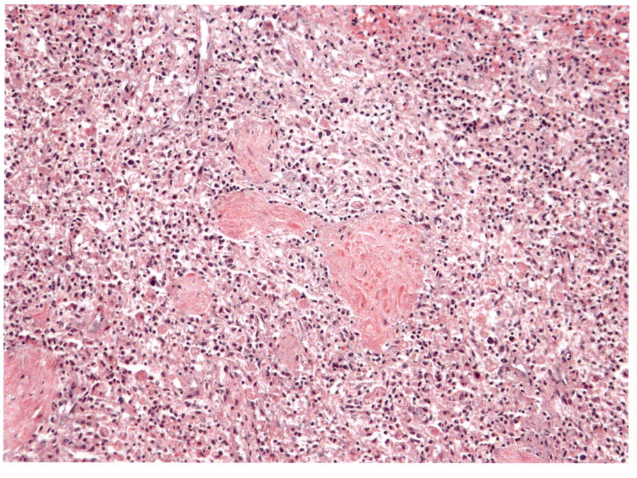

Fig. 6.129 Nezelof syndrome. High-magnification image of Nezelof syndrome, showing a near complete absence of lymphocytes in PALS regions. (*Slide courtesy of* T. George, Palo Alto, CA.)

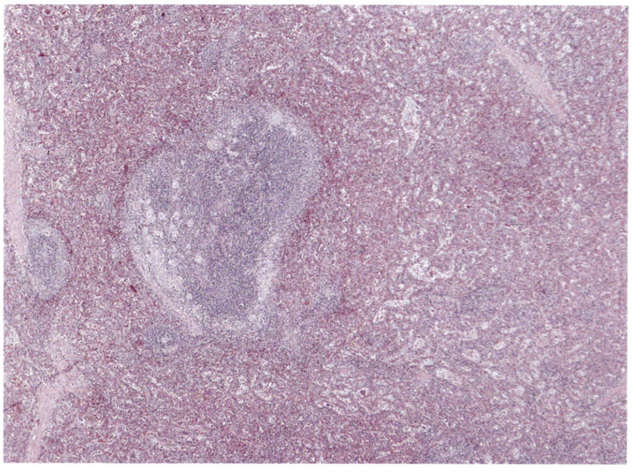

Fig. 6.130 DiGeorge syndrome. DiGeorge syndrome has thymic hypoplasia and associated severe T-cell defects. In this example, white pulp nodules (B-cell areas) are enlarged but PALS regions (T-cell areas) are absent. Note that there are granulomas present in the perifollicular area

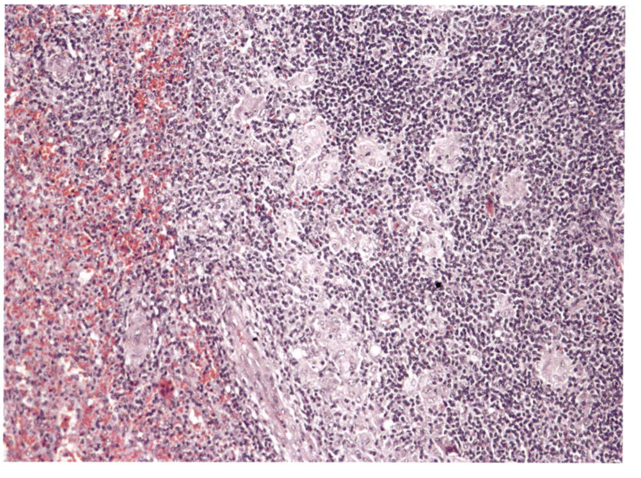

Fig. 6.131 DiGeorge syndrome. Granulomas in the spleen of a patient with DiGeorge syndrome may be primary or may be due to infections. Because of the severe immunologic defect, these patients may be susceptible to opportunistic infections

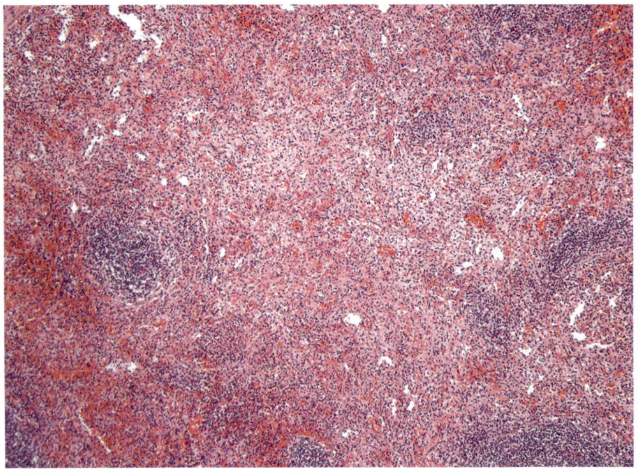

Fig. 6.132 Omenn syndrome. Low-magnification image of Omenn syndrome, another type of severe immunodeficiency. In Omenn syndrome, there is a reduction in white pulp elements, as seen in this case

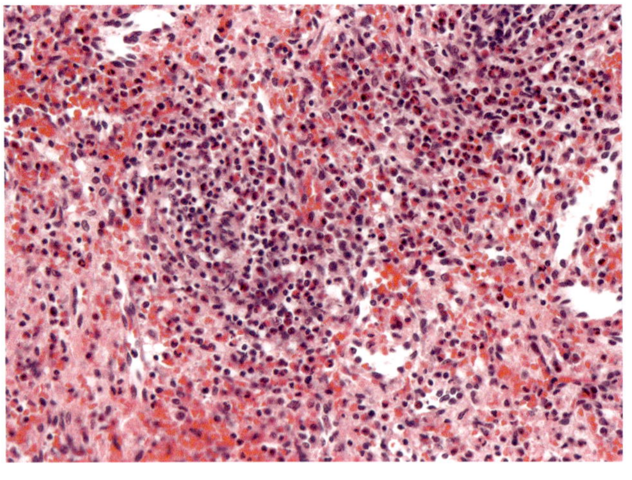

Fig. 6.133 Omenn syndrome. In addition to a general depletion of lymphocytes, there is an increase in eosinophils seen in Omenn syndrome. This is accompanied by elevations in serum IgE

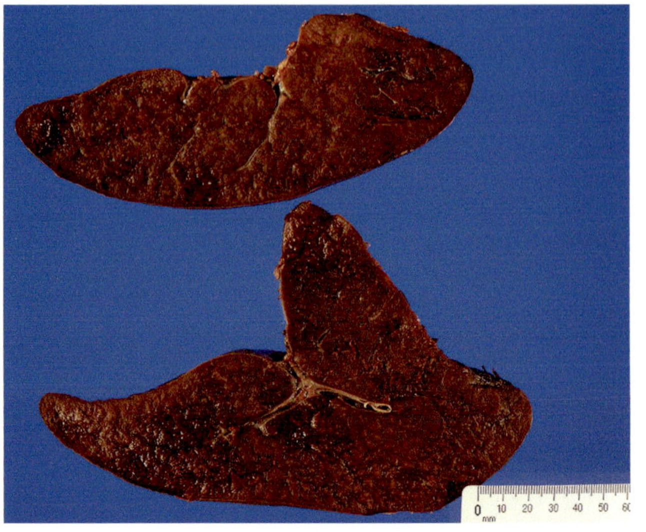

Fig. 6.134 Autoimmune lymphoproliferative syndrome (ALPS). Gross image of spleen in patient with ALPS. Although relatively normal in size, there is a subtle enlargement of white pulp structures. Prominent follicular hyperplasia is one feature that can be seen in ALPS. (*Image courtesy of* J. Sidhu, Johnson City, NY.)

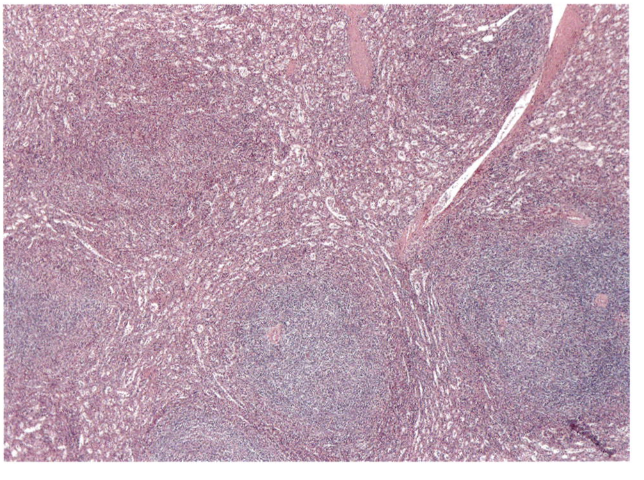

Fig. 6.135 Autoimmune lymphoproliferative syndrome. ALPS is associated with splenomegaly. They will often have a hyperplasia of lymphoid components in the spleen. In this case, germinal centers are enlarged. By flow cytometry, T cells that lack both CD4 and CD8 can be identified

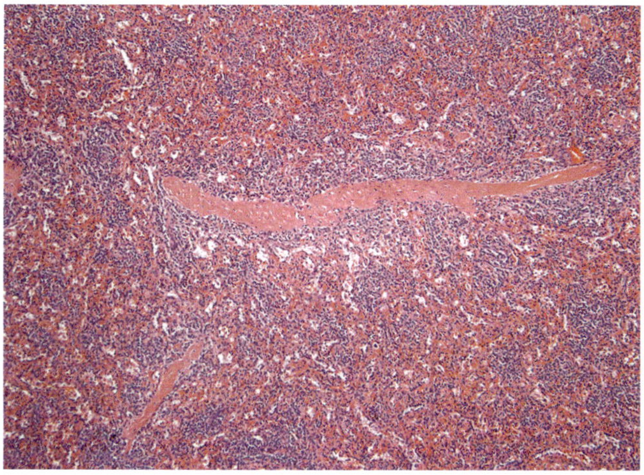

Fig. 6.136 Autoimmune lymphoproliferative syndrome. Another example of ALPS involving the spleen. There is an overall increase in white pulp structures. In this case, there is a marked increase in the amounts and size of PALS regions (T-cell areas)

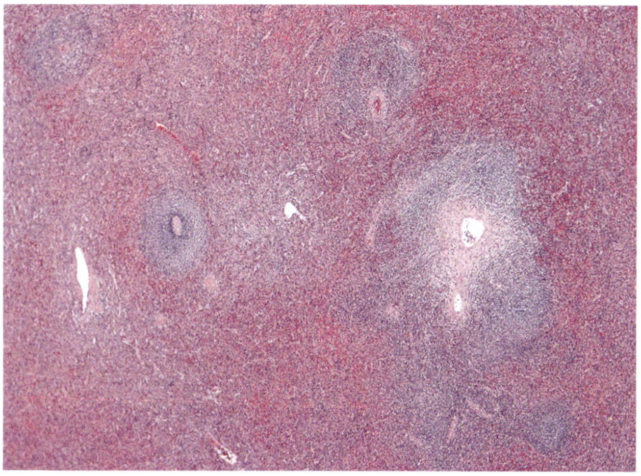

Fig. 6.138 Common variable immunodeficiency (CVID). As suggested by the name, the findings in CVID can differ in individual cases. In this example, there is a relatively normal appearance of the splenic histology

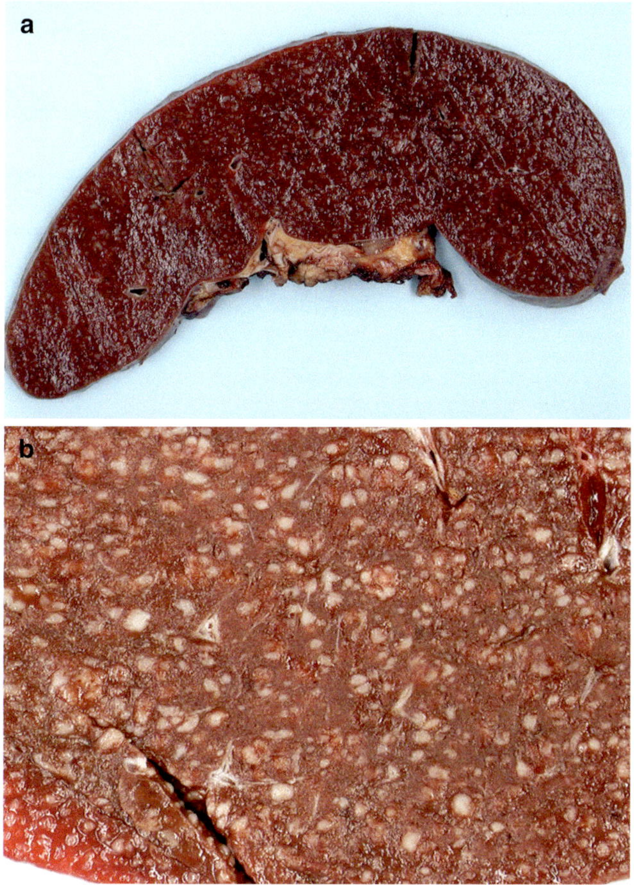

Fig. 6.137 Common variable immunodeficiency. In this gross image, the spleen of common variable immunodeficiency is markedly enlarged (**a**). Higher-magnification image of the cut surface (**b**) shows an overall increase in white pulp structures, imparting a miliary appearance and mimicking low-grade lymphoma. (*From* O'Malley et al. [1]; with permission.)

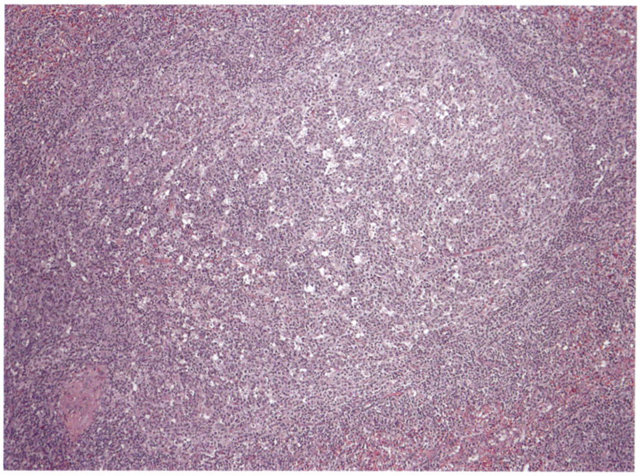

Fig. 6.139 Common variable immunodeficiency. In some cases of CVID there can be prominent follicular hyperplasias, as is seen in this case

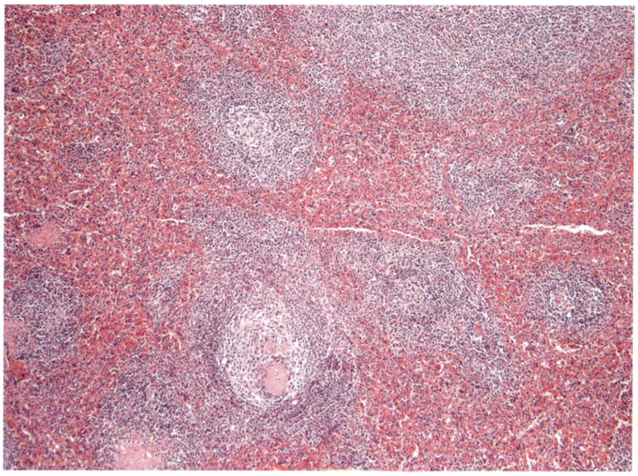

Fig. 6.140 Common variable immunodeficiency. In this example of CVID, there is expansion of both lymphoid follicles (B cell) and increased PALS regions (T-cell areas)

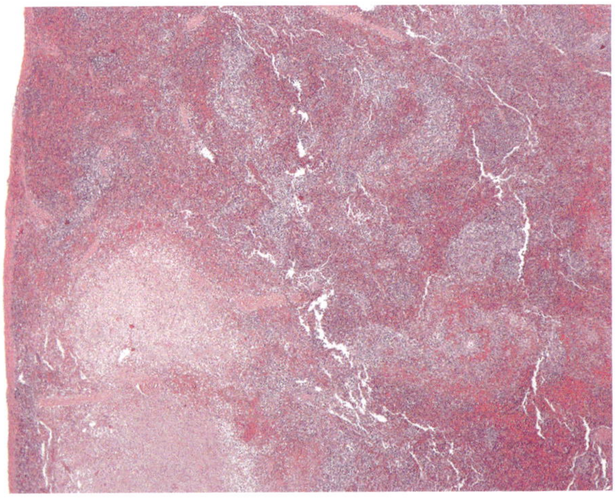

Fig. 6.141 X-linked lymphoproliferative syndrome (XLP). XLP, also known as Duncan disease, is a rare immunodeficiency associated with a lack of appropriate responses to EBV infection. In this example, the spleen shows large areas of necrosis, with lymphoid depletion and vascular congestion

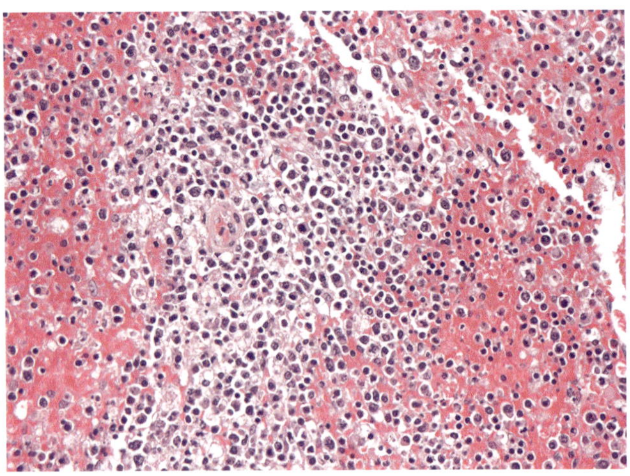

Fig. 6.142 X-linked lymphoproliferative syndrome. High-magnification image of the spleen in a patient with XLP. There are many plasma cells present in lymphoid areas, with admixed B- and T-cells

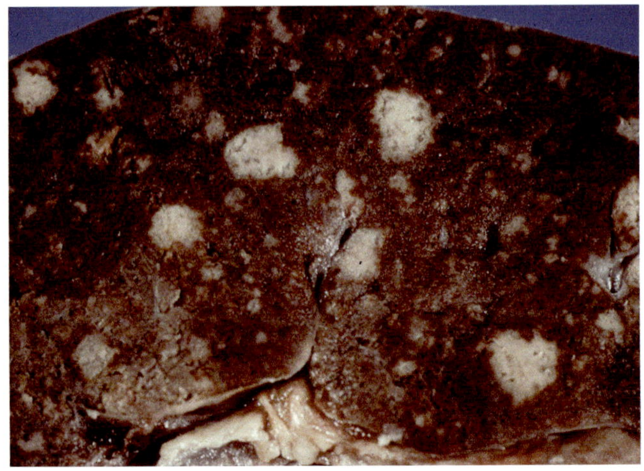

Fig. 6.143 Chronic granulomatous disease. Gross image of chronic granulomatous disease. A primary immunodeficiency, these patients have an impaired neutrophils response to certain infections. There is often a combination of both follicular hyperplasia (as highlighted by the increase in small lymphoid nodules) and a granulomatous response. The granulomas are variable in size, including the larger structures seen, with irregular borders and a pale appearance. (*Image courtesy of* D. Farhi, Atlanta, GA.)

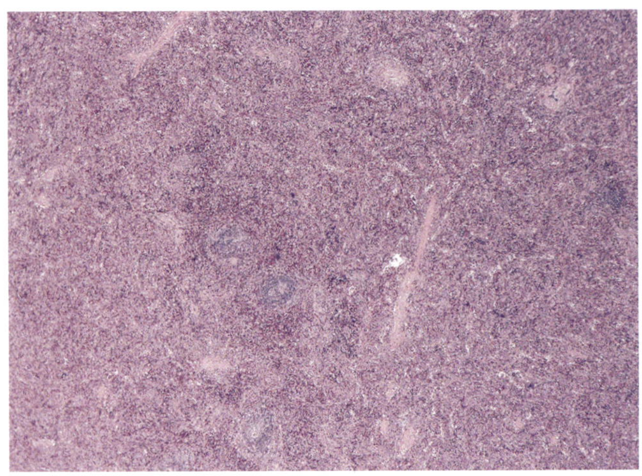

Fig. 6.144 Wiskott-Aldrich Syndrome. Wiskott-Aldrich syndrome is an X-linked immunodeficiency associated with increased development of lymphoproliferative disorders. In this low-magnification image, there is a depletion of white pulp structures in the spleen

Reference

1. O'Malley DP, George TI, Orazi A, Abbondanzo S. Armed Forces Institute of Pathology: Benign and reactive conditions of lymph node and spleen [book]. Silver Spring, MD: American Registry of Pathology Press; 2009.

7 Infections

In this chapter, the broad range of infections and their histologic effects in the spleen are illustrated. A broad variety of infections occur in the spleen representative of bacterial, mycobacterial, fungal, viral, and protozoal/parasitic infections. The variety of bacterial and mycobacterial effects illustrated in the spleen includes generalized bacterial abscess and infections of the family of mycobacteria including tuberculosis, leprosy, and *Mycobacterium avium-intracellulare*. The relatively common features of infectious mononucleosis, due to Epstein-Barr virus (EBV), are illustrated. Other viruses, including *Cytomegalovirus*, Herpes simplex virus, and HIV, are illustrated. A variety of fungal and protozoal infections, most often seen in immunocompromised patients, are illustrated, including *Pneumocystis* and *Cryptococcus*. Other fungal infections are also illustrated. Finally, protozoal infections common in the underdeveloped world, malaria and leishmaniasis, are illustrated.

D.P. O'Malley, *Atlas of Spleen Pathology*, Atlas of Anatomic Pathology,
DOI 10.1007/978-1-4614-4672-9_7,

7.1 Bacterial

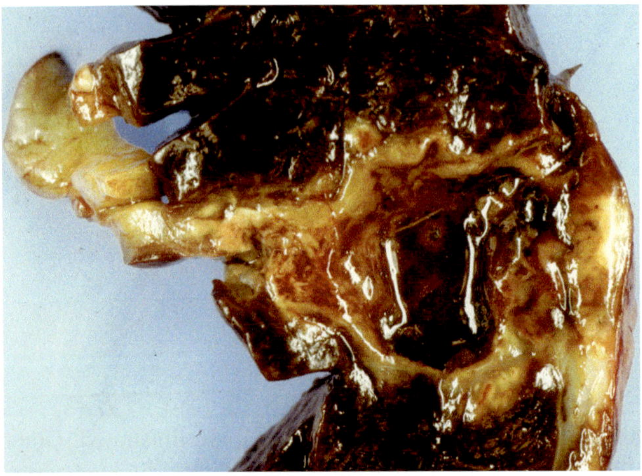

Fig. 7.1 Bacterial abscess. Gross image of a sectioned spleen with an *Escherichia coli* abscess. The abscess has a liquefied center with a rim of necrotic yellow material. (*Image courtesy of* D. Farhi, Atlanta, GA.)

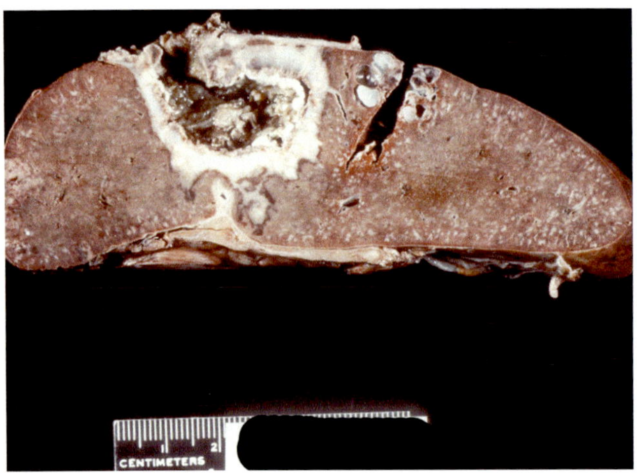

Fig. 7.2 Bacterial abscess. Gross image of a splenic abscess with subsequent infarction. In this case, the spleen has been fixed. There is a core of necrotic debris. In contrast to a recent infection, there is a fibrous capsule formed around the abscess. (*Image courtesy of* D. Farhi, Atlanta, GA.)

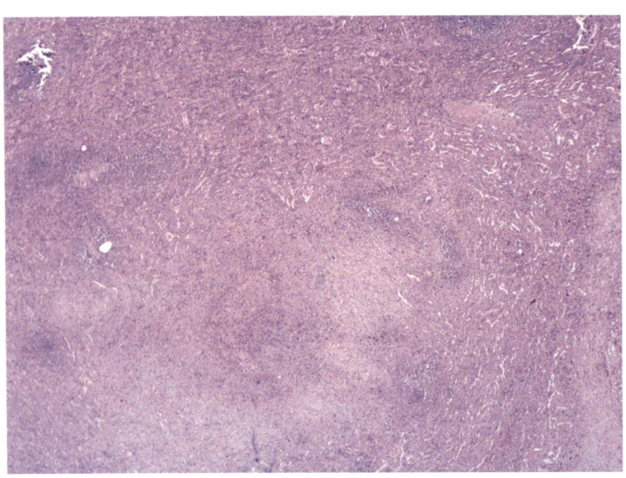

Fig. 7.3 Actinomyces. Splenic infection with *Actinomyces.* An area of necrosis associated with abscess is seen

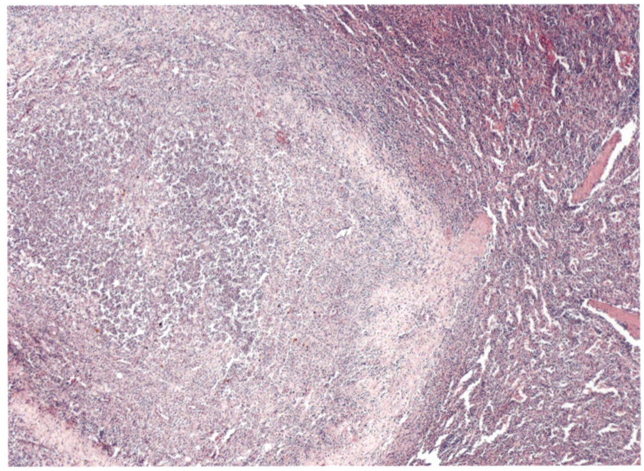

Fig. 7.4 Typhoid. Infection of the spleen in a patient with *Salmonella typhi.* A large area of necrosis with neutrophils (abscess formation) is identified

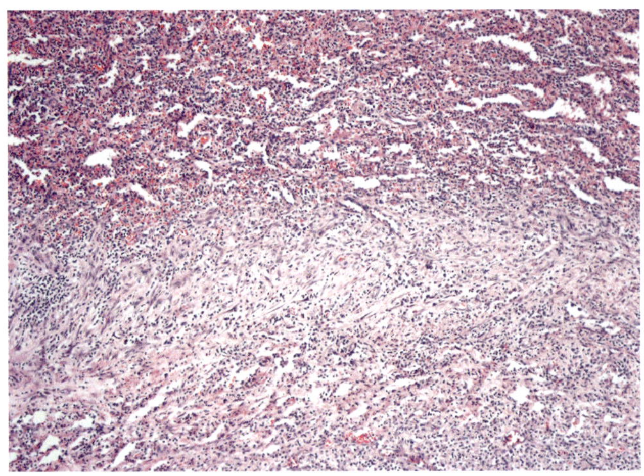

Fig. 7.5 Typhoid. In this spleen infected with typhoid, the area of necrosis is filled with neutrophils and cellular debris. Individual bacteria are not recognizable

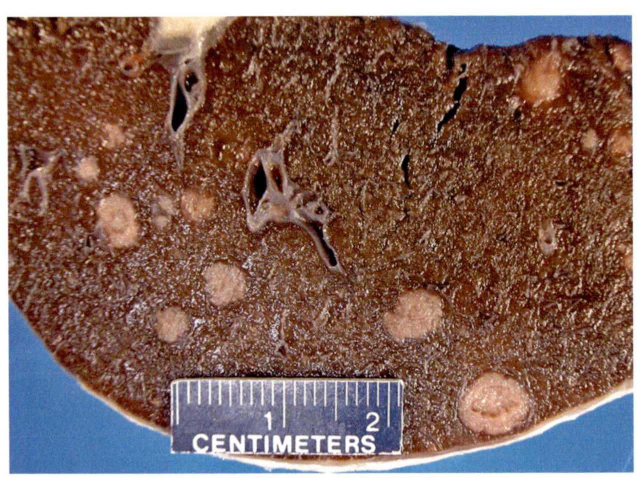

Fig. 7.6 Tularemia. Tularemia caused by *Francisella tularensis* in the spleen. In this case, granulomatous nodules are present. (*From* O'Malley et al. [1]; with permission.)

7.2 Mycobacteria

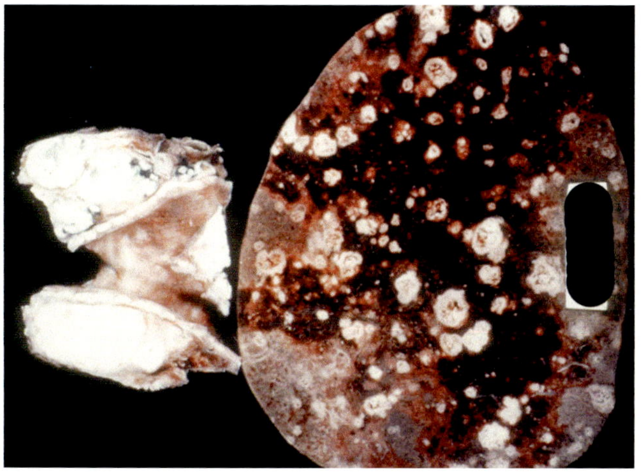

Fig. 7.7 Tuberculosis. Gross image of a spleen with massive involvement by *Mycobacterium tuberculosis*. The granulomas seen show central caseation and necrosis. Multiple nodules are present, suggesting severe systemic infections. (*Image courtesy of* D. Farhi, Atlanta, GA.)

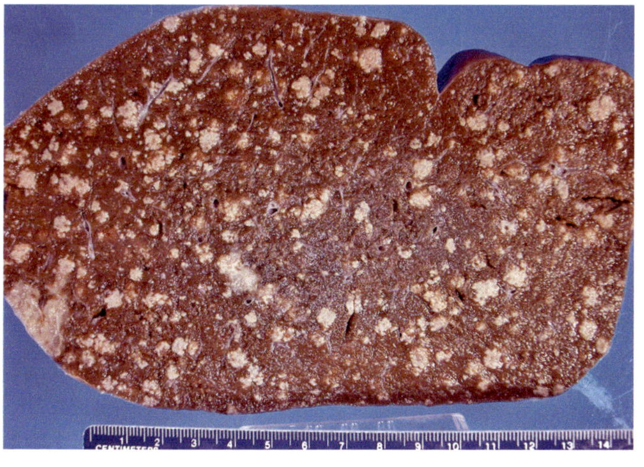

Fig. 7.8 Tuberculosis. Gross image of tuberculosis involving the spleen. Multiple granulomatous nodules are present and some have central necrosis or caseation. (*From* O'Malley et al. [1]; with permission.)

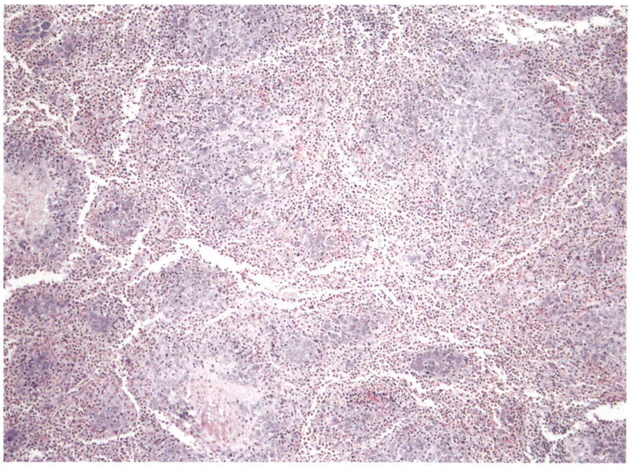

Fig. 7.9 *Mycobacterium avium-intracellulare* (*M-ai*). *M-ai* can be seen in the spleen as part of disseminated disease. In this case, there is severe and diffuse infection by *M-ai*. There are areas of necrosis present surrounded by gray-blue aggregates of organisms. Adjacent areas are composed of many neutrophils

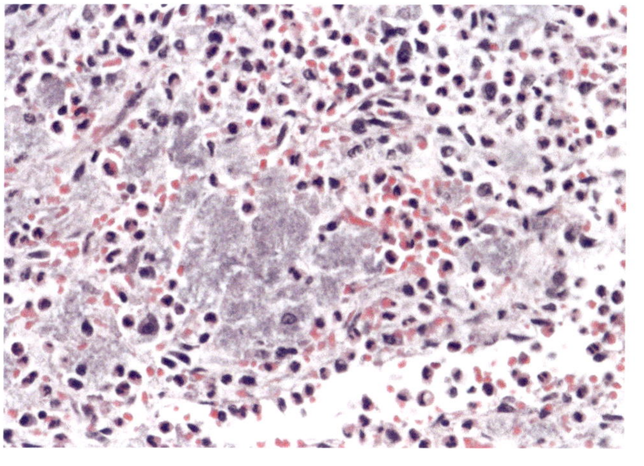

Fig. 7.10 *Mycobacterium avium-intracellulare*. In this high-magnification image, large clouds of mycobacteria are seen. The organisms are gray-blue in color. These would be positive for both acid-fast bacillus (AFB) stains and periodic acid-Schiff (PAS). *M. tuberculosis* would not be positive for PAS, but would be positive for AFB

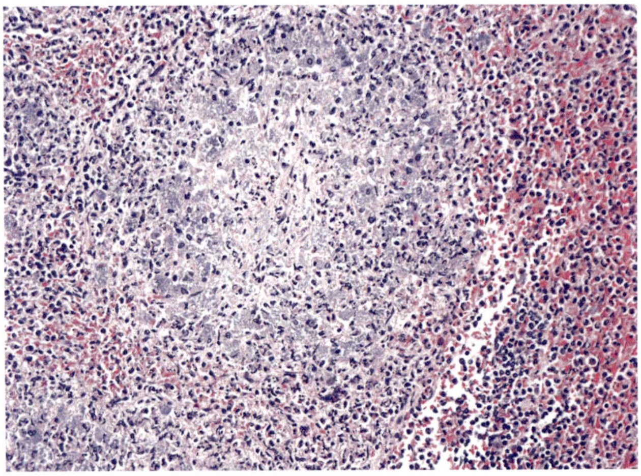

Fig. 7.11 *Mycobacterium avium-intracellulare*. A nodular aggregate of organisms with necrosis and partial granuloma formation within splenic red pulp

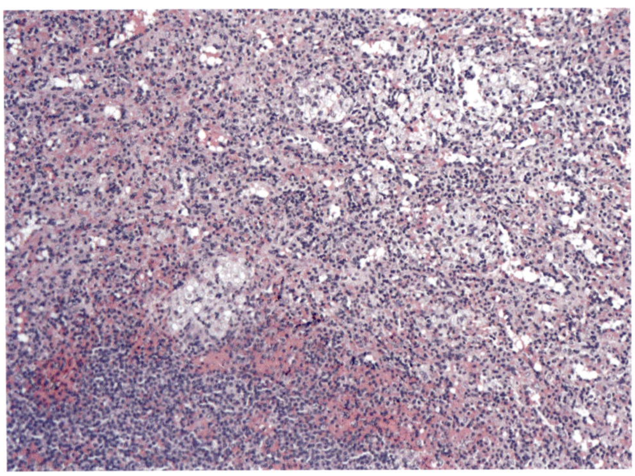

Fig. 7.12 Leprosy. Intermediate-magnification image of a spleen with involvement by leprosy. Granulomas and collections of foamy histiocytes (*leprae cells)* are seen. Although the location is not specific, granulomas are most often seen adjacent to vascular elements. The foamy macrophages contain many *Mycobacteria leprae* organisms

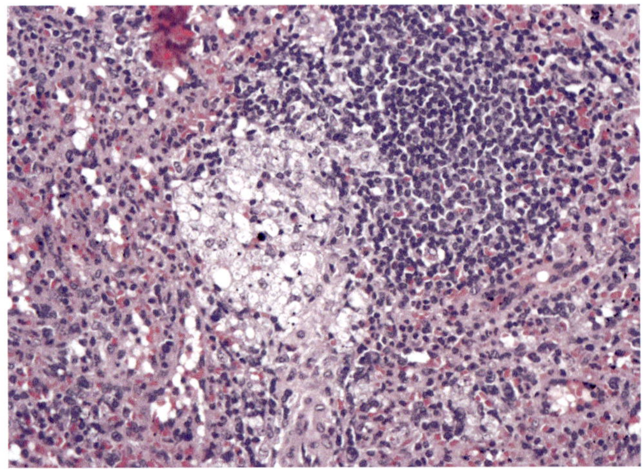

Fig. 7.13 Leprosy. High-magnification image of granulomas of leprosy in a spleen. The macrophages present have foamy or granular cytoplasm. A blood vessel is seen projecting into the bottom portion of the image

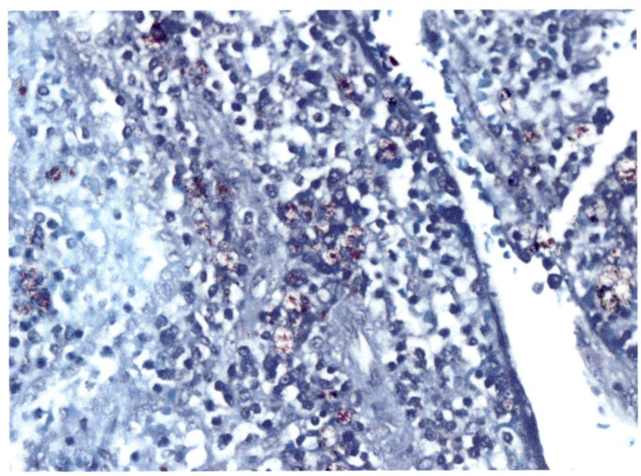

Fig. 7.14 Leprosy, AFB stain. AFB stain highlighting *Mycobacterium leprae* organisms

7.3 Viruses

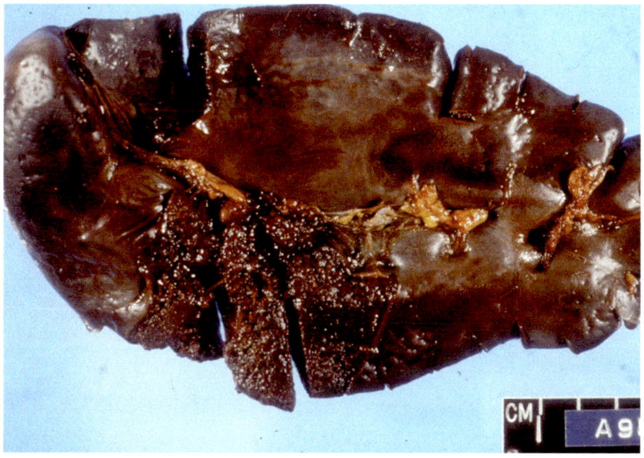

Fig. 7.15 Infectious mononucleosis. Gross image of a spleen with acute Epstein-Barr virus (EBV) infection. In this case, there is evidence of splenic rupture, which contributed to the patient's demise. Also, note the deep indentations along the edges of the spleen. These are a rare but normal congenital variation, where there is only partial fusion of the primordial lobes of the spleen, suggesting a sort of "incomplete" polysplenia. (*Image courtesy of* D. Farhi, Atlanta, GA.)

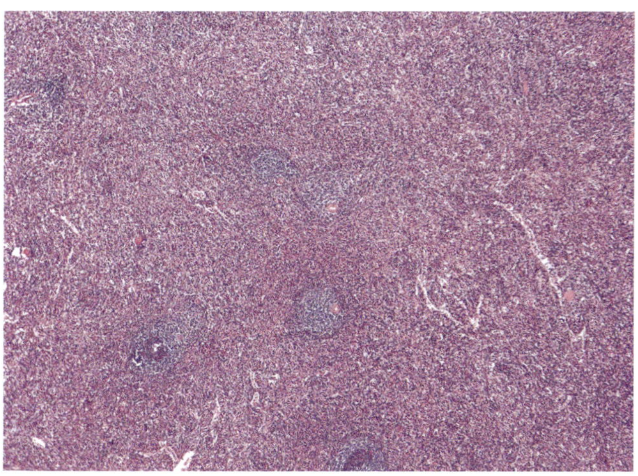

Fig. 7.16 Infectious mononucleosis. Low-magnification image of acute infection by EBV, also known as infectious mononucleosis. The overall splenic red and white pulp architecture is maintained. However, there is an overall increase in lymphocytes in the red pulp

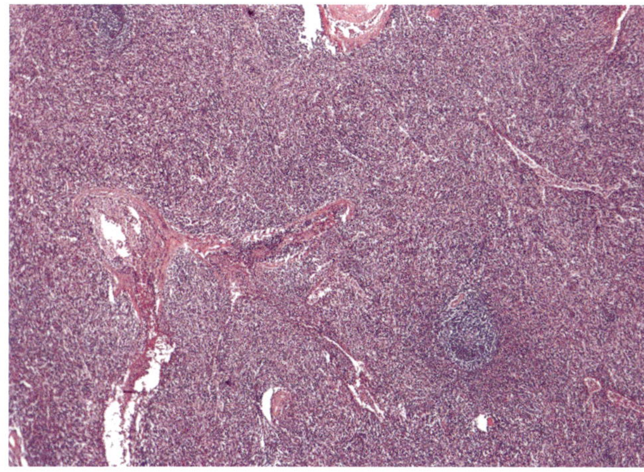

Fig. 7.17 Infectious mononucleosis. Another low-magnification image of infectious mononucleosis in the spleen. In this case, large caliber vessels are seen, and there is notable vascular invasion by lymphocytes

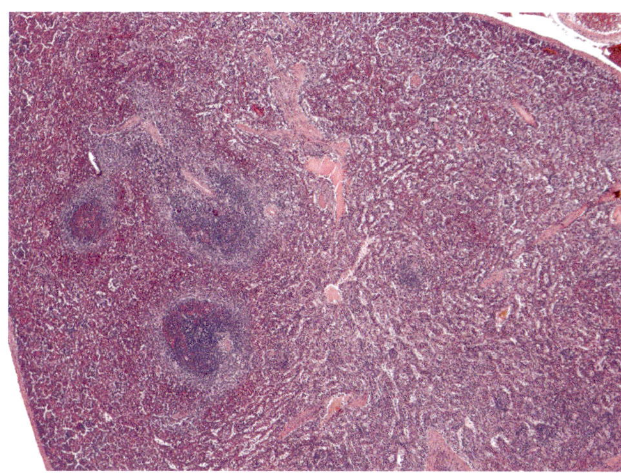

Fig. 7.18 Infectious mononucleosis. In this example of infectious mononucleosis, there is some follicular and marginal zone hyperplasia present, which is not always seen

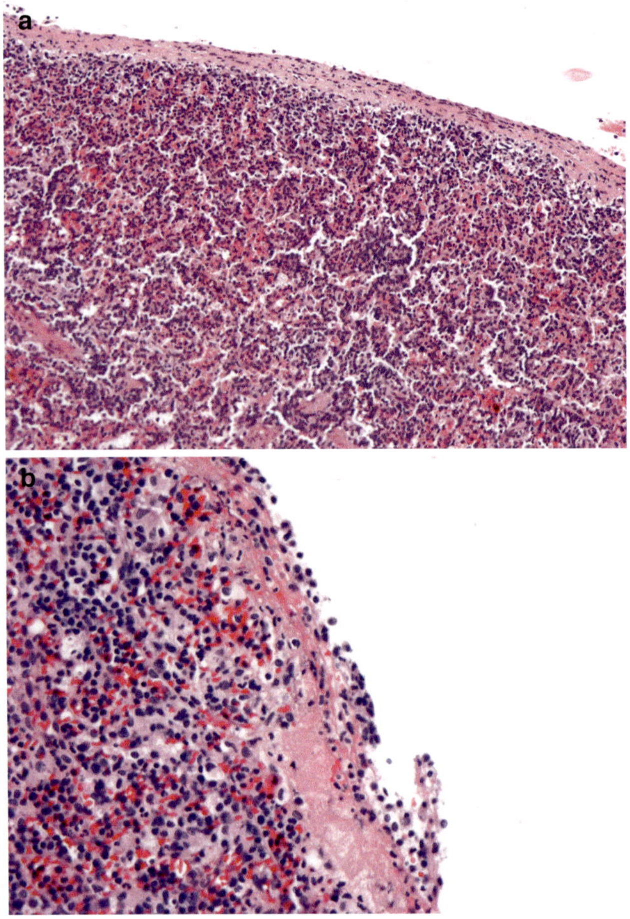

Fig. 7.19 Infectious mononucleosis. An underappreciated finding in infectious mononucleosis in the spleen, the capsule is thinned (**a** and **b**), with some infiltration by lymphocytes. This finding is the underlying cause of splenic ruptures associated with infectious mononucleosis. The thinned and weakened capsule is more prone to rupture after minimal trauma

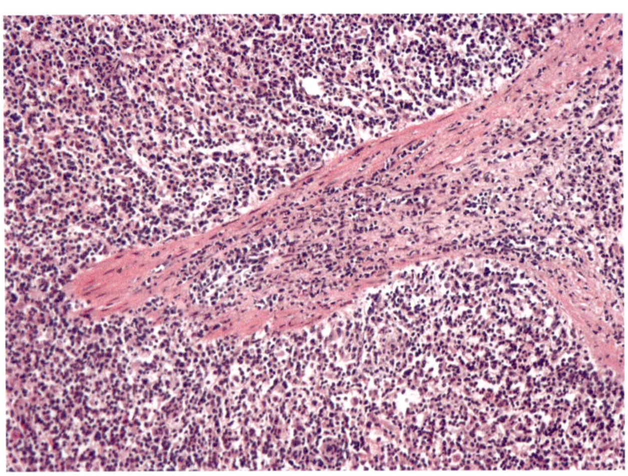

Fig. 7.20 Infectious mononucleosis. One almost pathognomonic finding in splenic infectious mononucleosis is a prominent vasculitis in large caliber vessels, usually arteries or large arterioles. The infiltrating cells are small and large lymphocytes, as well as some rare plasma cells

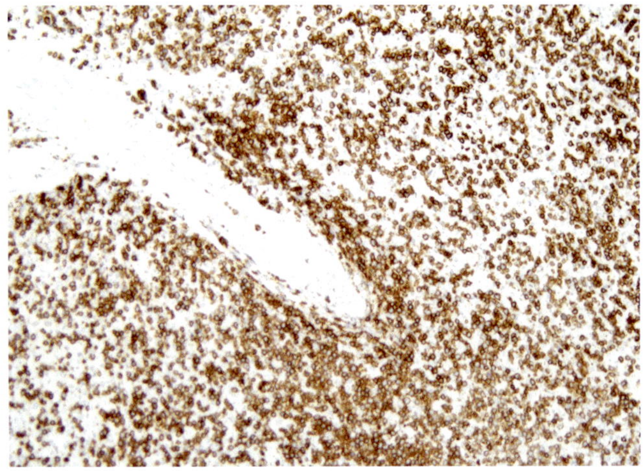

Fig. 7.21 Infectious mononucleosis, CD3 stain. This CD3 stain highlights the marked increase in red pulp T-cells, as well as cells infiltrating blood vessels. This finding is not specific for infectious mononucleosis but it is seen in almost all cases

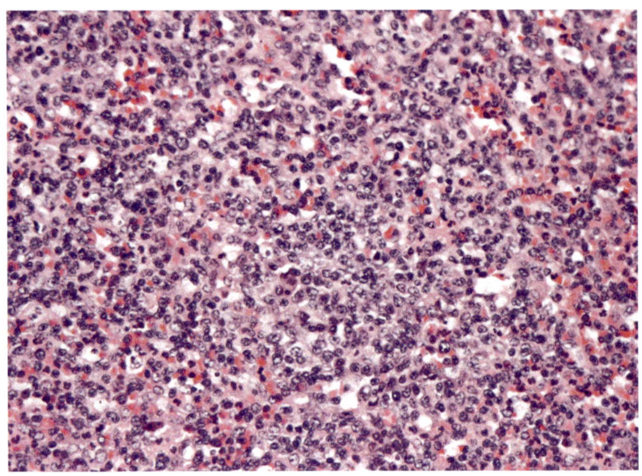

Fig. 7.22 Infectious mononucleosis. High-magnification image of red pulp in infectious mononucleosis. There is a marked increase in lymphocytes including large transformed cells (immunoblasts). These are predominantly of T-cell lineage. Rarely, Hodgkin-like lymphocytes can be seen in infectious mononucleosis

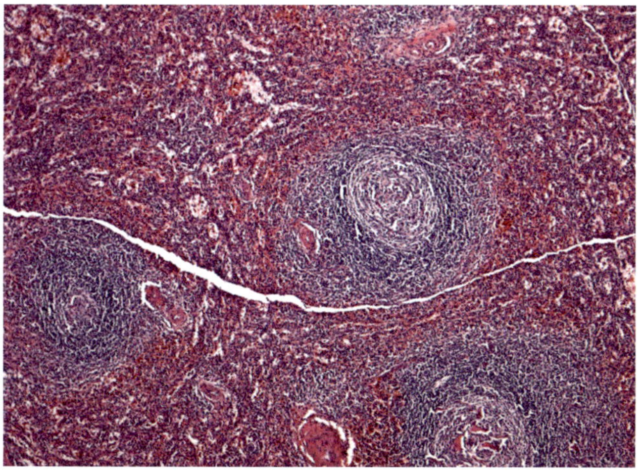

Fig. 7.23 Human herpesvirus-8 (HHV-8) infection, plasma cell Castleman disease (PC-CD). Lymphoid hyperplasia with marked increase in red pulp plasma cells, compatible with PC-CD. Most cases of PC-CD are positive for HHV-8

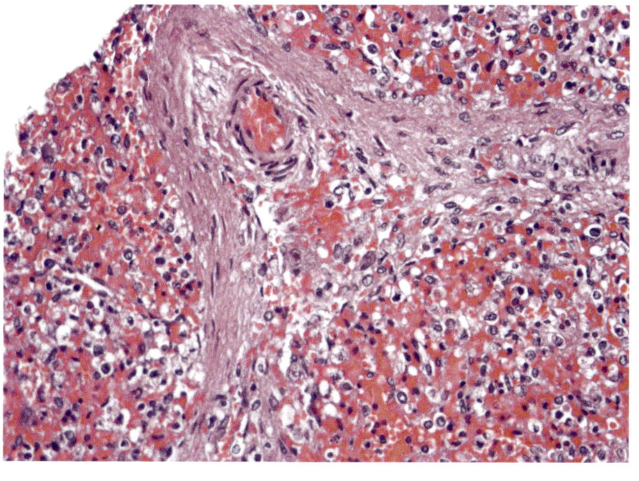

Fig. 7.24 Cytomegalovirus (CMV). Spleen from an autopsy specimen of perinatal CMV infection. A large, red intranuclear inclusion can be seen in the large atypical cell (*center*)

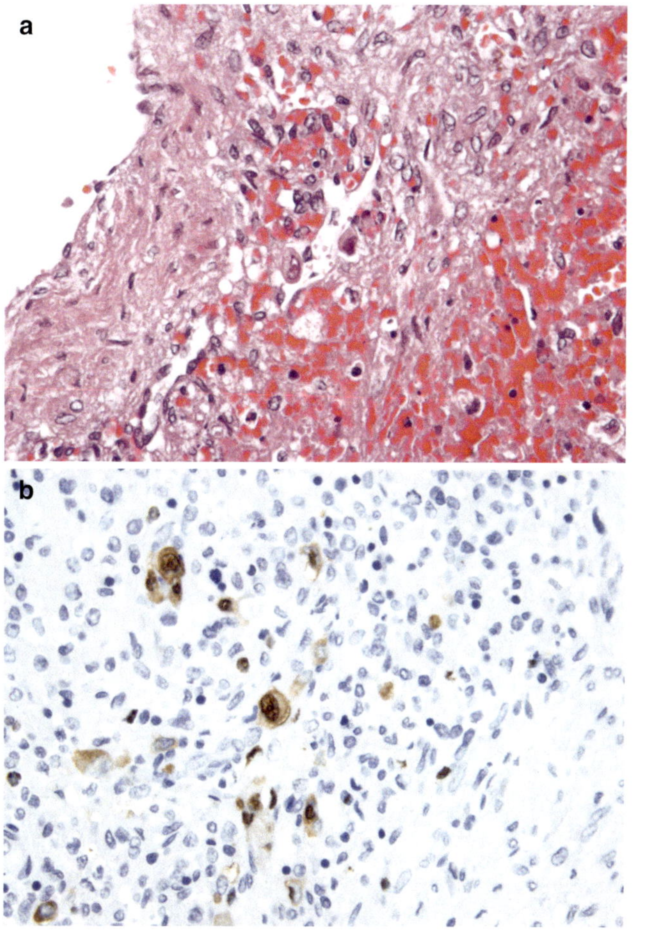

Fig. 7.25 Cytomegalovirus, hematoxylin and eosin (H&E), CMV stain. The inclusion of CMV is usually intranuclear and red in color (**a**). Identity of the virus can be accomplished by immunohistochemical staining (**b**)

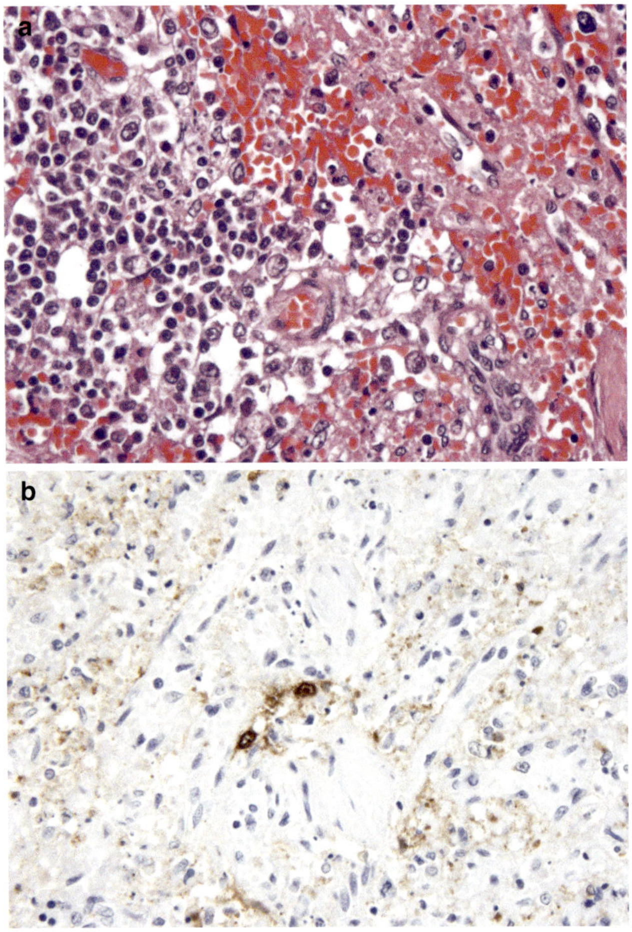

Fig. 7.26 Herpes simplex virus. In this example of a perinatal herpes simplex virus infection, lymphocytes with enlarged, cleared nuclei are seen (**a**). There is an overall increase in histiocytes and many transformed lymphocytes. An immunohistochemical stain (**b**) for herpes simplex virus, highlighting viral inclusions in infected lymphocytes

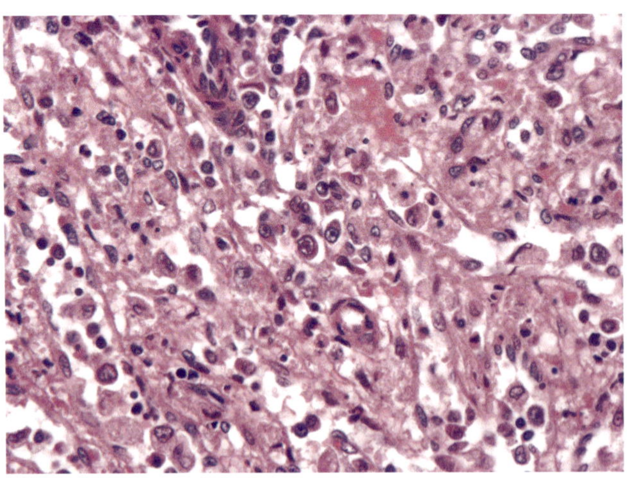

Fig. 7.27 Herpes simplex virus. Splenic red pulp in a case of herpes simplex virus infection. Note the large intracellular viral inclusion (*center*)

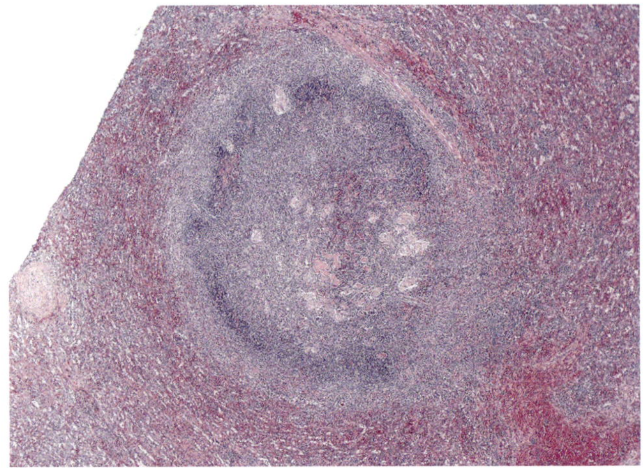

Fig. 7.28 Human immunodeficiency virus (HIV). Low-magnification image of a spleen in a patient with HIV infection. In this case, there is prominent white pulp hyperplasia. Within the white pulp nodule there are small epithelioid granulomas. Although spleen changes can be nonspecific, HIV has an increased incidence of both opportunistic infections and lymphoproliferative disorders

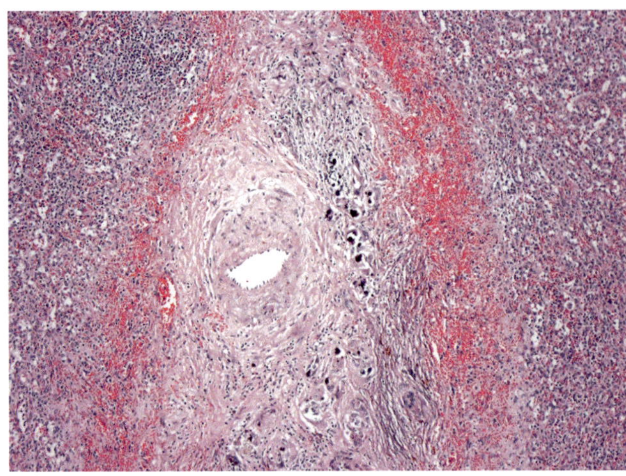

Fig. 7.29 Human immunodeficiency virus. Intermediate magnification of splenic arteriole. There is evidence of previous hemorrhage with iron and pigment deposition (Gamna-Gandy body)

7.4 Fungi

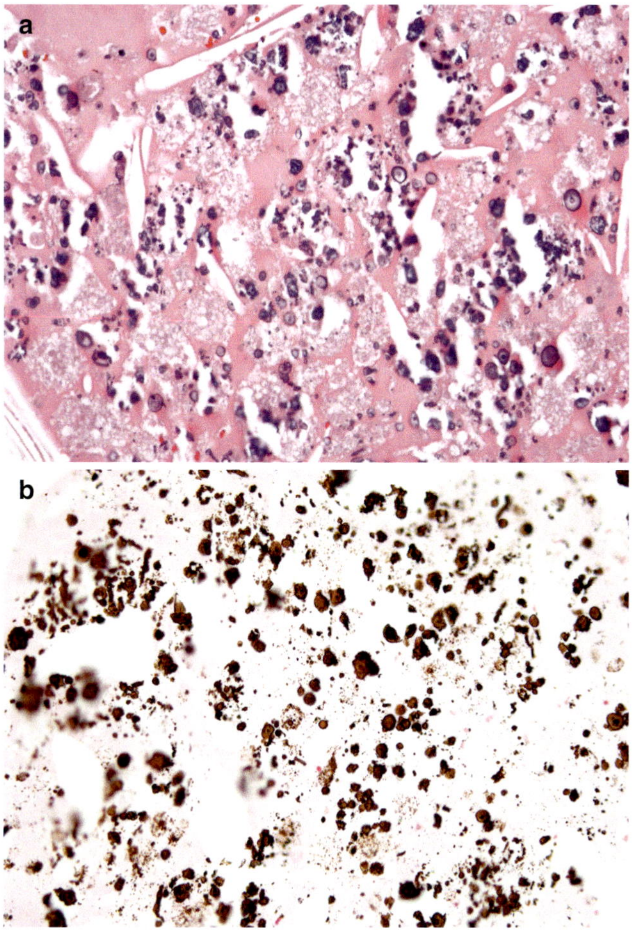

Fig. 7.30 Pseudofungus, H&E, von Kossa stain. Areas of calcification within a cyst of the spleen. The small spherules could be mistaken for fungal forms (**a**). A von Kossa stain highlights these small spherules as calcium (**b**)

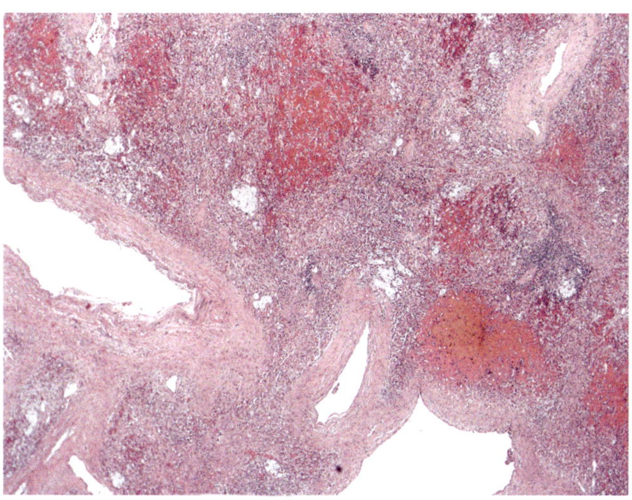

Fig. 7.31 *Cryptococcus*. Low-magnification image of the spleen shows clusters of pale macrophages present in red pulp and in perivascular locations

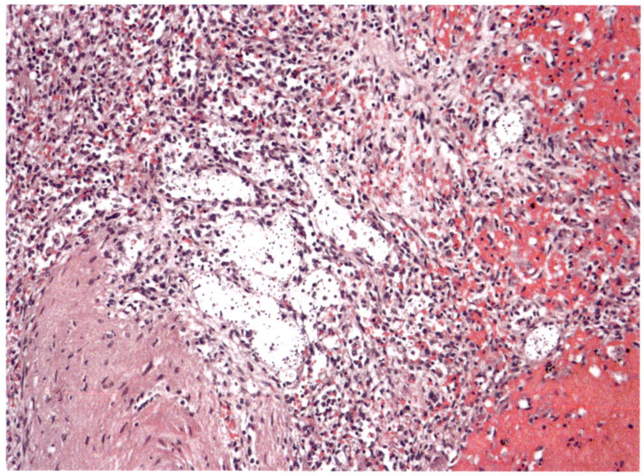

Fig. 7.32 *Cryptococcus*. Higher-magnification image showing a cluster of macrophages and organisms. *Cryptococcus* has a thick mucinous coating, which is clear on H&E stain. The small central organisms can be seen within large clear spaces

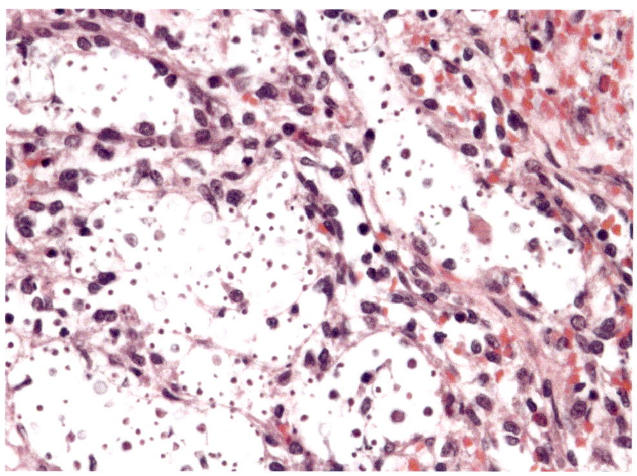

Fig. 7.33 *Cryptococcus*. The cryptococcal yeast forms can be seen as pale spheres within large areas of pale space. The pale spaces are the thick polysaccharide capsule of the organism

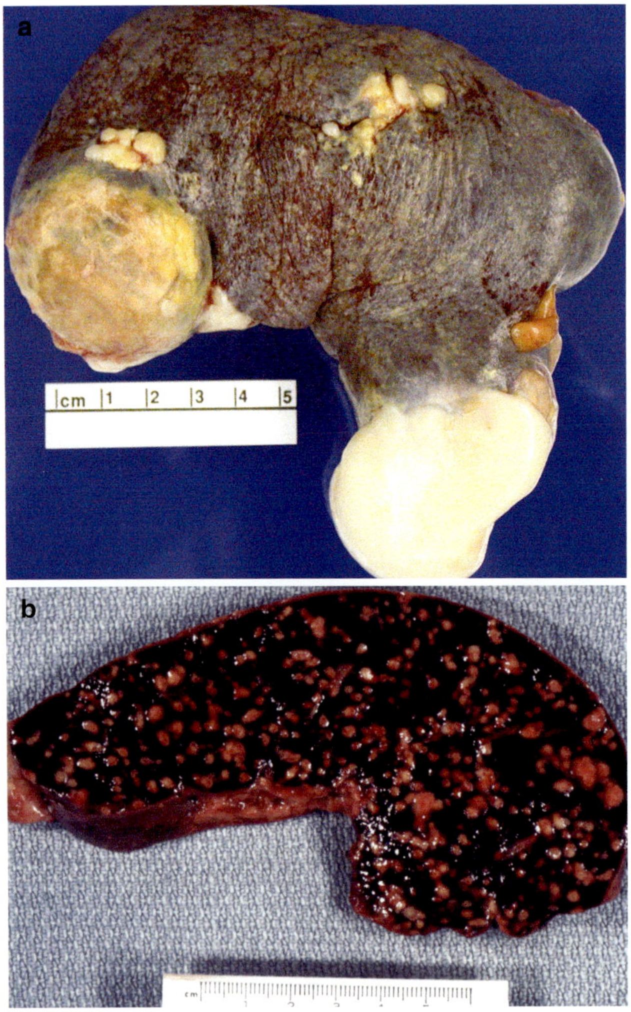

Fig. 7.34 Histoplasmosis. Unsectioned (**a**) and sectioned spleen (**b**) with massive involvement by histoplasmosis. Multiple granulomatous nodules are present that represent nodules of fungi and associated granulomatous inflammation

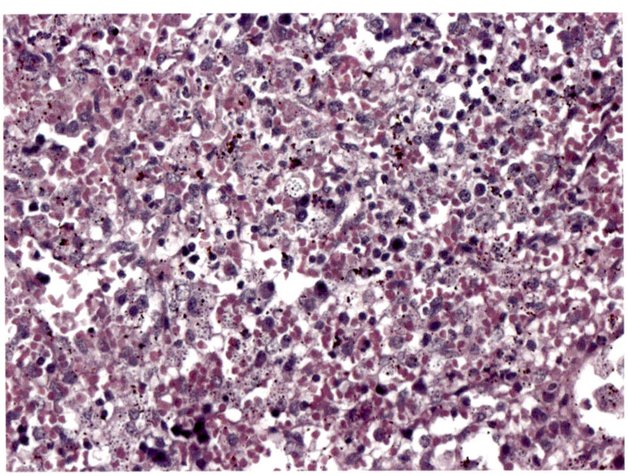

Fig. 7.35 Histoplasmosis. Individual yeast forms of *Histoplasma capsulatum* are seen within macrophages of the spleen. Splenic involvement is usually seen in disseminated cases. Granuloma formation can also be seen in some cases

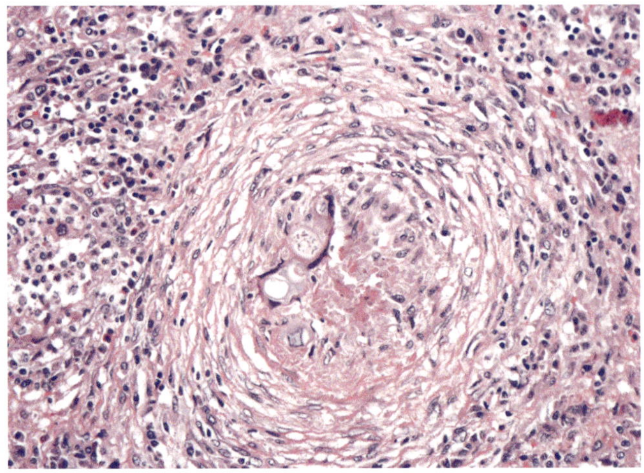

Fig. 7.36 Coccidioidomycosis. Caused by the dimorphic fungus *Coccidioides immitis*, coccidioidomycosis is a rare systemic fungal infection that can be seen in the spleen. In most cases, the fungal organisms are associated with well-formed granulomas, sometimes with central necrosis. Larger spherules (10–60 μm) are packed with small endospores (2–4 μm)

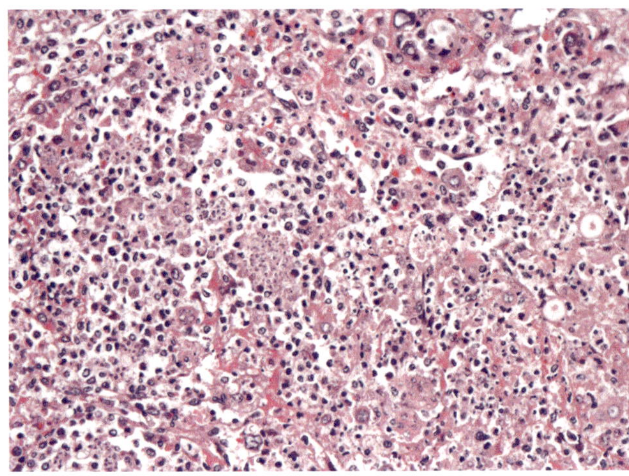

Fig. 7.37 Coccidioidomycosis. In more severe cases, the entire splenic parenchyma may have evidence of coccidioidomycosis spherules and endospores

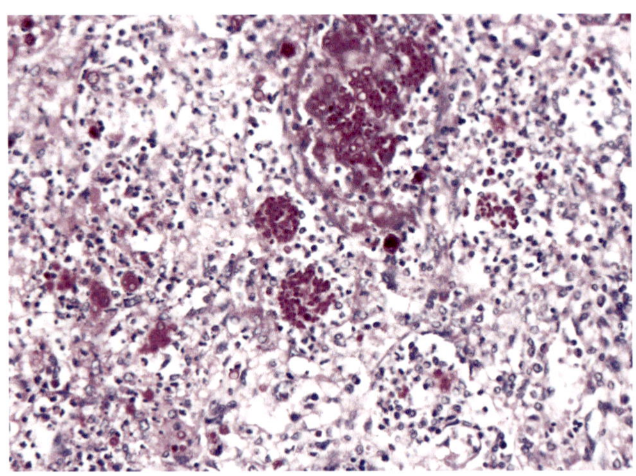

Fig. 7.38 Coccidioidomycosis. A PAS stain will highlight the organisms of coccidioidomycosis

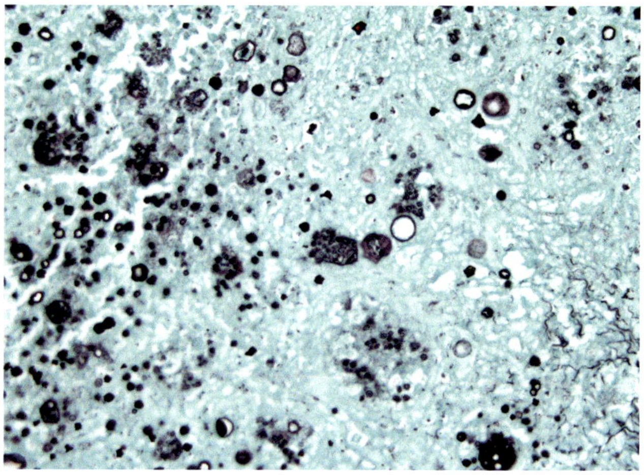

Fig. 7.39 Coccidioidomycosis. The organisms of coccidioidomycosis can also be highlighted with a silver stain such as Gomori methenamine silver (GMS)

Fig. 7.40 *Candida* infection. Gross image of the spleen with massive *Candida* infection. The distribution in small nodules is referred to as "miliary," after the appearance of millet seeds. Fungal infections such as these are most commonly seen in patients with severe immunosuppression and systemic infections. (*Image courtesy of* D. Farhi, Atlanta, GA.)

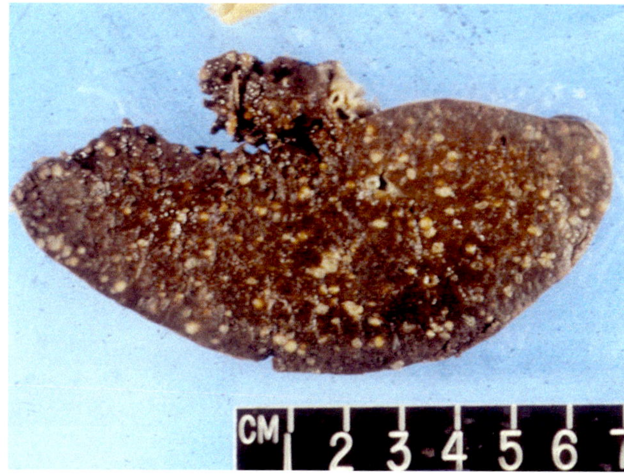

Fig. 7.41 *Candida* infection. Another image of *Candida* infection in the spleen. In this case, there are numerous aggregates. (*Image courtesy of* D. Farhi, Atlanta, GA.)

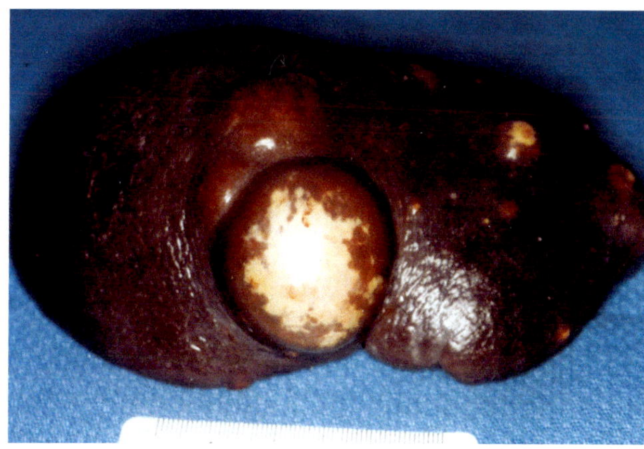

Fig. 7.42 Pneumocystis. Multiple subcapsular bulging nodules in this removed spleen. Although not apparent, these are abscesses filled with inflammatory elements and pneumocystis. (*Image courtesy of* J. Sidhu, Johnson City, NY.)

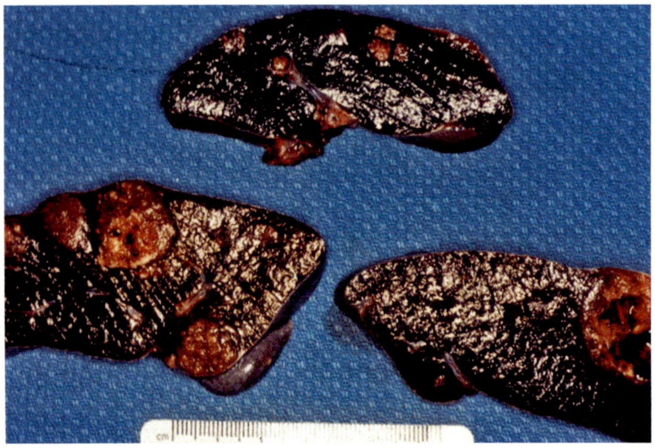

Fig. 7.43 Pneumocystis. Cut surface of above case showing multiple nodules with a variegated appearance. Some show areas of necrosis (more yellow) with some cystic degeneration. (*Image courtesy of* J. Sidhu, Johnson City, NY.)

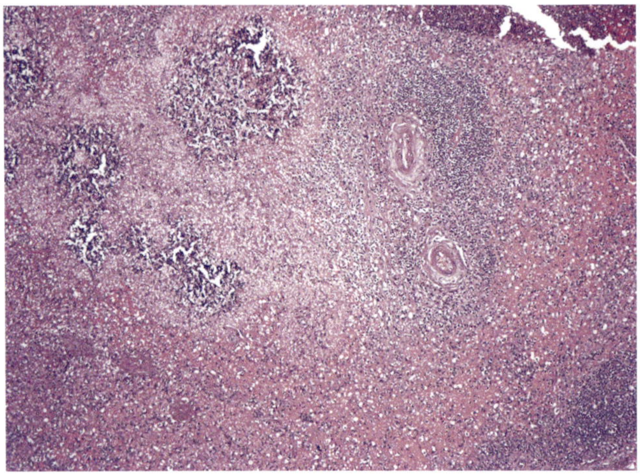

Fig. 7.44 Pneumocystis. Low-magnification image of a spleen with involvement by *Pneumocystis jiroveci*. Red pulp areas are present surrounded by histiocytes (pale pink) and with central areas of blue-black material (necrosis and cellular debris)

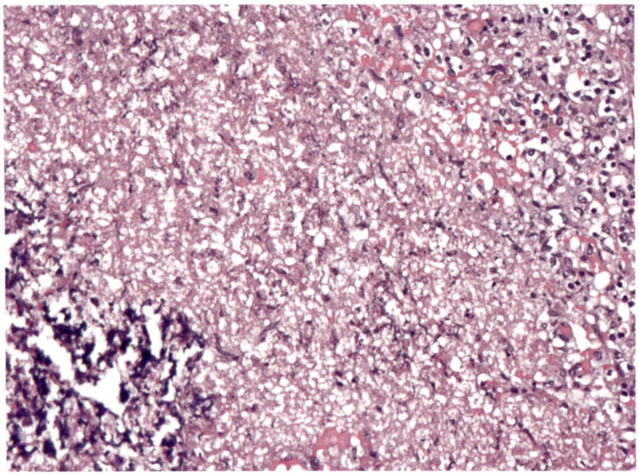

Fig. 7.45 Pneumocystis. High-magnification image of pale areas shows the foamy material that contains many *Pneumocystis* organisms

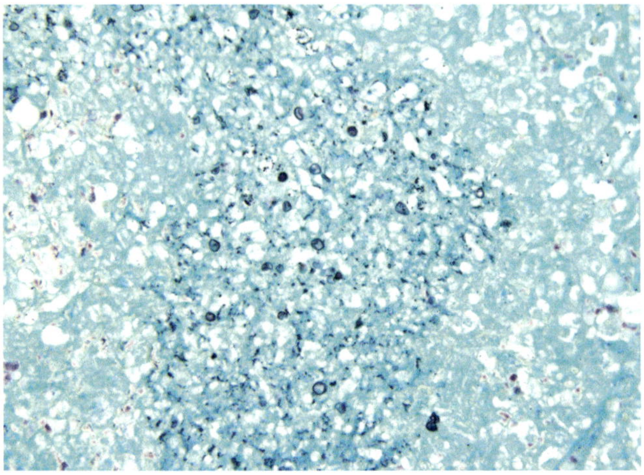

Fig. 7.46 Pneumocystis, GMS stain. High-magnification image of GMS (silver) stain outlines the *Pneumocystis* organisms. They are described as having a "crushed ping-pong ball" appearance

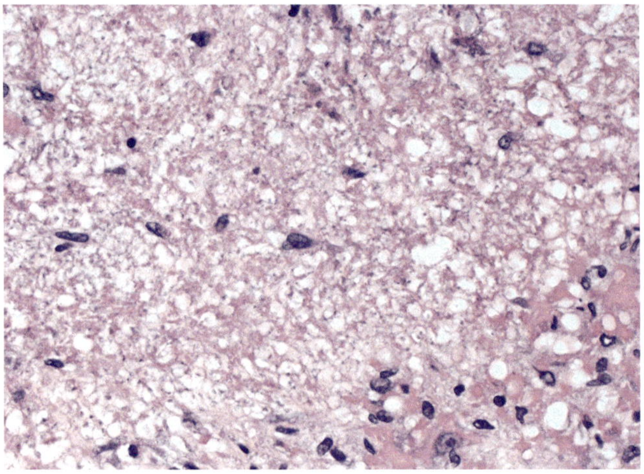

Fig. 7.47 Pneumocystis. Another high-magnification image of frothy paucicellular material. Although difficult, careful inspection of H&E sections does reveal some organisms. (*Image courtesy of* L. Weiss, Aliso Viejo, CA.)

7.5 Parasites and Protozoa

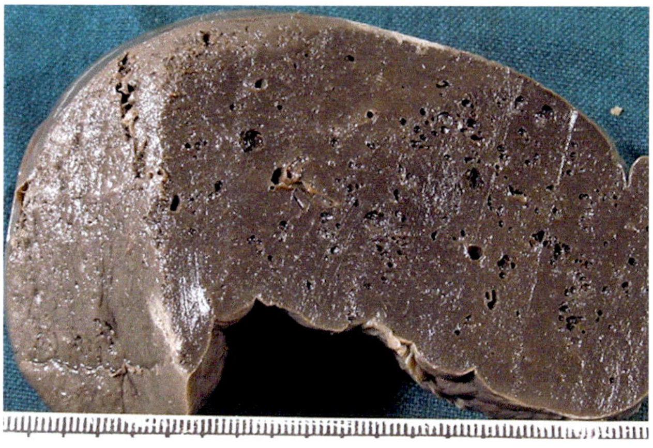

Fig. 7.48 Malaria. Gross image of a spleen with long-term involvement by malaria. The slate-gray color is typical and is due to long-term deposition of "malarial pigment." (*From* O'Malley et al. [1]; with permission.)

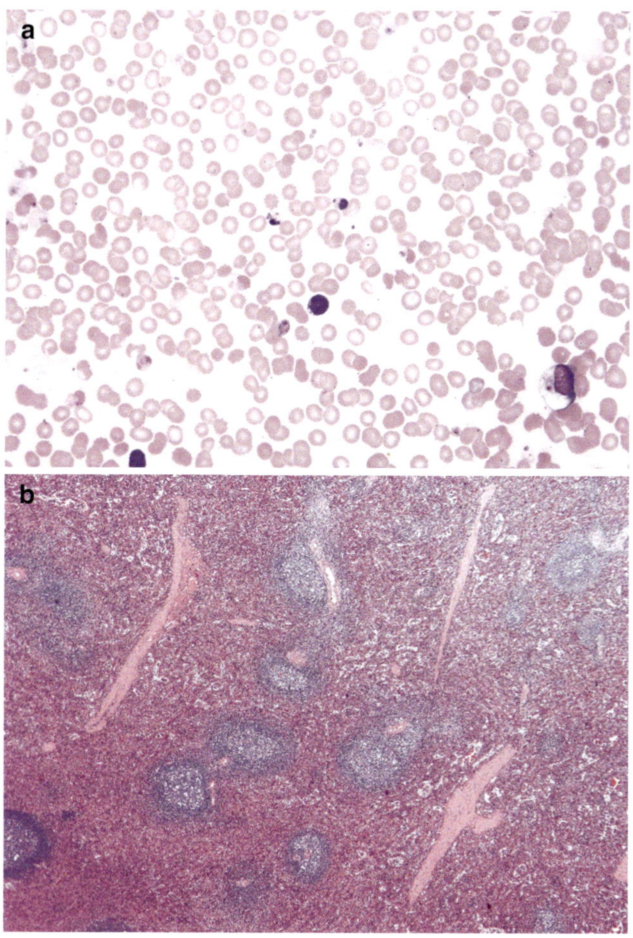

Fig. 7.49 Malaria, peripheral blood smear, spleen. This peripheral blood smear shows malarial forms in red blood cells (contrast with small lymphocyte in central portion of image) (**a**). This patient had significant splenomegaly (**b**). There is subtle expansion of red pulp

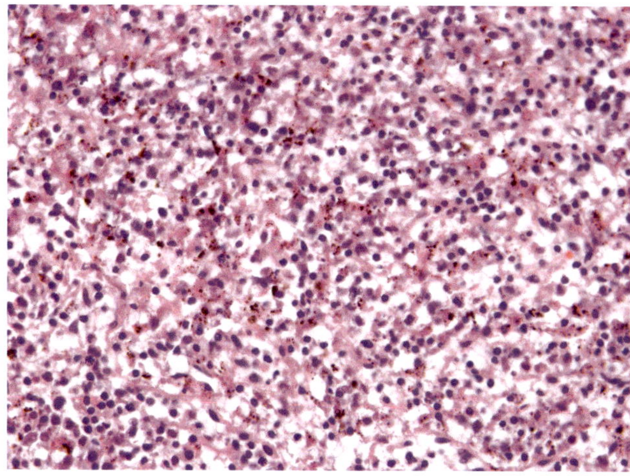

Fig. 7.50 Malaria. There is deposition of "malarial pigment" within the red pulp. Although difficult without oil immersion, rare organisms can be seen within red blood cells of the spleen in patients with malaria

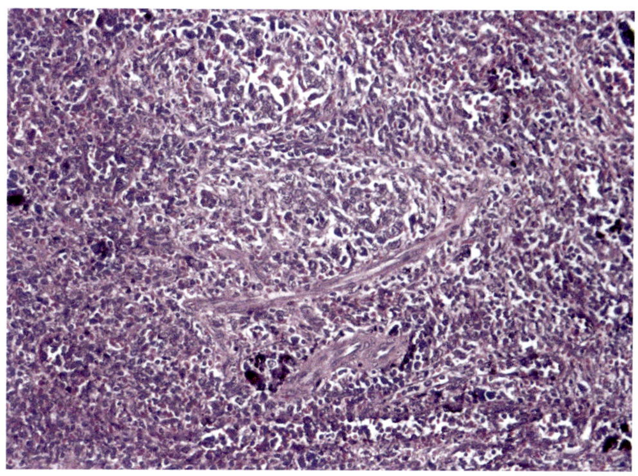

Fig. 7.51 Leishmaniasis. Intermediate-magnification image of a spleen with *Leishmania donovani*. The organisms are present within macrophages in the red pulp

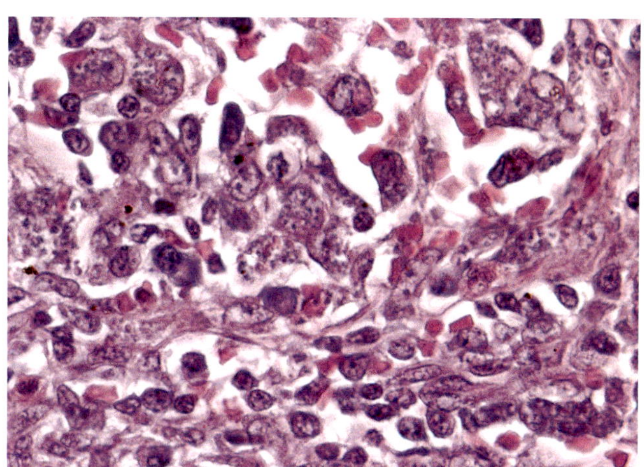

Fig. 7.52 Leishmaniasis. High-magnification image of splenic macrophages in a spleen with leishmaniasis. Individual organisms are called amastigotes. Leishmaniasis is a relatively common protozoal infection in the Middle East and Indian subcontinent

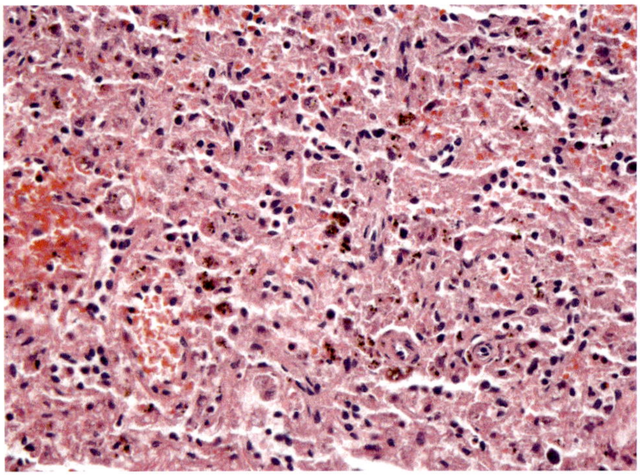

Fig. 7.53 Toxoplasmosis. Histologic image of toxoplasmosis in the spleen. Cordal macrophages show large coarse brown granules. Hemosiderin deposition is difficult to distinguish from *Toxoplasma gondii* cysts forms

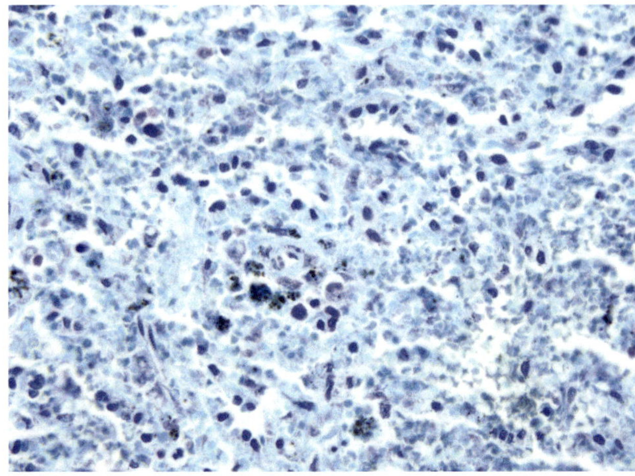

Fig. 7.54 Toxoplasmosis. Another image of *Toxoplasma gondii* in the spleen. In this Giemsa stain, the structures appear more green-black

Reference

1. O'Malley DP, George TI, Orazi A, Abbondanzo S. Armed Forces Institute of Pathology: Benign and reactive conditions of lymph node and spleen [book]. Silver Spring, MD: American Registry of Pathology Press; 2009.

Index

D.P. O'Malley, *Atlas of Spleen Pathology*, Atlas of Anatomic Pathology,
DOI 10.1007/978-1-4614-4672-9, © Springer Science+Business Media New York 2013

MIX
Papier aus verantwortungsvollen Quellen
Paper from responsible sources
FSC® C105338

If you have any concerns about our products,
you can contact us on
ProductSafety@springernature.com

In case Publisher is established outside the EU,
the EU authorized representative is:
Springer Nature Customer Service Center GmbH
Europaplatz 3, 69115 Heidelberg, Germany

Printed by Libri Plureos GmbH
in Hamburg, Germany